THE NEW NATURALIST LIBRARY

A SURVEY OF BRITISH NATURAL HISTORY

PEAK DISTRICT

ROBERT GILLMOR

In 1986, with the publication of *British Warblers* by Eric Simms, Robert Gillmor
was unveiled as the new artist for the New Naturalist Library, replacing the iconic
Clifford and Rosemary Ellis. The last 35 years have seen Robert produce some of
the most beautiful book jackets in British publishing, redefining the series and
book jackets in general. This year Robert painted the jacket for *Ecology and Natural
History* by David M. Wilkinson – his 72nd jacket. Robert has been the most creative,
wonderful, amusing, patient and knowledgeable artist for the series: we simply
could not think of anyone better. Thirty-five years since he originally vehemently
said 'no' to the offer of being the series artist, and after a long period of ill health,
he has decided it is time to step down. Robert's contribution to the visual appeal
of the New Naturalist series will be hugely missed. His 72 jackets will stand as one
of the great contributions to British book design.

THE NEW NATURALIST LIBRARY

PEAK DISTRICT

PENNY ANDERSON

WILLIAM
COLLINS

This edition published in 2021 by William Collins,
An imprint of HarperCollins*Publishers*

HarperCollins*Publishers*
1 London Bridge Street
London SE1 9GF

WilliamCollinsBooks.com

First published 2021

© Penny Anderson, 2021

Photographs © Individual copyright holders

A CIP catalogue record for this book is available
from the British Library.

Set in FF Nexus, designed and produced by
Tom Cabot/ketchup

Printed in Bosnia and Herzegovina by GPS Group

Hardback
ISBN 978-0-00-825737-8

Paperback
ISBN 978-0-00-825739-2

To Derek Yalden –
Mr Peak District Naturalist

Contents

Editors' Preface

ALMOST 60 YEARS AGO PROFESSOR K. C. Edwards, then Professor of Geography at the University of Nottingham, wrote in the preface to his *The Peak District* (New Naturalist 44): 'So many books have been written about Derbyshire and the Peak District that there would seem little excuse for adding to them'. Much of the justification, as we and the author went on to explain, lay in the declaration, ten years earlier, of the area as Britain's first National Park.

Much the same comment might well be made now, but the environmental and social changes that have unfolded since the early days of the National Parks are both profound and challenging and the need to bring their stories up to date is compelling. In the case of the Peak District, we were exceptionally fortunate in that two distinguished authors had planned to collaborate in such a project: Penny Anderson, a professional ecologist and botanist who has lived and worked in the Peak District for fifty years; and Derek Yalden, who spent the whole of his professional life as an academic zoologist in the Department of Biology at the University of Manchester, studying the birds and mammals of the Pennine uplands, as well as being a leading authority on the history of the country's mammal fauna. Coincidentally, both authors benefitted from an ecology and conservation training at University College London – Derek as an undergraduate and Penny from the post-graduate MSc in Conservation. Sadly, Derek died in 2013 and Penny assumed sole authorship.

Penny Anderson is uniquely qualified to undertake the task. In 1971 she established her own ecological consultancy, Penny Anderson Associates, specialising in habitat creation, management and restoration. At the same time, she is a naturalist in the finest tradition with an intimate familiarity with the plants and birds of the Peak District's richly varied landscapes and has carried out many voluntary surveys of the area's hay meadows and moorlands. One of the severest consequences of industrialisation on upland moorland is the erosion and desiccation of upland blanket bog leading to severe gully erosion, one of the most challenging of all landscape restoration problems, and of which Penny Anderson is the acknowledged authority. Since retiring she has been actively involved in a number of conservation organisations including membership of the Peak District National Park Authority and the preparation of the Peak District's State of Nature report as well as surveys for Natural England and the local Wildlife Trusts. On the national scene she served as a member of the National Trust's Conservation Panel and was elected president of the Chartered Institute of Ecology and Environmental Management. In 2015, in recognition of her significant career in conservation, she was awarded its prestigious Medal for Outstanding Lifelong Contribution to Promoting High Standards of Ecological Consultancy and Habitat Management.

Few people can have had such a single-handed influence on a major landscape and it is with pleasure that we welcome Penny Anderson's account of the Peak District as a significant contribution to the New Naturalists continued commitment to its regional titles.

Author's Foreword and Acknowledgements

D
EREK YALDEN AND I HOPED to write this book together, but very sadly he died suddenly before we could approach the publishers. I felt an imperative then to continue in his memory and hence dedicate this book to him. Derek was 'Mr Peak District Naturalist', a Manchester University zoologist and avid Peak District explorer. He surveyed and published on moorland waders, mountain hares, deer, pigmy shrews, wallabies and other mammals, and some invertebrate groups. He collected carcasses to provide a database for identifying archaeological remains. He (and I) developed an interest in the effects of outdoor recreation on wildlife, especially breeding birds, and he supervised research into disturbance effects on Golden Plover on the Snake summit blanket bogs. We worked together on the Peak District National Park's Moorland Restoration Project for many years. He was a true friend and an unending source of information on the Peak District.

With Derek's loss, I have been fortunate in having equally able support from other specialists. I am a botanist and ecologist and, following a move to Manchester from the south in 1969, the Peak District became a regular weekend escape – gradually becoming deeply immersed. I ran an ecological consultancy based in the Peak District all my working life, undertaking numerous surveys and prepared management and restoration plans for many Peak District sites. An early commission, by the National Park Authority (NPA), was a management plan for the Longshaw Estate, followed, in 1979, by vegetation mapping of the Longdendale catchment (about 78 sq km) for the former Nature Conservancy Council. I worked my way up all the cloughs (steep, deep valleys in the moorlands), across blanket peats and heather moors and through the flushes, recording plants and habitats. It was magical – and hard work. This led to work for the NPA on the Moorland

Restoration Project (1979–97), which identified the extent of damaged ground, worked out the causes and then started the restoration ball rolling. This provided the foundation for the 'Moors for the Future' programme.

My consultancy provided ecological support to the NPA in the 1980s before it appointed its first in-house ecologist. We surveyed all kinds of places, finding the rare and the common. Further commissions included preparing management plans for some of the NPA's large Estates – the Eastern Moors, North Lees, and the Harpur-Crewe Estate in the South West Peak. I co-wrote, with Dave Shimwell, *Wild Flowers and Other Plants of the Peak District* (1981), updating a seminal Peak District volume by the Cambridge University botanist, C. E. Moss (1913).

I was Chair of the Peak Park Wildlife Advisory Group (PPWAG) for some 21 years, co-ordinating and reporting issues of local concern to the NPA. Derek Yalden was a founder member. The PPWAG was superseded by the Biodiversity Action Plan development in the 1990s, but before that time, I learnt a huge amount from its members regarding the Peak District's wider wildlife interests. I have been involved in the Peak District ever since.

My specialist interest in moorlands has led to the production and implementation, with my colleagues, of major restoration plans for our many large upland estates, commissioned largely by water companies, but also by Sheffield City Council. I have also been closely involved in the limestone areas, where the greatest floristic diversity lies. From 1977 on, I began monitoring orchids and other vegetation in Derbyshire Wildlife Trust's Miller's Dale Nature Reserve. I include some of the findings here. Since I retired I have been monitoring hay meadows and moorlands as a National Trust volunteer – from which I have also learnt much. As a Member of the Local Nature Partnership (Nature Peak District) I wrote the *State of Nature Report* for the Peak District (2016) and apply those findings in this book. It does not always make for happy reading, showing declines in common with the country as a whole – but there are bright spots, such as the massive amount of moorland restoration being achieved.

I hope this long association qualifies me to deliver this Peak District volume. I want to share the wonder and magic of our fantastic wildlife – and by wildlife, I mean everything that is alive and wild – but, at the same time, reveal its ecological foundations, show how it has fared over the centuries and project what the future might hold. To do this, I weave in elements of vegetation history starting after the last ice age and moving through the centuries of human occupation – a major factor in shaping what we see today. I end by considering the future, particularly in respect of possible climate change. I slip in stories, myths and legends where they embellish understanding and add some insight into the past uses of plants, linking them back to human cultural associations.

I try to distil scientific endeavours and recent findings (the sources are in the bibliography), embellished by my experience and observations. I use the scientific names of species the first time they are given in the habitat descriptions, but avoid cluttering the text with them thereafter, except where vernacular names are lacking, or not widely recognised. Needless to say, limited space prevents inclusion of many species and, in any case, exhaustive lists would not make for joyful reading. My selections illustrate the different habitats, but you will need to search the local atlases, or the internet, for more information on species that interest you.

My perspective is an ecological one, viewing the habitats and their species in an integrated fashion. This means avoiding separate chapters on birds or mammals and so on. This works well when the animals occupy specific habitats but less so when they are widespread. Forgive me, therefore, for some species featuring where they are first mentioned, rather than where you may have seen them!

The OS maps of the Peak District (1:25,000 is the best scale) are essential companions and Appendix 1 provides locational data. Although these maps are unsurpassed, the best visual impression comes from overlaying local maps with aerial photography on the internet. Zoom in and out to see blanket bog gully patterns, or old lead rakes marching across the hills and dales, for example. However, maps need checking for open access land, or rights of way to locations mentioned. I have tried to suggest accessible places as far as possible, for your enjoyment, and some of these are purposefully chosen to offer easy access.

If you are interested in looking for specific locations for species, or want to see where they occur, you will need to consult the county floras, bird books and other atlases. The recent, magnificent *The Flora of Derbyshire*, as the main Peak District county, is excellent. *The Flora of Staffordshire* and *South Yorkshire Plant Atlas* are equally good. *The Birds of Derbyshire* and South Yorkshire are obtainable as books, and the Birds of Cheshire and the Wirral is freely accessible online (www.cheshireandwirralbirdatlas.org). *The Mammals of Derbyshire* and *The Mammals of Cheshire* are a must if you are interested. There is *The Butterflies of the Peak District* and Sorby Natural History Society publishes several regional compendiums of, for example, dragonflies, butterflies, freshwater beetles and bugs as well as others. Staffordshire Ecological Record has a searchable atlas online.

Vascular plant scientific names here follow Stace (2019) and may differ from some of the older atlases. Mosses and liverworts follow Atherton *et al.* (2010) and other names match those provided by specialist groups on the internet.

This book has benefited enormously from other specialists to whom I owe a huge debt of gratitude and thanks. I hope I have remembered everyone. Rhodri Thomas, National Park Ecologist, has provided general support and unearthed all sorts of gems. Derek Whiteley, local invertebrate and mammal specialist and Sorby Natural History Society stalwart, has filled many of my gaps on these groups, supported by Peter Tattersfield on molluscs and Rob Foster on hoverflies. Rob also boosted my waxcap knowledge hugely, supported by Carol Hobart on fungi in general, and Steve Clements and the Sorby fungi group have educated me well. John Gunn's expertise has helped shape the sections on limestone geology and cave development, with Ken Smith undertaking a similar role for all things archaeological and historic. I am indebted to Don Stazicker and Nick Everall for sharing their vast knowledge of fishing and freshwater invertebrates. Steve Price kindly supplemented my lichen knowledge. I am indebted to many of these enthusiasts plus Guy Badham, John Barnatt, Buxton Museum and Art Gallery, Geoff Carr, Bob Croxton, Kev Dunnington, Thomas Eccles, Alex Hyde, Andy Keen, John Leach, Steve Orridge, National Trust, Sorby Natural History Society, Treak Cavern and Gary Ridley, who have also supplied photographs or drawings, mostly just for the pleasure of sharing them with others, and not least, my husband who has converted my graphs to something publishable and analysed the local climate and plant data.

The maps contain various free-to-use Ordnance Survey Opendata products. The climate graphs and tables use data from Met Office (2006): UK Daily Temperature Data, part of the Met Office Integrated Data Archive System (MIDAS); NCAS British Atmospheric Data Centre; Met Office (2006): MIDAS: UK Daily Rainfall Data; and NCAS British Atmospheric Data Centre.

The responsibility for integrating all the information provided into this book though, remains mine alone.

Penny Anderson, 2021

The Peak District. (Contains OS data © Crown copyright and database right, 2020)

CHAPTER 1

Introduction

THE NAME 'PEAK DISTRICT' IS misleading if it conjures up a land of sharp peaks. There are, in fact, few and they are essentially small. However, the name may have been derived from an ancient Anglo-Saxon tribe, the Pecsaetan or peaklanders who inhabited the central part of the area from the 6th century CE when the land lay within the kingdom of Mercia.

Our Peak District is a land of great natural beauty derived from the contrasts of its rocks, soils and history. Imagine a horseshoe, stretched wider in the northern middle, placed on a map over the Peak District. This horseshoe is

FIG 1. General view across the Dark Peak from Blackden Edge (Kinder) and tors east across Woodlands Valley to the Upper Derwent Moors.

FIG 2. Chee Dale with steep slopes, limestone outcrops, woodland and grassland and the Monsal Trail in the centre.

occupied by the gritstones and shales that account for the high hill tops, often fringed with striking rocky edges, overlooking deep, steep-sided valleys, locally called cloughs, in which streams tumble over rocky courses (Fig. 1). The central core tucked into the surrounding horseshoe is the limestone country; lower than most high moors, but this time falling into deep, sometimes craggy dales, with whitish rock exposures and screes, but where streams are scarce (Fig. 2).

The contrasts between acidic gritstones and shales and the calcareous limestone rocks are dramatic. The core shapes, habitats and species are quite different: the windswept, wide vistas of blanket bog and dwarf-shrub heath, interspersed with springs and grasslands in the moorlands, contrasting the grey-coloured walls, limestone grasslands, or ash rather than oak woodlands in the dales. Yet there are subtle similarities to surprise you.

Before venturing into detail, our area needs definition. If you know the Peak District, the National Park will be familiar: the country's first, declared in 1951. The region for this book covers all of it, but the National Park boundary is sometimes oddly-shaped (see frontispiece map, p. xv). The void arm, coinciding with the A515 southeast from Buxton and the A6 to the north, is lined by several large limestone quarries. Buxton, too, is excluded, yet is the District's main town and intimately linked throughout history. Some exclusions were made on planning grounds. The National Park is designed to protect its landscape, flora and fauna, which would have been difficult in the face of internationally important quarry resources, particularly of limestone. That is not to say that quarries are absent from the National Park, but mostly not the giants that occur just outside. Indeed, many inside are now abandoned to wildlife and have become excellent habitats with plenty of orchids and other treasures.

In recent decades, effort has been made to describe and classify the landscape of the whole country into 'Character Areas'. These combine geology, landform, habitats, landscapes, settlement patterns, industrial and cultural history as well as current land-uses and vernacular architecture. The Peak District has three key ones: the Dark Peak, the South West Peak and the White Peak (Fig. 3). The Peak District Biodiversity Action Plan (BAP, Appendix 2) is underpinned by these three zones, as there is a clear integrity of environmental dynamics, which anyone can see. It is these three areas that are the subject of this book – with the National Park boundary firmly ensconced across the centre. There are slivers of some neighbouring Character Areas that have been included to smooth the boundary. These do not change the key features of the area, but if you live in any of them, I hope you are not offended.

Our Peak District is essentially the southern end of the Pennines' broad mountain backbone, starting at the Cheviot Hills and dwindling to low hills along our southern boundary. It is the first upland you reach travelling north from South or South East England. It is exciting because of the upland species that cannot survive in the often intensively managed adjacent lowlands – Golden Plover, Red Grouse, Dunlin, Curlew, Short-eared Owl and plants like Crowberry, Cranberry and Bog Rosemary.

It is also where southern species often peter out and northern ones reach their southern-most limits. We are, therefore, at the crossroads for these, as well as some that are more abundant in the west. Some can be very rare at their extreme locations and more common in their core 'homeland'. Finding them and seeing what lives alongside them, gives new insights into associations. Occasionally, these 'strangers' meet up and hybridise, as with the Northern and Southern Marsh Orchid and the Bilberry/Cowberry hybrid.

The Peak District is around 88 km from north to south and 53 km across the middle, west to east. It stops rather abruptly in the north, at the A62 cross-Pennine road, only because the Pennines are continuous and we need an artificial boundary somewhere. This boundary is not too precious since similar moorland stretches north as far as you can see. The National Park's total area is 143,700 ha (or 1,437 km²), whilst the whole Peak District as we define it is 185,686 ha.

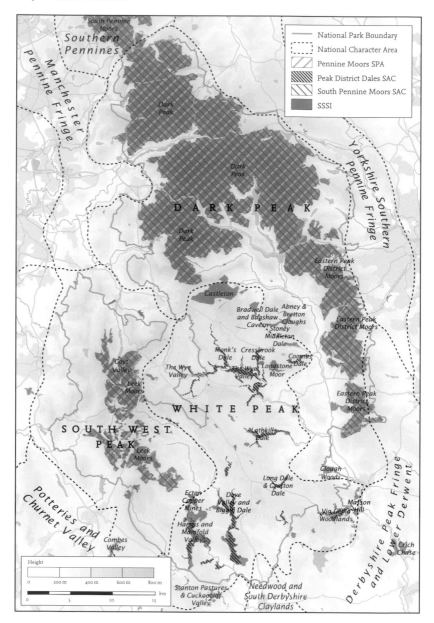

FIG 3. The National Park boundary, Dark Peak, South West and White Peak Character Area boundaries together with the Nature Conservation designations. (Contains OS data © Crown copyright and database right, 2020)

The landscape character areas cut across multiple boundaries (Fig. 3). Cheshire used to include a pan-handle that embraced the moorlands north of Longdendale up to the Yorkshire and Lancashire boundaries. Its odd shape dates back to pre-Domesday when Edward the Elder added modern-day Longdendale to the Mercian kingdom, possibly to control important routes through the valley to Yorkshire. Certainly, the monks from Basingwerk Abbey (Holywell, now North Wales, but within Cheshire in the 12th century) controlled the route for moving salt across the Longdendale valley east to Yorkshire. This pan-handle lasted until the 1974 re-organisation of counties, when it was passed to Derbyshire. This multi-county situation adds complexity to the National Park's management and results in regional divisions, often with different boundaries, for various agencies. Partnership working ensures equal attention across these borders.

This Peak District New Naturalist edition follows in the illustrious footsteps of K. C. Edwards' (and friends') 1962 volume. Our knowledge and understanding of the Peak District in the intervening period has improved significantly. Considerable research has been undertaken on the moorlands, particularly on peat, its origins, the effects of human activity, the extent of degradation and how best to restore it. Recent research, carried out at Sheffield University's outdoor laboratory in Buxton, reveals some potential climate-change effects on our limestone flora. The Peak District's BAP (see Appendix 2) has been responsible for new findings, following the disappearance and restoration of species and habitats, and planned new multimillion-pound projects are set to take us to new heights of habitat conservation and restoration.

The area is no less important to humans. The National Park's permanent population is some 38,000, but this expands significantly in good weather and especially at weekends – the area is still probably the best known of the National Parks. It is within a half-day's travel of nearly 50 per cent England's population – a place for day visits to all its wondrous corners. Until the South Downs National Park was established, the Peak District National Park revelled in being the second most visited in the world (with only Mount Fuji in Japan, capped by a shrine, with understandably higher numbers). Our area has been an escape ground for many working in the former grim conditions of the mills of Manchester, Sheffield and other nearby towns for many years, with the notorious Kinder Scout Mass Trespass in 1932 part of recreational and open-access history. The expansion of the freedom to roam on the windswept high moors and now to many of the dales as well, has been an integral part of that history.

Tourism, though, is not new in the Peak District. The area was on the visitors' circuit back in 1586 when William Camden first listed the Wonders of the Peak. Others added to his list, but none included the Dark Peak moorlands. Rather, caverns, Chatsworth and natural holes, and the wells in Buxton were more fashionable destinations. In this now well-travelled and loved landscape, it is hard to appreciate Daniel Defoe's dismissive description of the High Peak as a 'howling wilderness' which 'is the most desolate, wild, abandoned country in all England' (Defoe, 1726). How views have changed.

Before access was even considered, though, more local people would have been working in the landscape, with mineral and coal extraction, farming and woodland management and all the services needed to support a rural economy. There are records of carpenters, bakers, shoemakers, butchers and many other services in small villages that can barely support a general store nowadays. Ours is a working landscape – there is nothing completely natural about it, but, rather, it has evolved under the effects of diverse human activity from at least Palaeolithic times. Many naturally occurring plants and animals have made it their home, sometimes reacting positively to human activity, whilst others come and go as management and work patterns change, as we shall see.

This book, then, links the wildlife and habitats to the development of the landscape through its geology and geomorphological processes, whilst simultaneously incorporating stories of human activities past and present, as well as some of the legends and old wives' tales to bring to life the evolution of the Peak District and the special habitats and species that we call our own. We have a splendidly rich wildlife and this book gives a flavour of its diversity and value. However, there is more to wildlife than this: it provides the textures, sounds, smells, colours and beauty that clothe the landscape, all contributing to the spirit of place, something we can all love and cherish.

To tempt you to read further, we start by glimpsing the Peak District landscape, its main features and character. An account of the rocks follows – how they were laid down and then fashioned into today's landforms. Woven into this is the history of our wildlife integrated with that of humans in the landscape, drawing the connections that affect evolution of these plant and animal communities. A short account of our particular climate, soils and geographical position follows, before we dive into the Dark and South West Peak habitats: the peatlands, heather moorlands and associated wetlands and grasslands, the woodlands and hedges. A connecting chapter on neutral and farmed grasslands where we can delight in flowery meadows is followed by the limestone habitats in the dales and on the plateau tops. Finally, we muse over where the Peak District might be in the coming decades. Enjoy.

CHAPTER 2

The Spirit of Place that is the Peak District

L ANDSCAPE HERE DOES NOT JUST mean what you see, but more the spirit of place; the characteristics that, together, form the rich tapestry of what we feel with all our senses. A quick sketch of the rich variation of this landscape across the Peak District illustrates how the geology, landform, soils and hydrology all play a massive part in helping understand the forces that have combined to give the impressive and often dramatic upland landscape on which the wildlife depends. But these have not been the only forces shaping the landscape. Human activities have also been critical. Exploitation of the geological resources has been significant – sometimes of international importance and some of great antiquity; differing uses of wood have shaped the woodlands; and agriculture has moulded and fashioned the open land, striving to produce crops and meat under challenging climatic conditions.

The Peak District is not, however, isolated from outside pressures. The industrial development of surrounding areas played a huge part in generating all-pervasive air pollutants that have had major impacts on upland vegetation and peat; demand for water led to damming many moorland streams to create reservoirs; water was the energy source for the many mills, those in the Middle and Lower Derwent Valley helping to drive the beginning of the Industrial Revolution; and the beauty of the area gave welcome relief to those living in grim urban conditions seeking some 'fresh' air and exercise. All these factors have combined, along with others, to produce the Peak District we know and love today.

THE DARK PEAK

Expansive wilderness and weather-beaten extremes personify the higher, more remote landscape. Extensive, elevated plateaus with seemingly never-ending views across heather and moorland blanket bog, lie over black, deep peat. Deeply incised tributary valleys, punctuated by rocky outcrops, tumbling rocky streams and small waterfalls hide amongst the sweeping moors (Fig 1). These steep, often deep, cloughs drain into the main river valleys that are wider, less frenetic, but no less beautiful.

Most of the Peak District lies above 200 m, outside of a few river valley bottoms. Kinder Scout boasts the highest point at 636 m, just above Edale Head, but the whole plateau is over or around 600 m. Its very distinctive shape, approximating a squashed parallelogram from the air, sits on the west side of the Peak District above Hayfield, northeast of which is the more rounded Bleaklow ridge – also regularly exceeding 600 m along its roughly 9 km length. Compared with other upland areas, these are dwarfs amongst the giants of Snowdonia and the Lake District, but they make up for this in grandeur and magnificent views out across lowland Cheshire and Greater Manchester, the central Pennines or the eastern hills. Equally, they stand out in views from the surrounding regions – their visibility informing local people about local weather when they are obscured by cloud, or snow-clad.

These elevated hills are composed of hard-wearing gritstones, sometimes with rocky, abrupt, dramatic edges surmounted with suggestively shaped tors of massive boulders (Fig. 1), many with local names. The lower, more rounded slopes and surrounding valleys tend to be gouged out of softer shales, often with crumpled margins, as in the Longdendale or Woodlands Valleys. This crumpling represents a wealth of landslips, where the softer shales slip over lubricated bases, sometimes on an impressive scale, into a jumble of hillocks, boulders and hollows.

The upper horseshoe-shape of the moorlands is draped round the northern, western and eastern edges of the Peak District. The inflated northern section represents the southern end of the Pennine Chain where Kinder Scout and Bleaklow lie, but arms of similar character also flank the Peak District into the South West Peak on the west side and Matlock on the east. This gives stunning views coming into the Peak District from either side, with the rumpled edges giving long ridges that alternate with valleys stretching from north to south, especially on the west side. There are striking views of this from the top of Monks Road that looks south towards Hayfield (Fig. 4), and from the top of the Goyt Valley.

FIG 4. View south from Monks Road towards Hayfield showing folded rocks creating long, deep valleys giving striking views.

The Dark Peak spawns a plethora of streams that flow eventually to join the Trent, the Calder, the Humber Estuary or the Mersey. The frontispiece map (p. xv) shows how closely together the headwaters originate, with rain drops falling just metres away from each other at the watersheds destined for seas to the west or east. Many of these rivers have been dammed to provide drinking-water reservoirs – a recurring feature of the Dark Peak. Strings of reservoirs fill the Longdendale and Upper Derwent Valleys (Fig. 5), but there are many others. Although benign in the landscape now, and often a focus for visitors, flooding some of the villages and farms was controversial at the time. Former mill ponds still also survive here and there. Although most large reservoirs were established by the water companies, some were created as feeder reservoirs for the canals, like Black Moss and Swellands Reservoirs on high moorland near Marsden.

Within the valleys, woodland is generally scarce, covering only about 10 per cent of the area, with broadleaved oak woodlands being characteristic and concentrated in the Middle Derwent Valley. These semi-natural woodlands, however, are in the minority, with modern plantations predominating, especially fringing the reservoirs (Fig. 5). More widespread afforestation, as achieved in some other upland areas such as the Lake District and in Galloway, was constrained here by high levels of air pollution in the 19th century, which stunted tree growth and rendered wider planting futile.

FIG 5. (above) Derwent
Reservoir, part of the Upper
Derwent chain of reservoirs,
with bilberry and heather
moorland in the foreground,
extensive conifer plantations
around the reservoir and, beyond,
distant heather moorland.

FIG 6. (right) Hare's-tail
Cottongrass in summer
abundance, on blanket bog at the
upper end of the Goyt Valley.

It is on the high, peat-covered moorlands where you will find expanses of blanket bog mostly dominated by cottongrasses, whose fluffy white seed heads can simulate a summer snow storm (Fig.6). On drier soils, sweeps of heather moorland cascade into the more intimate cloughs, which boast seepages, springs and flushes (where water generally flows just below the surface), beds of Bracken and upland woodland as well as the streams and rivers themselves. The floristic interest is concentrated in the cloughs. Red Grouse (Fig. 104) and waders like Golden Plover (Fig. 88) and Dunlin grace the windswept moorland tops, while Ring Ouzels (Fig. 119), Wrens and Stonechat (Fig. 113) favour the shelter of the edges and cloughs, where Green Hairstreak (Fig. 124) and Bilberry Bumblebees (Fig. 125) dance amongst the flowering Bilberry and Marsh Thistles. You can still find some flower-rich hay meadows and pastures in the inbye fields further downslope, as we shall see.

Sturdy gritstone walls, still sometimes blackened by soot from the industrial revolution, are typical enclosures, separating rough grazing from more intensively used fields in the broader valley bottoms. At the upper edges, enclosures are large and often irregular, perhaps reflecting gradual intake of land into more intensive agricultural management over time. Hedges are rare and most seem fairly recent in origin. There are more trees and a few hedges in the lower valleys, such as the Edale and Middle Derwent Valleys and on the edges of the Dark Peak. Fields in these valleys and around farmsteads tend to be smaller and often irregular in shape compared with those in the White Peak. Many have been agriculturally improved with little wildlife interest remaining.

The Dark Peak is critically important for nature conservation. A very high 45 per cent is designated for its importance in a national or European context (see Appendix 2 and Figure 3) – especially for its blanket bogs and dwarf-shrub heaths, its bird assemblages and its geology. Many plants and animals are special – the breeding wading birds, some birds of prey, upland oakwood birds, Mountain Hare (Fig. 87, the only area outside Scotland where it is found), plant species at the edges of their range, some of the grassland fungi and a range of invertebrates.

As you might expect, land-use in the Dark Peak is dominated by agriculture and grouse moor management. Farming is predominantly stock-based, mostly sheep, but often with small numbers of cattle. There are negligible amounts of arable cropping, but this would have been different in the past as evidenced by old ridge and furrow and traces of medieval strip fields in enclosure patterns around some valley settlements. Management of the uplands for grouse shooting is focused on heather moorland where small-scale burning or cutting provides young shoots on which Red Grouse depend for food, with taller plants adjacent

FIG 7. Heather moor in the Goyt Valley showing patchwork quilt effect of regular cutting to right, contrasting with acid grassland to left and moorland not managed for grouse in the foreground. Plantations in the valley below.

for cover and nesting. The division of the heather moorlands into mostly small, irregular blocks of different-aged heather provides a very distinctive, patchwork quilted-landscape (Fig. 7).

The remote, wilderness appeal of the upland moorland is accentuated by the general lack of buildings and man-made structures. This openness and tranquillity is crucial for people, enabling them to leave their busy lives behind and contributing to their mental wellbeing. The lack of lighting in this expansiveness emphasises its seclusion and supports the Peak District's importance for dark night skies, where you can see constellations of stars or shooting stars twinkle and flash as they shower the landscape (on a clear night of course).

Farmsteads reflect the development of agriculture over centuries and are a characteristic feature of the Dark Peak (Lake & Edwards, 2017). There is a high density of dispersed settlements in the larger valleys, most lying on medieval to 17th-century sites. Upland farms once supported more cattle than currently, so barns for housing these and working horses were essential, along with hay and fodder storage, usually above the stalls. The farmstead pattern is often one of linear buildings, the farmhouse at one end and barns occupying the other (Fig. 9), or in clusters on two sides of a central yard, although more modern buildings now confuse the layout. Nevertheless, there is a high preservation of the farmsteads' historic style. Construction is in local gritstone, with lovely large gritstone slates on the older rooves. All are often still streaked with soot that epitomises the industrial past.

Villages are scarce in the Dark Peak. They are mostly nucleated and situated in the lower, more expansive valleys like Edale, Hathersage, Bamford and Grindleford. The main population is confined to the edges as in Chapel-en-le-Frith, Glossop and New Mills to the west and Marsden in the north. Traditional buildings also exhibit local gritstone construction and late-16th and 17th-century gentry's houses are a notable feature. These local building materials and styles give the Dark Peak a sense of place and a high degree of visual unity between towns, villages and farmsteads. Altogether there are 1,239 Listed Buildings here reflecting the strong local character and its long term conservation (Natural England, 2012).

There is rich evidence of a long history of human activity in the area since prehistoric times. There are Mesolithic remains and evidence for activity underneath the deep peat. Investigations on the Eastern Moors (Bamford to Beeley), particularly on the gritstone shelves overlooking the Derwent Valley, have revealed extensive Bronze Age field systems and cairnfields with some pot burials of cremated remains (Barnatt & Smith, 2004). Iron Age relics are well illustrated by the Iron Age fort crowning Mam Tor (Fig. 49), with its surviving ditches and ramparts and magnificent views south to the White Peak across Castleton and north onto Kinder Scout. This is just one of 127 Scheduled Monuments in the Dark Peak.

The continuity of human use is evidenced from peat cutting, Bracken harvesting, quarrying and the former hunting forests. Gritstone for walls and buildings was excavated, but the most iconic product were the millstones and grindstones. Adopted as the Peak District National Park emblem welcoming you along its major roads, these were fashioned in quarries mostly in the east above Hathersage and Baslow. Crafted since at least the 13th century, their heyday was in the 16th and 17th centuries. The industry collapsed suddenly in the 18th century as continental stones of superior quality became available (Barnatt & Smith, 2004). Many unfinished stones still litter the former quarries and extraction areas, visible to today's explorers (Fig. 64). At the height of the lead mining industry, wood was made into charcoal for smelting and many burning platforms abound in the Upper Derwent Valley especially. The impact of the industrial revolution was felt more in the towns on the western fringes of the area, but stretched too along the Middle Derwent Valley down to Matlock in the southeast.

The linkages in this landscape are varied and derived from how people needed to move around. The old packhorse routes are often still visible, such as at Edale Cross which marks one passing south of the Kinder Scout massive, and another possibly following the Long Causeway, probably of medieval origin, between Stanage Edge and Buck Stone. The Longdendale cross-park main road follows in part the ancient saltway to Penistone from Cheshire and others cross Big Moor to the east. This was at a time when there were no turnpikes in the Peak – the first appearing in the first half of the 1700s. Visitors to the area needed guides to find their way around. Now furnished with the excellent quality of the OS maps, GPS and Satnavs we can more easily negotiate the plethora of public footpaths and areas of open access, providing multiple opportunities and choices for access to the Dark Peak, not to mention the Pennine Way (starting in Edale) and the Trans Pennine Trail.

THE SOUTH WEST PEAK

Although composed of the same suite of rocks as the Dark Peak, the South West Peak boasts a lower, more intimate landscape, encompassing smaller expanses of peat-based blanket bog and heathland in a fragmented mosaic with more woodland and wet rushy fields compared with the Dark Peak (Fig. 137). Still upland in flavour, the scenery is often striking and distinctly diverse. There are strong north-south ridges, continuing the theme from the Dark Peak. Shining Tor above the Goyt Valley at 559 m is the highest point along the northern-most ridge between Cats Tor and Whetstone Ridge. Axe Edge to Morridge between Quarnford and Onecote (pronounced 'Oncot') forms the backbone of the central/southern area.

The views out of the Peak District from these higher points provide southern connections to the lower land in the Potteries and Churnet Valley and west towards Cheshire. In contrast, easterly views reveal the massive limestone hills along the lower Manifold Valley, with Ecton and Wetton Hills towering over the river hidden in the deep dale. At the northern end of the South West Peak, there are more distant views of the Kinder Scout plateau, Rushup Edge and the eastern moors. The ridge and valley landscape gives it a distinctive pattern and character. There are isolated gritstone ridges and tors such as Ramshaw Rocks, The Roaches and Windgather Rocks, which provide dramatic contrasts with their nearby green, walled fields. Combs Moss is different in being a large saucer-shape depression with rocky edges towering above Buxton.

Like the Dark Peak, the watersheds in the South West Peak determine to which sea any raindrop passes (Frontispiece map). The Goyt and Dane Rivers accentuate the north–south valleys between the ridges, the latter flowing south through Three Shires Head, where Cheshire, Staffordshire and Derbyshire meet marked by an 18th-century pack horse bridge listed Grade II by Historic England (Fig. 66). In contrast, the Rivers Manifold and Dove originate off Axe Edge and Morridge and flow off to the east and then south where they transform into iconic limestone scenery rivers in the White Peak. The River Dove runs along the limestone/shales boundary, defining the eastern edge of the South West Peak along some of its upper course, as well as the Staffordshire/Derbyshire county boundaries for much of its length. As in the Dark Peak, some of these rivers are dammed to form reservoirs for drinking water in their lower reaches. The Goyt Valley and Tittesworth Reservoirs are amongst the largest, but others occur in several valleys. Combs and Toddbrook reservoirs near Whaley Bridge feed the local canals.

Cottongrasses, mixed more with Heather and other dwarf shrubs than in the Dark Peak, typify the blanket bog core at the higher elevations. On the drier slopes heathlands provide drifts of pink colour humming with bees in summer. As in the Dark Peak, the wide range of flushes, springs and marshes are a feature of the cloughs, often with species that occur less in the Dark Peak, reflecting the lower altitudes and different conditions.

Breeding Curlew is one of the South West Peak's flagship birds (Fig. 118). Other specialities include Snipe, more often in the rushy fields, and Short-eared Owls that swoop silently over the higher moorlands (Fig. 120). There are a number of plants, such as Cloudberry, and invertebrates including some northern ground beetles that reach their southern-most locations in the Peak District here, as we reach the uplands' southern edge.

Below the moorlands, there is a greater preponderance of pastoral, enclosed fields than in the Dark Peak (Fig. 8), more of them agriculturally improved in the Manifold and upper Dove Valleys, plus many rushy pastures and hay meadows, some of which still boast a plethora of colourful flowers and associated invertebrates. Some fields are critical for fungal groups like the colourful waxcaps. Gritstone drystone walls dominate as in the Dark Peak, but there are more trees, many mature or older, associated with this landscape, giving it a more wooded feel. The increased incidence of hedges at the lower elevations adds to this, which is accentuated by an abundance of semi-natural woodland cover in the broader

FIG 8. More intimate and varied landscape of the South West Peak, looking north from Back Forest across fields, woodland and heather moorland.

valleys, particularly along the Dane and Shell Brook valleys in the western half of the landscape area (Fig. 8). These woods tend to be richer in species than those in the Dark Peak, with more Bluebells for example. This is where to find important upland woodland birds like Pied Flycatchers, Redstarts, Wood Warbler and Tree Pipits (Chapter 10). Some woodlands are clothed in ferns and drool with different mosses. There are also large conifer plantations near Macclesfield and in the Goyt Valley (Fig. 7), mostly associated again with reservoirs.

As in the Dark Peak, the most important habitats and species are protected by nationally or European-level nature conservation designations which cover some 13 per cent of the South West Peak (Fig. 3). These focus principally on the blanket bogs, heathland, rush pastures and hay meadows important for their breeding birds and habitats. These areas are supplemented by a wide range of local wildlife sites.

There is some archaeological evidence for widespread occupation from prehistoric times, for example from the Neolithic settlement found at Lismore Fields, Buxton. There are 57 scheduled monuments of national importance in the South West Peak, including Bronze Age barrows and an Iron Age promontory fort on Castle Naze, above Chapel-en-le-Frith. Add to these many more features including medieval packhorse routes, field systems and settlements, post-medieval turnpike roads, gritstone quarries, coal mines, lime kilns and disused mills, and it is clear that the influence of past human activities is extensively reflected in the landscape. Much of the land would have been used for upland grazing, with more cultivation over time in the lower areas. There are some enclosures which could be Roman and others that date from the medieval period. Some open field farming still survives near Warslow and Butterton and old ridge and furrow has persisted in the Manifold catchment at lower elevations. There were parts of three medieval hunting forests, Macclesfield Forest, Malbanc Frith in Staffordshire and the Royal Forest of the High Peak in Derbyshire that would have limited settlement until later medieval times.

As in the Dark Peak, a feeling of remoteness is engendered by the scarcity of buildings and manmade structures in the upper moors, which contrasts with a high density of dispersed farmsteads associated with inbye land below this. These are mostly medieval to 17th-century relating to irregular fields enclosed from woodland or on a piecemeal basis over the same period. The farmsteads are similar to those in the Dark Peak (Fig. 9), typically small-scale, reflecting the importance of housing cattle and storing hay. Buildings were arranged around or on one side of a yard used for stock management and where manure was stored for spreading on the fields. Field barns are also prevalent and a feature of the uplands in general (Lake & Edwards, 2017).

FIG 9. A typical linear moorland farmstead, built of gritstone with gritstone slate rooves, in the South West Peak.

Nucleated villages are rare in the South West Peak, with Longnor (Fig. 53) being the largest, perched on the south-facing side of a ridge along a spring line. Warslow to the south is the second largest, but others are smaller – Sheen, Upper Elkestone, Wincle, Flash, Hollinsclough and Onecote for example. Many of these are located above flood plains along spring lines. The grandest building is Lyme Park, Disley, a National Trust property with a registered historic park and garden now also famous for its portrayal as Pemberley in the 1990s BBC's filming of Jane Austen's *Pride and Prejudice*.

Older buildings are generally constructed of local gritstone, but there are subtle differences between those in the South West and Dark Peaks. Some are richer in iron giving a rusty colour, whilst greater use of the local pink-toned grits for quoins, walls or whole buildings (Fig. 10) give a more local vernacular character. More rooves in this area are Staffordshire blue slates, although thick gritstone slates are still evident, possibly replacing thatch of an earlier period. More red brick appears towards the southwest of the area, influenced by the proximity to brick works. Altogether, there are 463 listed buildings and structures in the South West Peak, including barns, mileposts, crosses and bridges (Natural England, 2013).

Mineral extraction was much more important in the South West than the Dark Peak. Coal mining was first recorded in 1401 near Goldsitch Moss south of Flash, with the main seams being exploited on Goyt's Moss and around Axe Edge, southwest of Buxton. The visible remains are everywhere in this landscape if you know where to look – waste heaps, shafts and bell pits (depressions often now full of water), especially from Flash (claimed as the highest village in England) to the Goyt Valley. Lime-burning was fuelled by much of this coal and was widespread

FIG 10. The pink tones of Chatsworth Grit in a derelict farmhouse's stonework near the Roaches, South West Peak.

in the South West Peak where access to limestone was easier. Small quarries probably mostly provided for walls and building stone. Few are extensive and most are now re-vegetated, although some provide valuable geological cross sections. These extractive industries, combined with farming, engendered a higher density of habitation than in the Dark Peak.

A number of packhorse routes developed in this area in order to move these extracted materials as well as products from the lowlands across the Peak District. The main routes can still be traced in places through hollow-ways – usually multiple, roughly parallel, narrow sunken lanes worn down through time where packhorse leaders chose the driest route on the day. Old routes include the drovers 'Great Road' from Congleton to Nottingham, which passed north of Leek across Gun Moor, over Revidge and down to Hulme End; the Manchester to Derby Roman road; more saltways from Cheshire to Chesterfield and Sheffield; and, later, the silk route from Macclesfield to Nottingham. Four packhorse routes meet at Three Shires Head (Fig. 66).

The South West Peak especially is the place for stories, myths and legends – part of the intangible cultural heritage of the area and indelibly linked with the landscape. There are mermaids in pools, white knights on horseback, bottomless pits, peculiar happenings and rituals. The rift in the gritstone that forms Lud's Church (Back Forest) is a special place with a unique ambience linked to *Sir Gawain and the Green Knight* and the Lollards that sheltered there, as we will see in Chapter 4.

The South West Peak tends to be the hidden gem in the Peak District. Although there is a high density of footpaths and tracks and open access on the higher hills, visitor numbers are generally lower. There are exceptions of course, with the Roaches being one of the most popular destinations at its southern edge; a mecca for walkers and, especially, climbers.

THE WHITE PEAK

Sitting snugly within the horseshoe of the gritstones and shales of the South West and Dark Peak landscapes is the White Peak; so named for its limestones. These form an elevated plateau, gently stepped from north to south, dissected by dramatic, deep dales and supporting occasional prominent steep hills, knolls and cliff faces. The highest areas are in the north, just south and west of Castleton, with hills and knolls regularly exceeding 400 m. The dales are varied in shape and character, tending to drain east or south. They sometimes start as a gentle broad valley, before often narrowing, sometimes becoming gorge-like as in the

FIG 11. Lathkill Dale showing typical deep valley, steep slopes and limestone rock outcrops with dry floor.

Wye Valley and Dovedale. There can be towering cliffs and tors, whilst the dale slopes are often very steep and punctuated by rock outcrops and screes (Fig. 11). The lower ends of the dales may be wider, with a flatter alluvial bottom as in the lower stretches of the River Wye upstream of Bakewell.

As in other limestone areas, surface water can be permanent, seasonal, or missing altogether. The only rivers to flow permanently throughout the dales are the Wye and Dove, both of which emanate from the moorlands. Other dales have short water courses, some of which extend in wet seasons as water tables rises. The Manifold disappears in summer once it enters the White Peak. Small streams appear from caves and springs (Lathkill Dale) or just seepages (Monk's Dale), but disappear in dry conditions. Some dales are not usually wet at all, like Woo Dale and Long Dale (north of Hartington – there are two Long Dales). This general lack of water in the White Peak is replaced by thousands of dew ponds created for stock, but for the local people, it is the well dressings in the villages that were once a plea to keep the wells flowing in the dry season.

The limestone area is famous for its caves – the underground caverns are a mecca for the caving fraternity, but there are also show caves full of stalagmites, stalactites and special rock features. Besides these, there are many hillside rock openings that usually require some scrambling to access, many of which have yielded ancient bones or other evidence of the region's past.

The Dales are justifiably also renowned for their richly diverse limestone grasslands (Fig. 12). This is where to search out our rarest plants, but also simply to immerse yourself in the splendour of colours and scents on a fine day, to admire the plethora of bugs, butterflies and bees buzzing around the flowers and absorb the accompanying wildlife orchestra. Limestone woodland is another dales feature, mostly Ash-dominated, but the canopy protects splendid ground floras, the spring flowers being the most rewarding. The richness of these woodlands contrasts with those in the South West and Dark Peaks. There is little woodland on the limestone plateau, but scattered small plantations and shelterbelts prevail. The total woodland cover is small – only about 6 per cent of the area, giving a much less wooded feel compared with the South West Peak. You can still revel in some flower-rich hayfields and pastures, more so here than in the other landscape regions. There are also fragments of limestone heathland, once much more widespread, and the remains of the lead mining industry that supports very special species.

FIG 12. Flower-filled limestone grassland with Harebells, Wild Thyme, Eyebright, Bird's-foot-trefoil and many other flowers providing colour and scent on Priestcliffe Lees, above Miller's Dale.

Protected habitats in the White Peak are much more limited than in the other two landscape areas, owing to them being concentrated in the dales (Fig. 3). Only 6 per cent is under national or European designations, although these include the best of the grasslands, scrub, woodlands, limestone heaths and streams. White-clawed Crayfish (now very rare in the area owing to disease), Lamprey, Dipper and Water Vole are wetland animals of high importance, whilst woodland birds are abundant and Peregrine Falcon and Raven breed in the main dales or old quarries as in Miller's Dale. Several sites are also valued for their geology and underground caverns where bats can also be a feature.

From an agricultural perspective, the majority of the plateau is occupied by improved pasture and silage fields, there being more, better quality agricultural land here than elsewhere, which has been exploited by more intensive farming practices. The main enterprises, which tend to average over 100 ha in size, are dairy and beef cattle, sheep and pigs, with more sheep than anything else (Fig. 13). Dairy farming was more widespread but has been suffering owing to cost pressures in the 1990s. There is much less dairy farming in the South West Peak and little in the Dark Peak in comparison.

FIG 13. Limestone walls around fields of agriculturally improved grassland supporting sheep farming adjacent to the upper end of Hay Dale with more limestone outcrops and steep slopes.

As in the other landscape areas, the White Peak has a rich archaeological heritage, especially of Neolithic remains, as in the henge and stone circle at Arbor Low (Fig. 47), and the Bronze Age combined with the later Saxon burial place at Wigber Low, south of Bradbourne. There are some very well preserved historic landscapes such as the strip field patterns near Chelmorton (Fig. 56) and areas of ridge and furrow like those near Ilam and south and west of Bakewell. The Romans established a settlement in Buxton as well as a fort at Brough on the northern edge of this landscape (Barnatt & Smith, 2004).

The mineral riches of the limestone have been exploited for generations. The most important, on an international scale, has been lead mining, but copper, zinc, baryte and fluorspar have all played their part in the region's prosperity. The effect on the landscape of mineral extraction has been enormous, with humps and bumps or long rakes: a special feature of mineral workings. Associated with this are waste tips and heaps, buildings and infrastructure and smelting facilities. Limestone was extensively extracted for agricultural use, much of it being burnt in kilns at different scales. Again, waste heaps and the remains of old kilns are everywhere when you know where to look. The White Peak is famous for its black marble and Blue John stone. The first was excavated near Bakewell and made into beautiful decorative pieces – ornaments, inlaid table tops and other items (Fig. 62). Small scale Blue John Stone working for jewellery and other ornamental pieces is still practiced (Fig. 61). Unique in the world, it justifiably draws crowds to see the rock *in situ* or the finished products in Castleton and nearby.

The Peak District limestone is of particularly high quality and has been extensively quarried, leaving both old abandoned quarries, some of which are now nature reserves (as in Miller's Dale), and many large-scale modern quarries that are striking features in the landscape, particularly around Buxton, at Stoney Middleton, in the Hope Valley on the northern limestone edge, in the south between Balidon and Cromford, and southwest at Cauldon in Staffordshire. Gazing into these quarries is awe-inspiring, but you can see why the National Park boundary excluded them. The industrial heritage from quarrying and mining activities in the White Peak plays a major role in the flora and fauna of the waste materials, with several specialities that tolerate heavy metal contamination. Moreover, the waste heaps from lime-burning have prevented agricultural improvements in places, so retaining their richness of plants and animals.

You always know when you pass into the White Peak as the walls change colour (Fig. 13). In contrast to the other Character Areas, the White Peak walls are much paler rather than brown and gritty. Sometimes stuffed with fossils and draped with mosses, these walls are often straight and rectangular indicating late enclosure or demarcating the narrow strip fields of medieval times. This planned enclosure contrasts with gradual intakes over time as in the Dark and South West Peaks (Lake & Edwards, 2017). Many of the village houses and scattered farms and outbuildings are built of limestone, although the more resistant gritstones often appear in the quoins, window and door frames. This building style is also displayed in some of the large old mills (as in the Wye Valley) that are prominent reminders of the importance of water power for past textile manufacturing. This contrasts with darker gritstone buildings in the rest of the Peak District. More grand buildings within the White Peak were still constructed of gritstone. However, Haddon Hall just east of Bakewell (the seat of the Dukes of Rutland) has an interesting juxta-positioning of limestone and gritstone blocks in some of its old walls.

The settlements in the White Peak are more nucleated than elsewhere. Old farmsteads tend to be more attached to villages and less dispersed, although their organisation as linear or L-shaped clusters round a yard is similar. However, more of the White Peak farmsteads originated more recently than elsewhere in the Peak District, with the enclosures of the open fields round the villages or the commons and waste lands. The main towns of the Peak District are in the White Peak – Buxton and Bakewell, with Matlock and Matlock Bath being the largest. Buxton and Matlock Bath were established as spa towns. Buxton attracted early visitors to its waters from St Ann's Well (one of the Wonders of the Peak that visitors were encouraged to visit). There are more villages in the White Peak than in the other landscape areas; many small, quiet, charming and historic, each with a unique character. Tideswell is the largest with a range of shops and other facilities.

Routes through the White Peak are numerous, although the footpath network is not as dense as in the South West Peak. However, there are a number of now disused railway lines, the Limestone Way and the Pennine Bridleway, all of which offer access for quiet recreation. The rock towers and cliffs attract climbers and the rivers are famous for their trout fishing (heralded by Izaak Walton in *The Compleat Angler*, 1653).

THE EXTENT OF PROTECTED LANDSCAPES

This quick glimpse across the diverse landscapes highlights the typical and characteristic, but the bigger picture is important too. We have already seen how the Peak District National Park covers the majority of our area. The two purposes of English and Welsh National Parks, as set out in law, are to conserve and enhance the natural beauty, wildlife and cultural heritage, complemented by the second which is to promote opportunities for understanding and enjoyment of the National Park special qualities. There is also a duty to foster the economic and social well-being of local communities within the Parks. A number of the projects and measures to achieve these aims will be show-cased as we pass through the different chapters, followed by a more critical examination at the end.

There are other protected landscapes apart from the National Park. Nature Conservation has its own procedures for identifying and protecting wildlife and geological features of importance (Appendix 2) and, compared with neighbouring counties, the Peak District is blessed with extensive habitats of national and international value. The best nationally important habitats and geological features are designated Sites of Special Scientific Interest (SSSIs, see Appendix 2), with 49,280 ha (26 per cent) selected as SSSIs in 85 sites across the Peak District, many also important for their earth science features. More are in the Dark Peak. We also have four Special Areas for Conservation (SACs), which represent habitats of European Importance, and one Special Protection Area (SPA) of European importance for birds. These are all also SSSIs (see Fig. 3). This is a comparative wealth of wildlife habitats with the highest level of protection compared with neighbouring lowlands where sites tend to be small, fragmented and disconnected, excepting large coastal estuaries. This emphasises how important the Peak District is for its wildlife.

In addition to this cream of the nature conservation sites, there are some 381 county wildlife sites totalling a further 12,672 ha, which is a significant addition to higher value habitats and species, although their quality will not generally be as good. These sites are less well protected by the planning system. New surveys are always ongoing, new habitats are created or develop naturally, some are also lost, so the area and number of lower value sites, in particular, fluctuates through time.

WHAT NEXT?

This, so far, has been a general overview of the key features of the Peak District landscapes and the contrasts between the Dark, South West and White Peaks. It is time now to delve into some of the detail: to understand better how the rocks were formed; how they were forced by mountain building and erosion into the shapes and contours we see today; and how they have been affected by some of the human activities outlined above. The next chapter goes back in time 360 million years to an equatorial sea that is hard to fit with our image of the Peak District today.

CHAPTER 3

How the Rocks Were Formed

HOW THE PEAK DISTRICT LANDSCAPE ORIGINATED

HAVING GIVEN A FLAVOUR OF the Peak District's appearance, it is clear that its character is largely derived from the rocks that provide its structure. Our story starts in the Carboniferous Period; a geological era that occurred between 300 and 360 million years ago that is divided into three major time periods: the Visean when the oldest rocks, the limestones, were laid down; the Namurian, marking the deposition of the Millstone Grit group and the Bowland Shales Formation (previously labelled Edale Shales); and the Westphalian when the Coal Measures formed. Greater detail is provided by Aitkenhead *et al.* (2002), but if interpreting geological features whilst out in the Peak District interests you, find Paul Gannon's or Trevor Ford's excellent books, The Geographical Association's volumes on the White or Dark Peaks, or the BCRA Cave Studies Series, all detailed in the bibliography.

In our area, only rock exposures from the Carboniferous Period are visible. There are older Silurian, Ordovician or even Cambrian (see Table 1) rocks beneath these, as testified from boreholes, but none are exposed or influence the physical landscape we see today.

THE LIMESTONE FORMATION

Living now in the Peak District, it is hard to imagine it sitting in tropical seas near the equator 360 million years ago. Nothing about Britain's familiar form or pattern would have been identifiable then. Geologists think that the Peak District was part of a continent called Laurasia sitting on a tectonic plate that was being stretched, resulting in a drift across the equator from south to north, thus taking

TABLE 1. The Geological Periods and their main deposits in the Peak District

Era	Period	Epochs	Approx. starting time relevant to Peak District	Principal rocks & deposits in Peak District
Cenozoic	Quaternary	Holocene	Current, starting after last ice advance, 11,700 years ago	Alluvium, head, peat
	Quaternary	Pleistocene	2.58 mya (includes ice ages)	Terraces, gravels, glacial till, silts and clays
	Neogene (formerly part of Tertiary)	Miocene and Pliocene	23.03 mya	Pocket deposits, sands, silts and clays in solution subsidence hollows
	Paleogene (formerly part of Tertiary)	Palaeocene, Eocene and Oligocene	66 mya	
Mesozoic	Cretaceous	Various	135 mya	Some deposited, then eroded away
	Jurassic	Various	192 mya	Some deposited, then eroded away
	Triassic	Various	251 mya	Midland Plain to south, SW edges, eg Churnet & Rudyard Valleys
Paleozoic	Permian	Various	298 mya	None
	Carboniferous	Westphalian, Namurian and Visean	347 mya	Coal measures, Millstone grit & Edale shales, Carboniferous Limestone
	Devonian			none
	Silurian	Various	443 mya	Basement rocks at depth
	Ordovician	Various	510 mya	Basement rocks at depth
	Cambrian/Pre-Cambrian	Various	4,500 to 500 mya	Below base-rocks

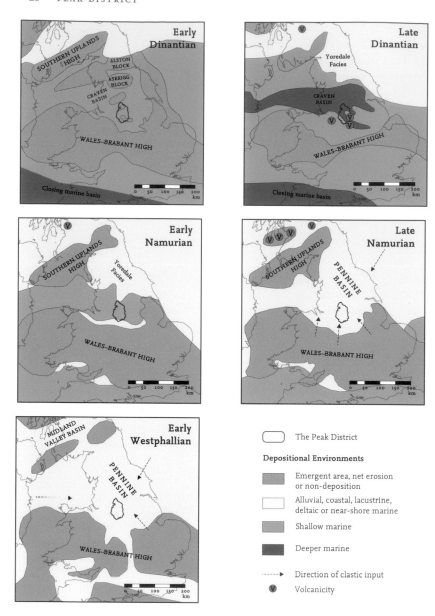

FIG 14. Sketch maps illustrating the generalised conditions of the Peak District in the Carboniferous Period showing the relative locations of emergent land and the marine environment. (Redrawn from Aitkenhead *et al.*, 2002, © British Geological Survey)

the land mass from a tropical dry zone that influenced erosion patterns in the earlier Devonian period to a wetter Carboniferous equatorial zone.

The Peak District was probably land at the end of the Devonian Period. However, the tectonic stresses resulted in major earth upheavals producing a very uneven surface topography of rifts creating fault-bounded high areas known as blocks and intervening basins. As these were actively developing, the land was continually sinking during the early Visean phase, resulting in a tropical sea transgressing our region and much of what is now the Yorkshire Dales, separating the Wales–Brabant High to the south from the Southern Uplands High to the north, visualised in Figure 14. The sea was at first fairly shallow, but deepened later in this period, overlying the very uneven surface giving shallow seas, ramp slopes and deep basins, resulting in local variation in depth, wave action and depositional conditions. This fragmented landmass continued to sink as materials were deposited on top of it for more than 50 million years.

Carbonate deposition was the dominant process with little sediment entering the sea giving rise to a clear, warm environment for a varied and abundant marine life. These calcium carbonate sediments ranged from muds to limestone pebbles to boulder beds deposited in different marine environments from shorelines, to shelves, to slopes and into deep basins. This is how the Pennine limestones were formed. There was a wide range of marine organisms, including those shown in Figure 15, nearly all of which have living relatives, but which were often more abundant and varied in the Carboniferous Period. The abundance and distribution from these fossilised creatures reveal much about changing conditions enabling re-construction of the Carboniferous environment. The seas and plates were not static and layers of sediment accumulated related to variations in the conditions. Shells or calcareous structures of marine creatures would have been discarded after predation or death, or partly digested by other organisms before being ground up by the wave and current action to produce a carbonate rain, termed the 'carbonate factory', which formed thick accumulations of carbonates on the sea bottom. These formed the 'platforms' – the massive building blocks of the limestone rocks. On top of these, reef knolls (mounds of fine-grained lime muds supporting many species in warm, shallow conditions) and apron reefs (at the edge of a drop-off) formed.

Where a warm, shallow sea covered low-lying areas, the sea bed would have been colonised by the Carboniferous equivalent of a coral reef and bottom dwelling creatures like the sea lilies (Crinoids). The sea lilies are one of the most abundant fossils you can find in the Peak District limestone. It is distinctive in form, either in cross or longitudinal section. Sea lily forests established on shallow sea bottoms, each long stalked individual up to 2.5–3.5 m tall, made by

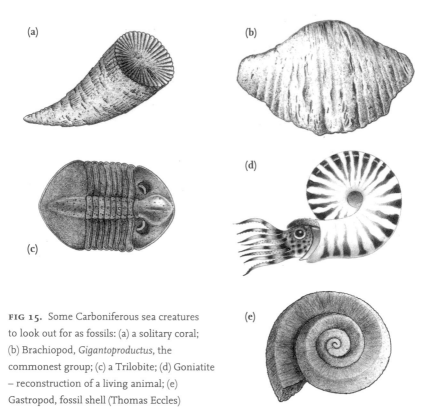

FIG 15. Some Carboniferous sea creatures to look out for as fossils: (a) a solitary coral; (b) Brachiopod, *Gigantoproductus*, the commonest group; (c) a Trilobite; (d) Goniatite – reconstruction of a living animal; (e) Gastropod, fossil shell (Thomas Eccles)

accumulating calcium plates as it grew. Five branching feeding arms, set on top of the stalk, like the fronds of a palm tree, filtered out small prey organisms, passing them to the central mouth. In the decayed state, the calcium rings disarticulated and were scattered on the sea floor. There are many examples of crinoid fossils exposed in rock faces, often with visible transverse plates about 0.5 cm across or tubular longitudinal sections (Fig. 16). Try scouring drystone walls or buildings to find some, for example Hartington Station buildings on the Tissington Trail, the stone steps to Thor's Cave in the Manifold Valley, or in natural rock outcrops as in Dovedale, too. Buxton Museum and Art Gallery has an interesting collection of Carboniferous fossils as well.

Brachiopod and bivalve shells are also common in many limestone outcrops. Most occur as fragments rather than as whole shells, sometimes in dense clusters as if dumped on the sea floor. Good examples can be seen at the National Stone

TABLE 2. The main Carboniferous warm-sea creatures

Organism	Type
Algae	Seaweeds.
Coral *(a)	Not unlike some modern-day corals in tropical seas.
Sponges	
Bryozoans	Very small, aquatic invertebrate animals filter feeding using a crown of tentacles, known as moss animals, often living colonially
Brachiopods *(b)	Lamp shells, with an upper and lower shell, hinged at the rear end, while the front opens for feeding
Gastropods *(e)	Snails, usually with coiled shells
Bivalves	A class of molluscs with compressed bodies enclosed in a two-hinged shell
Trilobites *(c)	Extinct marine arthropod (invertebrates with exoskeleton, segmented body and paired jointed legs), something like woodlice
Goniatites *(d)	Cephalopods with an external shell divided into chambers internally and gas-filled to give buoyancy, similar to Ammonites and current day Nautilus
Ostracods	Small, shrimp-like Crustacea, flattened side to side and protected by a calcareous shell-like valve.
Echinoderms, especially crinoids	Marine group with a five-point radial symmetry like starfish. Crinoid means 'lily', adults usually have a stem with a group of feeding arms attached.

* organism illustrated opposite

FIG 16. Crinoid fossils from 'Once a Week' Quarry, Sheldon (where workers were paid weekly rather than the normal fortnightly). Sea lilies visible as transverse plates and tubular longitudinal sections. Similar polished fossilised stone has been used at Chatsworth House and the new Coventry Cathedral. (Thomas Eccles)

Centre in Wirksworth, but look out for them in walls, by footpaths and in rock outcrops as well. A good range of fossils can be seen in stones during the walk from Monsal Head to the Dale bottom below some of the limestone cliffs. These Monsal Dale Limestones hold corals, brachiopods and crinoid stems. Best to leave any you find for the next visitor to enjoy as many lie in protected SSSIs. Where the limestone is purest, calcite crystals can form, which are six-sided polyhedrons with diamond-shaped faces, white (unless coloured with impurities), partly translucent and easily scratched. These crystals are very common and are visible in limestone screes and rock outcrops.

The detailed nature of the limestone rocks relates to their location in the shallow/slope or deep seas and to the abundance of reefs. Looking at different outcrops in the White Peak will reveal their variable nature with different colours and surface shapes. This is because the rock's character is linked directly to the environment in which it was deposited and to its associated animal life. There are thickly bedded, grey massive limestones such as in the Monsal Dale limestones, the Woo Dale limestone or the Bee Low limestones. From the ramp locations on the edges of the deep-sea basins, there are grey, thinly bedded limestones, often with chert nodules forming knoll reefs (or mud mounds) for example identified in the Eyam, Hopedale and Milldale limestones. Ramp limestones are more like conglomerates (mixtures of materials from different locations) often showing soft-sediment structures and chert nodules such as in the Ecton limestones, whilst basin limestones are darker, thinly bedded, cherty, fine-grained limestones in the Widmerpool Formation in the North Staffordshire Basin. Cherts are a silica-rich, hard, flinty rock that look like large variably sized and shaped pebbles embedded in the limestone. They were formed from silica-like sponges and can be seen in some railway cuttings (as west of Wirksworth), quarry faces and in some of the caverns.

Limestone exposures are generally arranged in layers, which can be clearly seen as bedding planes (the horizontal or slightly dipping lines between the layers akin to the cement between stone layers in walls) (Fig. 17). Where conditions remained stable for long periods, bedding planes are absent and the limestones are referred to as massive. The purity of some of these has resulted in their high demand in chemical and other industries and accounts for many of the major quarries in the Buxton region and at Longcliffe near Brassington in particular. Where conditions alter or internal pressures set up tensions, the beds vary from very thin to thicker, divided by bedding planes. In these exposures, vertical lines or joints are also clearly visible, which represent drying cracks as water was squeezed out as rock was formed from loose sediment. The exposures at the northern end of High Edge, south of Buxton, are a good example.

FIG 17. Bedding planes and vertical joints in limestone outcrop, Tideswell Dale. (Thomas Eccles)

Ancient reefs are another feature that are exposed in the White Peak where they form some of the more dramatic and prominent hills. For example, the 30 m high reef knoll forming High Tor dominates the River Derwent gorge near Matlock; while Gannon (2010) lists the Dovedale and Manifold valleys, including Wetton Hill, Narrowdale Hill, Thorpe Cloud and Bunster Hill as reef knolls that formed on the slope down into the Widmerpool Gulf. There is a very good Hamps and Manifold Geotrail that explains many of these limestone features from accessible routes in these valleys. Lathkill Dale, in the central limestone area, is also cut through a reef knoll in its narrowed section near Ricklow Quarry and the high, vertical limestone rock face in Alport Dale east of Lathkill Dale is a minor reef knoll (not to be confused with the moorland Alport Dale).

Apron reefs which developed around the rims of the Derbyshire Platform provide the dramatic scenery around Castleton and Earl Sterndale. Snels Low (opposite Eldon Hill) is one of a series of hills that are apron reefs west of the Winnats. Their shape mimics but is reduced compared with the shelly reefs of upper Dovedale, which form the spectacular crags of Chrome and Parkhouse

FIG 18. Chrome (left-hand side) and Parkhouse Hill, in Upper Dovedale formed from shelly reefs. They are Geological SSSIs.

Hills (the only significant peaks in the Peak District, Fig. 18). Fossils are often particularly abundant and diverse in these exposed reefs, set in a limestone cement matrix. Reefs do not generally exhibit bedding as in the limestone beds.

Although the picture drawn so far is of a lengthy, quiet period of sedimentation in which the limestones formed, volcanic activity disrupted this occasionally. The tensions between the tectonic plate edges south of the Peak District spilled over into the accumulating limestones as vents, lavas and tuffs most of which are basalts, locally called Toadstones. Interesting exposures of Toadstone occur in Tideswell Dale's quarry and the grassy slopes just to the north. Some of these erupted on the surface, sometimes when the seas had receded and dry land was temporarily present. In other situations, clouds of volcanic ash erupted and thence were deposited on the limestone floor or on islands within the reefs. These volcanic ash beds of tuff are called clay-wayboards, and occur sporadically across the limestone sequence. A good example where molten lava flows onto the earth's surface is on the Monsal Trail above Litton Mill, where an ancient lava flow is visible. This cooled and shattered as it

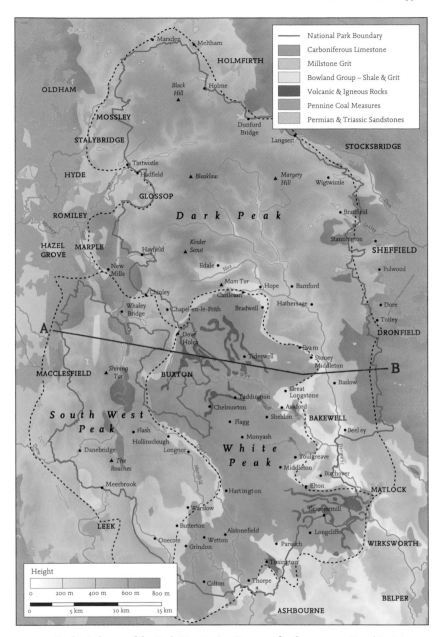

FIG 19. Geological map of the Peak District. See Figure 21 for the cross-section. (Contains British Geological Survey materials © UKRI 2020)

reached the sea, with the resultant explosions included giving rounded solid shapes called pillow lava, before being covered with more limestone deposits. An interpretation board explains the cutting and what you are seeing. The site is of national importance and protected by being part of the Wye Valley SSSI. Tuffs are produced by pyroclastic eruptions of volcanic ash: that is fragments of hot rock thrown out in volcanic explosions. Vents are where molten lava remains in the volcano's pipes without reaching the surface. Volcanic materials in our limestones are concentrated around two areas: the northern region east of Buxton north nearly to Castleton and south towards Monyash; and the southeast ore field from Wirksworth to Matlock, with a small outlier near Tissington (Fig. 19).

THE GRITSTONE FORMATION

Anyone who climbs in the Peak District – as on Stanage or the Roaches, will be familiar with the gritty texture of our Millstone Grits. Gritstone is merely a sandstone where the grains are particularly large. The same climber will also recognise that many of the rock faces vary in layers composed of finer or sandier sediments. These are the finer-grained sandstones and finer still mudstones and siltstones within this series.

There was a major change from the calm, clear tropical seas of the Visean as we entered the Namurian at around 326 million years ago, which lasted some 13 million years. Namurian is named after the Belgian city of Namur where similar rock strata occur. Continual subsidence combined with southwards movement of a major deltaic system from the north resulted in the deposition of silts, muds and sands over the top of the carbonate accumulations. A very large river system (something analogous to the Mississippi in scale) was draining mountains that would probably have been in an area that includes present day Scandinavia and Greenland. These were tectonically active mountains, being uplifted and eroding rapidly, generating huge quantities of sediment into a vast delta system, which repeatedly retreated and advanced as sea levels fluctuated. Sea ingressions gave rise to regular incursions of marine conditions along the delta and incoming river valleys, along with their marine organisms. The resulting black muds slowly accumulated 0.5 m to 3.0 m deep along with the remains of marine organisms, most particularly goniatites: a genus of now extinct cephalopods – ammonite-like, shelled-animals related to squid (Fig. 15). Sea levels fluctuated over several cycles, with marine deposits, then estuarine and deltaic again, before repeating the sequence. Depending on the location of the rivers in the delta (which moved

FIG 20. Cross-bedding in Roaches Grit on the edge of the Roaches showing the pattern of an ever-changing river channel.

over time as sediment blocked exits or storm surges eroded new channels), the sequence of muds and sandy or gritty sediments formed multiple layers, varying horizontally depending on the location of the delta.

Since rivers entering a delta drop their larger material first as water flows slow and their finer sediments last, often well out to sea, then the ebb and flow of sea level and delta in one spot will result in layers of marine muds, alternating with silts, muds, larger grained sediments and the largest gritty material and pebbles laid down in freshwater. The layer's thickness was controlled by the length of time conditions remained stable and the quantity of material brought downstream. Tides and strong currents moved and re-arranged some of the sediments producing distinctive layers of deposition visible in many outcrops today. The distinctive cross-bedding on the Roaches Grit (Fig. 20) shows the pattern of an ever-shifting river channel, possibly controlled by floods. In contrast to the limestone bedding planes and joints described already, cross-bedding runs in a series of curved lines, often truncated at the top by another series at a different angle. These are visible in many places as on Kinder Scout and Millstone, Froggatt and Stanage Edges.

When the delta dominated and the sea receded, tropical, fast-growing vegetation established on the sand banks and mud flats composed, for example, of *Lepidodendron* and *Sigillaria*, similar to modern-day clubmosses but growing tree-sized, unlike these now small herbaceous plants. These forests lasted long enough to build up deposits of organic matter and sometimes peat where conditions were waterlogged. When the marine transgression re-advanced, these forests and soils were covered with more sediment and degenerated into organic layers at the base of the marine muds. After compression over time, these now represent thin coal facies in the stratigraphical sequence, particularly towards the top of the column. So far geologists have recognised 46 marine bands, with three new ones recently found, each distinguished by a different goniatite species prevalent at the time, allowing a dated sequence to be constructed. Since there are more layers of these gritstones in the northwest end of the delta in the North and West Pennines and Bowland than in the Peak District, this dating helps identify the same layers in different locations.

Fossil identification in these marine bands is possible in several Peak District geological SSSIs, for example in Blake Brook, where the stream sections exhibit complementary exposures of a related series of shales and sandstones representing the middle and upper Namurian Series. The rocks contain many marine goniatite fossils, which together with other abundant micro-fossils, permit dating of the rocks and correlation of sequences. Another equally important exposure near Orchard Farm (adjacent to a public footpath) shows a sequence of shales on a stream bank outcrop above the Ringinglow Coal, which includes marine horizons with different goniatite faunas. The rock sequences at both these sites, which lie within the Leek Moors SSSI, have been proposed as geological reference points and are therefore internationally important for studying Carboniferous rocks.

The oldest layer of the Namurian series is the Bowland Shale Formation, representing the first inundation of fine muds in advance of the main delta which were dropped all round the Peak District, filling the deep basins around the Derbyshire Dome, eventually accumulating into the shales and finer sandstone layers of the Bowland Shales that are 260 m or more deep. There is some evidence that the materials in the sediment in the south of the region originated from the south and southwest, presumably off the Wales–Brabant High rather than the landmass to the north (Fig. 14). The upper layers of the Bowland Shales are characterised by tough, pale-grey siltstones and sandstones interbedded with dark, fissile mudstones. Different fossils define different layers, suggesting varying salinity, with abundant iron bands and nodules in some. Groups of these rock layers in the North Staffordshire Basin now outcrop as the tops of hills along some of the north–south ridges described in the last chapter for the South West Peak along Bosley Minn and petering out on Gun Hill to the south.

These Bowland Shales fringe the limestone core of the Peak District; they form the bottom of the Edale Valley between Kinder and the Losehill Ridge and the Castleton to Hope Valley south of this, they lap the bottom of Chrome and Parkhouse Hills, underlie Carsington and its reservoir on the southern border of the District and fringe the limestone around its west and eastern margins. Most famously, the Bowland Shales are exposed at the foot of Mam Tor, near Castleton, over which the Mam Tor beds lie. This is the largest and best example of a landslip in the country – more of which later – and shows the nearly horizontal, mostly shallow layers of shales and sandstones.

Above the Bowland Shales are the Mam Tor Beds, followed by bands of Shale Grits over which a number of distinct massive Millstone Grits lie separated by shale layers. The general sequence is shown in Table 3, and their interrelationships are clearer in the geological cross-section across the Peak District (Fig.21, overleaf), but not all layers occur everywhere. It is difficult to imagine, but the Kinderscout Grits were being laid down further north in Yorkshire at the same time as the Mam Tor Beds were forming in the Peak District – all related to where the different parts of the delta lay at any time. The particular layers of different beds can therefore be quite localised. You can see Mam Tor Beds at their type locality exposed on the Mam Tor landslip. They consist of alternating bands of sandstones, siltstones and shales, at least 150 m thick. They are also present in Edale, Alport, Ashop, Derwent, Perryfoot and Hope Valley, where they vary from 60 m thick in their northern and eastern

TABLE 3 A simplified sequence of the Namurian Formation

Youngest	Rough Rock
	Huddersfield White Rock and Chatsworth Grit
	Roaches Grit (in the west), Ashover Grit (in the east)
	Upper and Lower Kinderscout Grits
	Shale Grits
	Mam Tor beds
Oldest	Bowland shales (including Edale Shales)

Note: Owing to the ever-moving deltas, not all strata are evident in any location. Shale-grit layers also separate the massive gritstones.

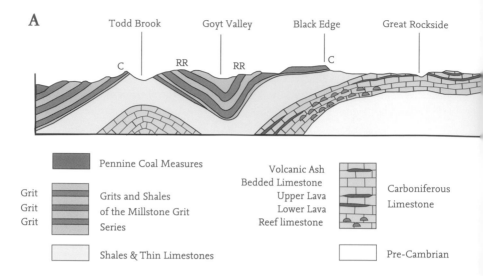

A

Todd Brook   Goyt Valley   Black Edge   Great Rockside

C   RR   RR   C

Pennine Coal Measures

Volcanic Ash
Bedded Limestone

Grit
Grit
Grit

Grits and Shales
of the Millstone Grit
Series

Upper Lava
Lower Lava
Reef limestone

Carboniferous
Limestone

Shales & Thin Limestones

Pre-Cambrian

exposures (north side of the Edale Valley for example), to their maximum at
Mam Tor, but they peter out beyond these areas to the southeast and southwest
(Stevenson & Gaunt, 1971). They were deposited in a cyclical sequence varying
from 1 to 3 m in thickness from shale and mudstone through laminated stone to
massive sandstone, often showing cross-bedding. These sediments were thought
to have been deposited in a deep marine environment at the distal ends of the
slopes that fronted the large delta lying to the north.

The Shale Grits lie on top of the Mam Tor Beds and comprise thick
sequences of sandstones, often massive – although not as massive as the
upper grit layers above them, with shale bands. They form the broad spread
of moorland flanking Kinder Scout, with strong escarpments but a less craggy
topography than the higher Kinderscout Grits, and extend into the upper
Derwent Valley, Hathersage and Offerton Moor. The thickness varies from
230 m around Hayfield to around 100 m near Hathersage. Within this depth, the
sandstones tend to be 16 to 20 m thick alternating with thinner shale beds up
to 3 m. The Shale Grits tend to disappear to the south and thin out in a southeasterly direction as the distance from their deltaic source of material increased,
leaving a single sandstone layer on which Eyam lies, but disappearing by the time
Curbar is reached to the south. These Shale Grits are interpreted as being laid
down within highly turbid currents, the upper layers in particular being closer to
the massive delta as it advanced more into our area.

Between the Shale Grits and Kinderscout Grit are shale layers exposed now

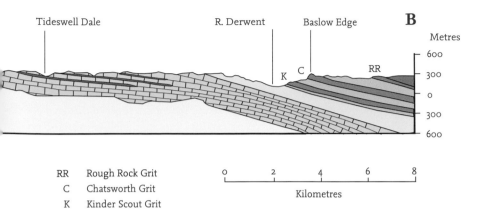

Tideswell Dale   R. Derwent   Baslow Edge   **B**

Metres

600

300

0

300

600

K   C   RR

RR   Rough Rock Grit
C   Chatsworth Grit
K   Kinder Scout Grit

0   2   4   6   8

Kilometres

FIG 21. Cross-section of the geology of the Peak District along the line shown on Figure 19.

in places like Grindslow Knoll and Grindsbrook on the southern flank of Kinder Scout above Edale. These layers are at their thickest north of Kinder Scout, but thin rapidly to the southwest before they could reach Dove Holes, and again peter out to the southeast on Abney. A series of shale bands again separates the Shale Grits from the more massive gritstones that form the magnificence of most of our higher hills.

The massive grits overlying the Shale Grit Beds are first the Kinderscout, then Chatsworth Grits, and finally a layer called Rough Rock, which is the youngest (Table 3). In between these are also the Roaches and Ashover Grits, which are believed to have been generated from sediment entering the area from the south rather than the north as for the other layers. This is evidenced by the greater gritty texture in the south as larger sized particles were dropped first in another delta, rather than to the north, as finer material reached further from its source northwards. Roaches Grit forms the whale-back of Axe Edge south of Buxton, the Roaches, Hen Cloud, Ramshaw Rocks, Back Forest and Hangingstone as well as the lower slopes of Shutlingsloe (the top is Chatsworth Grit), all in the South West Peak. The equivalent rock layer in the east, Ashover Grit, forms Stanton Moor and Harthill Moor. Much of the land around Edensor and Baslow and the area around Ashover is underlain and shaped by this rock.

In Blackden Brook, just below Seal Stones, where there are several waterfalls and rock outcrops, there are exposures of an excellent and fairly continuous sequence through these sandstones and shales, including the Shale Grit,

Grindslow Shales and Lower Kinderscout Grit, which provides accessible information to geologists on the various stages of the changes in the delta, especially the great variability in the slope depositional processes. This section is included as a geological feature in the Dark Peak SSSI.

The Kinderscout Grits are typically coarse and pebbly massive sandstones and give rise to many of our most striking topographical features in the District, especially the Kinder Scout plateau. Anyone walking round the edges of this area will be familiar with the many rock outcrops and tors, often moulded into weird and wonderful shapes by erosion (Fig. 1). Crowden Tower and Noe Stool are the most striking above Crowden Brook (the start of the first National Trail, the Pennine Way), with rounded rocks balancing on each other in layers, the largest being of the most resistant material as in the upper stone on Noe Stool. But there are many other equally dramatic features made of Kinderscout Grit including Kinder downfall on the west side of the hill, and others along Derwent Edge such as Wheel Stones and Salt Cellar, or Back Tor to the north. Kinderscout Grits also form much of The Edge on the northern edge of Kinder Scout, Derwent Edge and Bamford Edge and the western edges of the moors including Arnfield Moor, Chew Valley and Dovestones and on north to Standedge. Crowden Great and Little Brooks (tributaries of the Etherow) are flanked by Kinderscout Grits too.

The Chatsworth Grit forms one of the more constant sandstone horizons in the region and covers more ground than the other strata. It does, however, show marked variation in lithology (its general physical character and thickness) and depth. In the southern half of the area, south of a line from Kettleshulme to Moscar (north of Bamford), the Chatsworth Grit is coarse, pebbly, massive and full of feldspar (a group of minerals), often cross-bedded, and frequently with deep red staining due to an increased iron content. This red colouration is conspicuous where used for building stone in the South West Peak, especially in the Gradbach to Wincle area, showing preference for local stone in construction. (Fig. 10). It outcrops in the Goyt Valley, on the lower Roaches hills, Gradbach Hill and on Kerridge in the west.

The northern Chatsworth Grits are more yellow-brown in colour, a coarse to medium-grained sandstone and shallower, up to 27 m deep. They form the long line of impressive rocky edges on the eastern side of the Peak District stretching 15 km from Stanage in the north to Curbar Edge and beyond southwards (Fig. 22). On Hathersage Moor, Chatsworth Grit can be seen in its fullest development, together with shales containing a distinctive fossil fauna, where there is approximately 100 m of Chatsworth sandstones exposed that is interbedded with shales in the lower horizons. The grit here contains a lower strata of fine-grained flaggy sandstones with thin shale units and an upper layer

FIG 22. Part of Burbage Edge above Burbage Brook, showing massive Chatsworth Grit favoured by climbers, with numerous boulders tumbled below.

of rock that is fine-grained at the base but increasingly coarser upwards also showing large-scale cross-bedding. The lower sandstone layers 15 m thick are thought to represent deposits by a delta mouth bar, whilst the coarser upper cross-bedded sandstone represent the remains of distributary channels that migrated seaward as the delta accumulated. This stratification is of national significance and a feature of the Eastern Moors SSSI.

The last of the massive sandstones is Rough Rock, separated from the Chatsworth Grit by more shale bands, dated using the goniatite assemblage as in other marine layers, and present mostly on the western side of the Peak District. This too is a single massive coarse sandstone, often current-bedded and pebbly in places. There are also some flaggy beds in its lowest layers, which have been extracted for gritstone roof flags as on Chinley Churn, where the remains of an extensive quarry is visible. The thickness of Rough Rock is variable, from only 16 to 20 m between Glossop and Chinley, but deeper at 27 m in the Etherow Valley in Broadbottom fringing our area, where there is a deep gorge visible from the road bridge.

THE COAL MEASURES

The Coal Measures were deposited over a period of 6.5 to 13 million years, on top of the Namurian grits and shales, some 310 million years ago into a single depositional basin, the Pennine Basin, still between the Southern Uplands High to the north and the Wales–Brabant High to the south (Fig. 14). Altogether there was over 1,330 m of materials deposited in layers, although the depth varied with differential subsidence within the basin. The maximum depth was about 1,900 m near Manchester. Very few of the Coal Measures are actually coal. Rather, they are formed of a mix of interbedded mudstones, siltstones and sandstones interleaved by thin coal seams. Marine bands and the fossils used to date the shale layers between the gritstones are now rare and terrestrial fossils such as non-marine bivalves and plants take over. The layers tend to be grey in colour, laminated, sometimes with ironstone layers (Fig. 23).

FIG 23. Exposed Coal Measures at the upper end of the Goyt Valley beside the River Goyt showing laminated shale layers below some shallow bands of gritstones. The shales show vulnerability to erosion by the river.

Coal is formed from carbon, usually finely laminated, with a high proportion of cellular structure from bark and leaves or decomposed plant material mostly of species related to present day tree ferns, horsetails and clubmosses, none of which are flowering plants but which reproduced by spores. Flowering plants had not yet appeared in the fossil record at this stage. All these Carboniferous plants were tree-sized; 10 m high horsetails and clubmosses, and 30 m tall *Lepidodendron*. These trees did not have trunks like modern trees, but developed height by having a tough supportive bark formed of lignin (the main constituent of wood), which is also highly resistant to decay.

The animal fossils here are dominated by bivalves (similar to modern mussels), fish and ostracods. The habitat seemed to have been freshwater or brackish, with extensive waterlogged plains draining to the sea, still in equatorial latitudes in a hot, wet climate (similar perhaps to the current Amazon Basin, but with a completely different wildlife). Large shallow lakes and peatlands were separated by rivers crossing these plains. Peat consists of accumulations of partially decomposed plant remains where the conditions are anaerobic from waterlogging and stagnant water thus allowing undecomposed material to accumulate and then consolidate. To give some idea of the conditions for coal development, about 10 m of peat gives a 1 m coal seam through burial, but it takes some 7,000 years for this amount of peat to accumulate, depending on ambient conditions.

The Westphalian environment fluctuated at times as sea level was unstable and the location of coasts would have moved as sea levels changed on a world scale. This is indicated by shallow marine mudstones separating the sandstones and shales in the Lower Coal Measures. However, most of the Coal Measure rocks were laid down in freshwater and lake environments on these alluvial plains. Coal was formed, either from compressed peat in raised bogs where conditions are acidic and only fed by rainwater (giving low ash coals) or from peat developed in fen-like conditions which give rise to high-ash coals. The type of coal in the eventual seam depends also on the depth to which it is buried – which determines attendant pressure and temperature – and the duration of geological forces. Coal seams in the Peak District are generally fairly thin and form a minor part of the whole Coal Measure series, more particularly appearing in the Upper Coal Measures.

Although this is the time when the great coal fields developed in the fringing areas of Yorkshire, Staffordshire and Lancashire (and many more in Northern England, North America and Europe), coal was extracted in the Peak District, albeit on a much smaller scale than in the main coalfields. These Coal Measure rocks as a whole are concentrated at the surface in the west and southwest and along the eastern fringes of the Peak District (as shown in Fig. 19).

LATER ROCKS

Although the main rocks described cover over 90 per cent of the Peak District, there are minor intrusions of later rocks on the southern boundary where Permian and Triassic sandstones occur in the Ashbourne area. Evidence for this lies in Neogene deposits around 7 million years old, which, although minor in extent, are important for their silica sand quality. These sands and clays, derived from weathered Triassic Sandstones, probably formed a continuous surface over at least the southern half of the Peak District at a time when the land was still some 200 m lower than now. Ford (2002) surmises that the current extent of the Brassington Formation in numerous sinkholes or dolines (collapsed solution depressions in the limestone surface) developed as solution weathering of the underlying limestone resulted in collapse of unconsolidated sheets of sediments into the pocket hollows found today. These deposits can be tens of metres thick and have been worked extensively for refractory sand such as around Friden Brick works and near Brassington. All other Neogene deposits have been lost to erosion.

This geological setting is the basis for the development of the landscape we see today, with all its complexities of topography and processes, which have evolved over the intervening millennia, shaped by mountain building, water and ice. We need to turn to the next chapter to bring this landscape alive.

CHAPTER 4

The Shaping of the Scenery

WE HAVE SEEN HOW THE limestones, gritstones, shales and Coal Measures were laid down, how the environmental conditions determined the nature of the sediments deposited and how these were punctuated by volcanic activity, mostly into the developing limestone. However, most of these rocks were developing on top of each other in seas or deltaic environments, and we have noted numerous locations for seeing some of the associated features – fossils, outcrops, towering cliffs or enigmatically shaped stones. There is obviously broad divergence between these conditions 310 million years ago and our modern day scenery: the high plateau on Kinder Scout, which provides such a stimulating climb; the deep, craggy limestone dales and gorges; the splendid cloughs and the patterns of hills. As we have seen, the Kinderscout Grits were buried by up to 1,330 m of coal measures, so they were at least some 2,333 m below sea level compared with their present level. We now need to turn these rock layers into today's landscape and that is closely related to two major drivers: mountain building combined with the differential susceptibility to erosion of the rocks; and the effects of ice.

MOUNTAIN BUILDING, FAULTS AND FOLDS

We have already seen how the Peak District was located on the Laurasia tectonic plate in the Carboniferous Period, when the world's plates were gradually coalescing into a massive continent called Pangaea. It was the interaction of the plates in developing this super-continent that drove the general subsidence in our region in which the Carboniferous sediments were deposited. This type of super-continent has actually formed several times in earth's history, possibly driven by circulation dynamics in the mantle (the layer below the earth's thin crust). The collision, stretching and buckling of the plates continued during a major mountain building period called the Hercynian or Variscan orogeny for some 20 million years throughout the Permian following the Carboniferous Period (Table 1). The greatest effects were in what is now Belgium, northern France and southern Britain, but the Peak District was close enough to feel the effects of these forces. The prolonged subsidence in the Carboniferous Period ceased and a series of faults developed as rock stretching reached breaking point, with opposing sides sinking, rising or stretching sideways. Large areas were uplifted and rocks were folded producing synclines and anticlines. The cross section of the Peak District showing the marked folding on the west and to a lesser extent on the eastern sides of the gently mounded central limestone massive (Figs. 4 & 21).

As rocks were forced up into the terrestrial environment by tectonic activity, they were exposed to erosion: frost or heat shattering, wind action and water erosion, depending on the ambient climate. This was not a massive mountain erupting, but a slow uprising over millions of years simultaneously being eroded down. The Coal Measures, particularly the softer shales, are mostly more susceptible to erosion than the more solid gritstones, so these would have been stripped faster, especially on the uplifted anticlinal folds rather than in the relative protection of syncline bottoms. Similarly, within the gritstone layers, the intercalated softer shale bands were more vulnerable to erosion than the gritty sandstones. In this way the layers have been pealed back leaving the more resilient gritstones as exposed edges with the lower, more gently sloping landscape marking shales and shale-grit sequences.

Erosion of the limestones followed a different pattern once they emerged out of the sea. Rainwater contains carbon dioxide, which forms a weak carbonic acid that dissolves bare limestone, most particularly along the joints (Fig. 17). The stone itself is impermeable except along lines of weakness. The permeability of the volcanic Toadstones is less than the limestones, thus they have influenced the mineralisation and direction of percolating rainwater at various times. Unlike the limestones, gritstones are permeable and water can freely (slowly) penetrate them, although this reduces with depth due to the better preservation of cements binding sediments together. Where these become less permeable, or where an impermcable shale layer is reached, water is forced out as springs or wet oozes. This is very important, as we shall see later, in the lubrication of landslips and in determining vegetation composition.

Of outstanding importance to the Peak District's economy over the centuries has been the mineralisation related to the intense tectonic activity. Mineralisation takes place when super-heated fluids containing volatile elements circulating below the earth's crust under intense pressure, escape as faults judder and fracture allowing fluids to rush through fissures, bedding planes and any other structural weakness, heating the surrounding rock so that minerals leach out. As they cool, metal ores and other minerals crystallise out such as galena (lead sulphide from which lead is derived) surrounded by crystals of fluorspar (calcium fluoride) and calcite (calcium carbonate). Sometimes baryte (a barium ore) is found with the lead, whilst at Ecton and Mixton (in the Manifold valley), copper deposits occurred. This mineralisation has been more intense here than in other Carboniferous Limestone areas (South Wales, Yorkshire Dales for example), resulting in more extensive extraction over the centuries.

The main structures resulting from the tectonic activity and erosion are the Derbyshire limestone dome and more intensively folded younger rocks on either side, cut in different places by fault lines which jolt differently aged rocks side by side. The dome is a gentle anticline (Fig. 21) with horizontal or slightly dipping beds some 40 km long and 10–15 km wide. It is difficult to see these in the field as exposures are limited and dips mostly small. Gannon (2010) suggests a walk along the Wye Valley where strata at the western end dip to the west, before becoming more horizontal in the central part and then slope gently eastwards at the valley's eastern end. The whole anticline generally dips more steeply westwards than

FIG 24. Folding of impure, dark, thinly-bedded ramp limestone in the Ecton Formation at Apes Tor in the Manifold Valley.

eastwards, but this is complicated by the limestone forming more of a dome, dipping in all directions. The dome is also slightly stepped with three plateau areas, the highest in the north above Castleton at over 400 m, the central area some 20–40 m lower and the southern area generally another 20–30 m below this.

Folding is rarely visible in the limestone, but a good exposure along with some faults is evident in a small quarry at Apes Tor (Fig. 24) which is recognised as a Regionally Important Geological and Geomorphological Site (RIGGS) described in the Hamps and Manifold Geotrail. Longstone Edge (north of Bakewell) is a larger anticline on an east-west alignment, whilst others, like the Ashover and Crich anticlines, occur closer to Matlock.

Crumpled against the limestone edges are the grits and shales, with folding more pronounced in the southwest and western fringes (Fig. 21). The Goyt syncline is one of several aligned north-south and stretches from near Glossop south to the Roaches and Ramshaw Rocks, with a northwards dipping axis. Gannon (2010) offers a Roaches walk to see elements of the syncline. The imposing sculptured rock outcrops on the Roaches and Ramshaw Rocks in the Staffordshire Moorlands (Fig. 20) represent the more resistant upstanding Roaches and Chatsworth Grits on either side of the syncline, whilst the basin between, sheltering vestiges of the Coal Measures, is softer, lower scenery (Fig. 25).

FIG 25. A view of the synclinal basin behind Hen Cloud and the Roaches with Morridge in the distance, South West Peak.

FIG 26. Splendid view west from Millstone Edge above Hathersage up the Hope Valley. Kinder Scout plateau in the mid-right distance, the shale and grit Losehill Ridge stretching in front of Kinder to Mam Tor.

Folding tends to stretch the layers on the top of the anticlines, making these more vulnerable to being peeled back. This has given us the striking gritstone edges on both sides of the Peak District that provide such distinctive and impressive landscape features. Stanage, Millstone, Froggatt, Curbar and Baslow Edges form a more or less continuous, prominent outcrop from north to south facing west on the east side of the Peak District (Fig. 22). They are the scarp slopes of folded rocks composed of the massive Chatsworth Grit (Hathersage Edge) or Rough Rock (Froggatt Edge). There are footpaths above most, giving splendid, wide-ranging and diverse views, especially to the west, on a clear day (Fig. 26). Fine examples of cross bedding, the detailed layering of the grits, impressive tors of massive rocks, strangely shaped eroded giant rocks and peculiar holes of varying sizes can all be seen. The edges are characterised by their block form, large fissures between many of the blocks and heavily shattered sandstone debris and blocks below and between the opened joints (Fig. 22). Similar edges on the west side of the Peak District are generally discontinuous, lower and less impressive, as at Windgather Rocks, Turn Edge, Cracken Edge and Cown Edge, which is more spectacular but also magnified by a landslip. Most of these are west-facing except Cracken which faces east.

The geological evidence suggests thousands of metres of Carboniferous rocks have been denuded since their deposition (possibly up to 2 km). The uplift of the Pennines ridge may have formed a barrier between the adjacent Permian seas. Similar conditions are thought to have existed in the subsequent Triassic period with the Peak District mostly inferred as a rocky upland, feeding erosion sediments to seas eastwards and westwards. It is probable that further rock layers from the later Jurassic and Cretaceous Periods were deposited over our Coal Measures, mostly within a marine environment, and it was not until the beginning of the Cainozoic Era (Table 1) that more mountain building thrust us above sea level again, exposing rocks to further erosion. This resulted in the removal of post-Carboniferous rocks prior to the more recent Neogene Period apart from rare deposits preserved in solution features near Brassington.

THE CARVING OF THE MODERN SCENERY – RIVERS AND ICE

The landscape we see today has evolved from this eroded, uplifted surface, but is also shaped by ice ages. These together are responsible for the striking tors, jumbled rocks, caverns and landslides that are now so familiar. This picture is completed by the accumulation of sediments in the river valleys and disparate other deposits that add to the landscape's complexity. Peat development that clothes all our higher moorlands is described in the following chapter as it occurred later.

Water erosion and our river system

Adding flesh to these landscape bones, today's river systems established on the late Cretaceous surface, with water flowing with the dip of the rocks. This drainage was effectively superimposed on the then hidden Carboniferous rocks, but having once established, the rivers continued to erode down. Thus, rivers now flow through layers of hard and resistant rocks after stretches of softer ones, which would not have been the case if the drainage had originated on the Carboniferous surface. There are wider valleys interspersed with tight gorges, for example, which all add grandeur to the scenery. The Derwent Gorge

FIG 27. The dramatic limestone gorge in Chee Dale through which the River Wye flows. The footpath uses stepping stones along the left-hand side of the cliff.

at Matlock, the middle section of Bradwell Dale, parts of the lower Dove and Manifold Valleys and Chee Dale on the Wye are good examples (Fig. 27).

There were major sea-level changes during the warm and cold periods of the Quaternary. Lowered sea-levels would have accompanied ice formation and advancement. This, in turn, would result in lower base-levels for rivers, with increased energy as the hydraulic profile from source to sea steepened. Rivers would have contained more coarse sediment as ice sheets melted, with more bedload, which would have been deposited in lower reaches. Major sediment blockages also diverted rivers. Thus the Trent is known to have flowed to the Wash prior to the Anglian advance, and only turned north to the Humber during the last Devensian ice advance. This would have impacted our rivers by changing the gradient and hence their erosive powers, extending into tributaries like the Derwent, Wye and Dove (Banks *et al.*, 2012). There would have been renewed erosion in the upper reaches in particular when base-levels were lower, resulting in further incision of stream beds below terraces of elevated sediment. Such terraces are visible to a greater or lesser extent in most Peak District river valleys. Strata of more resistant rocks are marked in many of the gritstone

FIG 28. A pretty, small waterfall over a more resistant gritstone layer in Fair Brook Clough, east side of Kinder.

cloughs by waterfalls – nothing of fame, but many smaller features of interest and drama (Fig.28). Kinder Downfall has the longest drop (about 30 m), but even this can be dry in summer. It is more dramatic when strong westerly winds spray it backwards or when ice freezes it to a wonderland of icicles and ice flows: a mecca for ice climbers.

Drainage systems in the early Pleistocene were thought to be largely west-east, as in the Rivers Wye, Noe (Edale Valley), Lathkill and Via Gellia. However, the headward expansion of the River Derwent was faster, being on less resistant shales, which captured middle sections some of these west-east rivers. Additionally, the Lower Wye and Lower Lathkill rivers captured their upper stretches as a product of knick-point recession as a consequence of the Anglian or earlier glaciation plus land uplift (Banks *et al.*, 2012). The main stretches of the rivers Dove and Manifold have north-south courses and would appear to be young rivers compared with the others (Ford, 2002).

Ice Ages

The Pleistocene (Table 1), commencing about 2.5 million years ago, was characterised by alternating cold and relatively warm periods. An ice age is generally a group of several cold periods over a relatively short period of time. During 11 of the cold periods there were glacial events, plus other minor ones, with intervening warmer periods, some more so than today. However, generally evidence in the Peak District of different glacial periods is sparse as each can be superimposed on the last one. The earliest evidence for a glacial advance into the Peak District around 730,000 years ago comes from cave sediments at Matlock and Castleton, but little is known about this except that it could have been as extensive as the later Anglian advance (J. Gunn pers. com). It preceded three main advances for which there is more evidence. The Anglian was the most extreme in the middle Pleistocene, starting some 478,000 years ago, extending to southern Britain. There is some evidence of a second, the Wragby advance, which occurred in the late-middle Pleistocene, and the most recent, the Devensian Glaciation, starting some 80,000 years ago, peaking about 22,000 years ago.

Glacial erratics from the Lake District or other northern regions occur between the Rivers Wye and Dove and further east, some of which, together with river terrace analysis suggest a lobe of ice flowing south-eastwards into the Wye and Derwent Valleys interpreted as part of the Wragby glacial advance. This is when eastern and western ices sheets are envisaged circumventing the Peak District except for this one lobe crossing over a Dove Holes Pass and down the Wye Valley as far east as Matlock with a meltwater channel escaping down Winnats Pass (Fig. 35, Bridgland *et al.*, 2015). Wensley Dale is interpreted as an ice-marginal channel at this time at the limit of this lobe (Westaway, 2019).

The last Devensian Glaciation involved the growth, expansion and then decay of an ice sheet that extended from Scotland and Northern Ireland south to South Wales and southeast Ireland and which spread across the West Midlands, plus a lobe descending down the North Sea. The Peak District was close to this ice sheet, with possible minor incursions (Fig. 29), so periglacial processes played a major part in modifying the landscape (Johnson, 1985). There is a significant difference between big ice sheets and glaciers that grew in high mountains like the Lake District and Snowdonia and had enough power to flow and hence carve ice-eroded features like U-shaped valleys and corrie lakes. Here the ice cover emanated from elsewhere on all except the highest hills and did not exert strong erosive forces, meaning we have no strongly glaciated landscape.

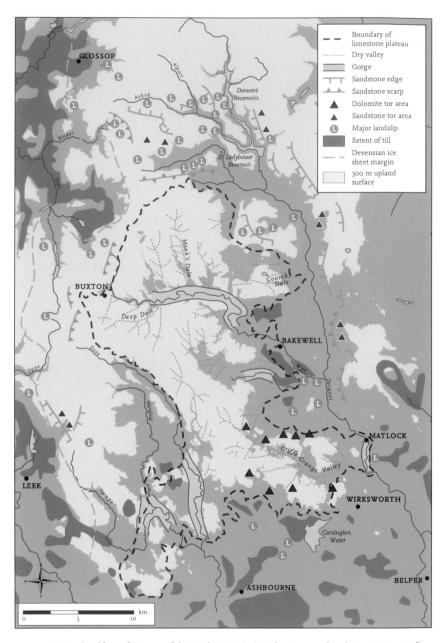

FIG 29. Major landform features of the Peak District. (© The Geographical Association Ltd)

Till, head and other deposits

We do have some post-glacial deposits though from these former cold phases. A material called 'till' – unsorted glacial sediment of stones, clay and sand derived from material that had been trapped in ice – has been left in the river valleys around Bakewell and in the lower Manifold valley around Butterton. Post-glacial conditions after these earlier ice advances might also have altered river courses as till (Figure 29 shows its extent) and other deposits filled valleys, forcing diversions, and copious powerful outwash meltwaters from the ice produced new rivers or new routes. Thus the till in the Wye and Derwent valleys upstream of Bakewell could be the remains of blockages. There is some evidence that the Derwent diverted from near Calver south to join the Wye upstream of Bakewell when its current valley was blocked by glacial debris. This would account for the over-wide Wye valley above Bakewell compared with its current river volume (Gannon, 2010). Dalton *et al.* (1999) point to the possibility of a further stream capture near the Roaches, where there may have been some interplay between Black Brook eroding eastwards and the Upper Hulme stream incising a course northwards, although there are other possible interpretations. Other stream captures occur on Howden Moors, at the upper end of Bull Clough Head for example. Here, the upper streams, which should have flowed to the Little Don, have been captured by the upper reaches of a Derwent River tributary through its greater erosive powers. This area too lies within the Dark Peak SSSI and is highlighted for its geological interest in the study of river landforms.

'Head' is the accumulated material that moves downslope through solifluction characteristic of saturated soils at high latitudes. With little vegetation to stabilise the surface, sediment movements would have been unimpeded. Head is extensive in our area beyond the last ice sheet, having accumulated particularly during the period of intense periglacial activity. It is generally unsorted and heterogeneous, consisting of stones in a sandy-clay to silty-sand matrix. It can accumulate to considerable thickness: a 10 m-thick bed occurs, for example, in many places in the Edale Valley. Other head deposits can be seen in the Ashop Valley, on Hathersage Moor and elsewhere and provides the foundation into which many of the block fields lie. Depending on the source of materials (shales or grits or both), the nature of the head materials that cover many of our hills and slopes produces some very variable soils (Dalton *et al.*, 1999).

Aeolian deposits called loess are also prevalent – these are windblown, fine silts, with some clay and small sand grains (visible in the lower part of Figure 71). They were transported mainly from unvegetated areas, glacial outwash plains and erosion of sandstones by the action of the prevalent strong, dry winds in the Pleistocene's colder periods. Much of the limestone plateau is blanketed with loess of variable depth, comprising an orange-brown, silty loam, mostly originating from the surrounding Millstone gritstones. This masks the underlying limestones, except where it outcrops, producing a more acidic soil.

Exposure of the edges and development of tors

Periglacial processes were instrumental in forming many gritstone features. The gritstone edges (shown on Figure 29) were subjected to frost wedging and freeze-thaw cycles, dislodging stones and generating many rock falls, thus producing the strident edges we see today along with boulder fields of large angular blocks still lying higgledy-piggledy below many edges, sometimes shrouded in Bracken or daubed with lichens or mosses (Fig. 22).

Tors are resistant, residual, isolated pinnacles of rock that have been exposed to severe weathering after removal of the surrounding rock. They are considered to have developed during periglacial periods through mass movement of shattered debris, although it is also possible that post-Cretaceous deep chemical

FIG 30. Gritstone tor on Froggatt Edge, one of many on the gritstone edges, overlooking the mid-Derwent Valley.

weathering in tropical conditions primed the strata for removal prior to the Pleistocene (J. Gunn, pers. com). The tors would have been battered by wind, rain and frost, which removed rotted rock and has helped form the many dramatic shapes we see today (Figs. 1, 20, & 30). Tors are common on the gritstone escarpments all along the eastern edges, in the Greenfield Valley east of Oldham, in the dolomitic limestone above Brassington at Black Rocks and Rainster Rocks, and above Matlock (Fig. 29). The Hathersage edges and tors are described along a walk with explanatory maps by Dalton *et al.* (1999).

Burbage Valley (Fig. 22) is a key geomorphological study site for upland periglacial landforms. Not only are there a wealth of tors, but there are also escarpments, structural benches, solifluction deposits, weathering features, blockfields, a variety of slope forms and a Devensian late-glacial soil. The variety of landforms and deposits demonstrate relationships between the features and show the long and complex evolution of landforms under periglacial conditions. These features are regarded as nationally important and are included within the Eastern Moors SSSI.

Landslides and bulges

Strong movements of frost-riven debris lubricated by water formed block-fields, mudslides and solifluction lobes all round the Dark Peak especially, producing a period of major landslides, mostly dating from after the last Devensian ice cover. Geomorphologists have identified over 600, largely associated with the shales and gritstone rocks (Fig. 29). Of the nationally important mass-movements in Carboniferous rocks, five out of seven are in the Peak District (Cooper, 2007). Many were initiated by ice or by periglacial conditions that left unstable slopes when ice melted, but the essential cause is the juxtaposition of shales and grits and the lubrication of the shale layers, particularly when dipping outwards. Water draining through the grits reaches the underlying impermeable shale surface and thence passes down the dip of the shale bed, oozing out at the outer edge forming a spring line. However, where the upper gritstones are too heavy and saturated, the wet shale layer can form a lubricated surface over which upper rocks slide. This is accentuated where the softer shales have eroded back, leaving overhanging sandstone beds. Gannon (2010) describes this process more fully and provides several walks to view the best.

It is worth scanning hillsides for evidence of landslides. They can be small-scale and clear-cut, or highly complex and massive. They tend to interrupt any hill side's smooth curve, leading to rumpling and crumpling of rocks. Sometimes massive blocks slide down in one piece without losing their integrity, as in Bretton Clough. Lud's Church's landslip in Back Forest shows minimal movement and a limited debris field below. It is as if a giant has cut the hillside with a machete leaving a deep, twisted wound 3–5 m wide and 220 m long, including all its side passages, with towering cliffs over 15 m high. (Fig. 31). It is easily accessible from nearby footpaths, but damp and muddy. The landslide has been identified as of national interest as a bed-on-bed translational sliding within a rotational mass (Cooper, 2007). The backface of the slip forms a short steepening slope about 25 m above the fissure, which therefore lies within the slipped mass. There are also hillside trenches downslope of the fissure, many hidden in the Bracken and woodland, and nearby Castle Cliff Rocks is a tor that may have originated as part of the landslip. The whole landslide features in the Leek Moors SSSI.

This highly atmospheric, dank, dark crevice, full of ferns and mosses, has spawned a wealth of stories and myths. Its hidden depths are reputed to have provided shelter for all sorts of renegades, even extending to stories of Robin Hood. It is more certain that the Lollards (the followers of John Wycliffe, an early church reformer) condemned as heretics, worshiped in it in the early 15th century. Their local leader, who along with his followers, was captured and

FIG 31. Lud's Church, Back Forest in the South West Peak showing a bed-on-bed translational slide, 3–5 m wide with towering rock faces.

slaughtered here, was thought to be Walter de Ludank or Walter de Lud-Auk, which may have given the chasm its name. Other suggestions are that is was named after the Celtic God, Llud.

Anyone interested in the medieval poem, *Sir Gawain and the Green Knight*, will know that links have been made between the Derbyshire/Cheshire/Staffordshire border dialect and the poem's language, suggesting its author (unknown) to be local. Some researchers also link the Castle of Hautdesert to Swythamley Grange (not far from Lud's Church) and consider that the Green Knight's Green Chapel to be modelled on Lud's Church (Armitage, 2007):

> *Then he strides forward and circles the feature,*
> *baffled as to what that bizarre hill could be:*
> *it had a hole at one end and at either side,*
> *and its walls, matted with weeds and moss,*
> *enclosed a cavity, like a kind of old cave*
> *or crevice in the crag – it was all too unclear to declare.*
> *'Green Church?' chunters the knight. More like the devil's lair.*

(p. 100, Armitage, 2007)

Look out for the supposed profile of the Green Knight's head covered in algae and mosses on the chasm's walls.

Leaving mythology and returning to landslides, others are deeper, rotational slides with little integrity surviving. Alport Castles, Longdendale Valley, Crowden Great Brook, the west-facing side of Kinder Scout and many in the Woodlands Valley are of this general type. The landslide on Cown Edge shows a series of even, stepped collapses, with Rough Rock overlying shales, that took place in phases over a two- to three-thousand year period starting around 5500 BCE. Dating techniques used pollen analysis from plants growing in the post-slide pools and peaty surrounds (Franks & Johnson, 1964). The Longdendale Valley seems to have been especially prone to landslides, with at least 10 identified. These are bedding plane slides often with non-circular rotational elements involved. It is the instability in the lower slopes that has often led to cambering and valley bulging so that slopes become deformed and earth flows develop at some sites as on Mam Tor (described below). Again using pollen analysis, the Longdendale slides are dated from between the early post-glacial years to as recently as 2,000 years ago, although they might have started earlier (Johnson & Walthall, 1979).

In contrast, Alport Castles is nearer 8,000 years old (Gannon, 2010). Located at the lower end of the Alport Valley, it is also of national importance, featuring in the Dark Peak SSSI. It illustrates mass-movement in Carboniferous rocks,

FIG 32. Alport Castle above the lower Alport Valley showing mass movement in the gritstones and shales in the largest landslide in England involving the whole valley side.

but this time showing composite mass rock creep, retrogressive rotational (on concave slopes) and translational (on a flat slope) movements including slides and flows of material (Cooper, 2007; Fig. 32). It is the largest inland landslide in England and consists of a single block movement, but one that involves the whole valley side. There is a high vertical face at the top, a tall pinnacle ridge and massive flat-topped detached sandstone mass. There is a complex range of features associated with this landslip exposing sections through the Carboniferous rocks, which can then be studied to understand better the conditions in the large delta apron of the time.

Our most famous landslide is Mam Tor (Fig. 33) and there are several others mostly on the north side of the same Losehill Ridge and Rushup Edge. Believed to be 3,500 to 4,000 years old, Mam Tor slippage is still moving at about 10 cm a year compared to others that are now stationary. Gannon (2010) provides details. Mam Tor is some 80 m high with a large area of exposed beds in the Shale Grit series overlying the weaker Bowland Shales, which, being crumpled, provide a very weak foundation. The landslide consists of sandstone material that lies on the surface and more finely fragmented shale beds from the lower Bowland Shales that rot into an amorphous mass. This can flow downhill as a slump-earthflow (Cooper, 2007), more so after extreme storms. Mam Tor's local name is Shivering Mountain owing to the shower of rocks that can tumble down. It is safer to observe it from the former road surface, which has been repeatedly disrupted, demonstrating its continuing downhill shift that is still being measured. You can see how the slippage is warping fences, stressing trees and pushing at electricity poles. The road was first built in 1810 and abandoned finally in 1979, but is accessible on foot. Indeed, it was the road's engineering requirements that drove more research, thus advancing knowledge on the landslide's processes and their management.

Another landslide site that shares some features with Mam Tor is Rowlee Bridge, which represents a complex valley bulge resulting in a spread of materials (Cooper, 2007). Like Mam Tor, the Edale Shales provide the weaker strata, which have been forced upwards by the weight of the overlying masses on either side. This results in intense folding (which is not, unusually, derived from tectonic pressures) from the compression and buckling, which can be seen in the River Ashop bank at Rowlee Bridge.

Landslips in the limestone area are much rarer. Hob's House is one example perched in Monsal Dale consisting of numerous irregular tors on a bed of softer, weaker mudstones with a range of screes behind it. It separated from and

FIG 33. Mam Tor or the Shivering Mountain, a landslide near Castleton, showing exposed Shale Grit beds which overlie weaker Bowland Shales. The former road crosses above the rushes in the middle.

slid down the slope and is a single rotational slip producing a slide of material (Cooper, 2007). It lies below a pre-Roman large hillfort on Fin Cop, where the precipitous landslide slopes down to the River Wye would have safeguarded it from attack. More dubiously, the story surrounding Hob's House is one of a giant, Hob, who emerged at night to thresh local farmer's corn in return for a bowl of cream. Hob was supposed to have lived in the caves and clefts in the landslide's rocks.

Peter's Stone, the famous landmark at the upper end of Cressbrook Dale, is another limestone landslip (Fig. 34). This too is believed to be a translational landslide, whereby Peter's Stone and a lower mass separated from the rock strata above it and moved, spinning as it did, as evidenced by the mismatch of mineral veins in the stone with the head strata. It lies on a 13 m thick bed of weaker volcanic ash called the Litton (the nearby village's name) Tuff, which itself is underlain by 5 m of basalt. It is believed to have originated from the higher limestone crags visible about 55 m south of the stone. There are apron screes that could be derived from periglacial activity after the landslide event. Research suggests that this landslide was initiated by a combination of processes, with undercutting and local

FIG 34. Peter's Stone, at the upper end of Cressbrook Dale, is a translational landslide which spun as it slipped. Anthony Lingard's body was hung on the stone in the early 1800s for eleven years, his skeleton blowing noisily in the wind until removed.

outward tilting of the limestone and the Litton Tuff on the inside of a high level meander (a river would have been flowing at the time, with the dale floor higher than currently). The landslide could be much older than others described in this section, possibly dating to nearer the beginning of the Pleistocene period, and moving only slowly over a long period (Brancaleoni *et al.*, 2016).

It is doubtful if Anthony Lingard would have appreciated Peter's Stone origins – it is also called Gibbet Rock and this 21-year-old Tideswell man was, according to local folklore, the last person to have been hung in chains on the rock having been put to death in Derby in 1815 after being convicted of murdering the toll-keeper at Wardlow Mires, just east along the main road. His bones were alleged to have been removed eleven years later after complaints about their clattering noise blowing in the wind (https://derbyshireheritage.co.uk).

Limestone features

There are other important limestone landscape features apart from landslides and tors. The White Peak is one of five major British areas of karst landforms, others being the Mendip Hills, South Wales, North Wales and the Yorkshire Dales (Gunn, 1992, Gunn *et al.*, 1998). Karst landscapes are those in which rock dissolution is the dominant process and most form on limestone bedrock. Rainwater contains a low, but steadily increasing, amount of carbon dioxide dissolved from the atmosphere. This forms a weak carbonic acid that dissolves limestone, but soils contain much more carbon dioxide so water percolating through them has an enhanced dissolutional ability. Water moves very slowly through small channels in the rock structure, but over time some of them reach a critical size (around 1 cm) after which the flow becomes turbulent resulting in faster expansion. Small channels are tributaries of larger conduits with turbulent flow and some conduits eventually become large enough for human entry and are then classified as caves.

The ability of limestone to absorb rainfall accounts for the lack of streams and rivers in the White Peak and only the Rivers Dove and Wye maintain a permanent surface flow through the limestone area (Fig. 29). At the upstream margins of the limestone, where water flows in from other rocks, there are sinking streams which drain to springs, sometimes passing through caves en route. Between sinks and streams there is a network of dry valleys and many closed depressions or dolines sometimes referred to as sinkholes. There are also a few potholes (narrow, largely vertical caves that are mostly relict) and open shafts. Nettle Pot on Eldon Hill is a good example of a pothole and Eldon Hole, is the best example of an open, unroofed, phreatic vein cavity in the Peak District. The entrance shaft is 30 m long, 5 m wide and 60 m deep and, although the original vein cavity has been enlarged by wall-collapse, it has no relation to present-day processes. The first reported descent was in 1770, although the authors report a descent by locals some 10 to 15 years earlier (Lloyd & King, 1771). They also mentioned a further shaft but a dig undertaken by cavers failed to locate it.

The number of streams starting life on other rocks and disappearing shortly after reaching the limestone is a striking feature of the White Peak. Good examples occur on Stanley Moor near Buxton, the Perryfoot Swallets at Rushup Edge's southern edge and east of Great Hucklow. The point where a stream goes underground is a swallow-hole or swallet, as for example Waterfall Swallet, east of Foolow, where a small stream originating on Eyam Edge's gritstones and supplemented by several ditches that run off the intervening fields, falls 10 m into a mostly sheer-walled depression. The water disappears at three points in the swallet base, but in wet periods inputs exceed the conduits' capacity forming a lake that can be over 5 m deep, the water eventually entering a small cave entrance in the northern cliff leading into the 43 m deep Waterfall Hole (Ford & Gunn, 2010). Waterfall Swallet is tree-lined and shaded, with a muddy, steep path providing access. Up to around 250 years ago the water emerged from Carlswark Cavern in Stoney Middleton Dale just below the junction with Eyam Dale, but the water was captured by a lead mine drainage level, Moorwood Sough, and now emerges in the private grounds of Stoney Middleton Hall.

The Perryfoot Swallets, stretching from Sparrowpit to Windy Knoll, are the finest collection of sinking streams in the Peak District. They are numbered P0 to P12 and most give access to caves that vary in length from a few tens of metres to Giants Hole (P12), with over 4000 m of passage descending over 120 m. With the exception of P0, all these streams originate on a large solifluction sheet extending from the lower slopes of the gritstones and shales off Rushup Edge. Before the swallets and the solifluction sheet were formed, water from Rushup Edge flowed down the now dry Perry Dale, on down Dam Dale to Monk's Dale and thence into the River Wye.

Not only do many small streams disappear underground, but so do two rivers: the Hamps and the Manifold. Starting below Flash in gritstone country, the Manifold flows south untroubled until it reaches the limestone at Hulme End. Downstream of here there is a complex underground conduit network into which water from the river leaks, although in most years there is continuous flow for about 4 km to Wetton Mill. Here there is a large sink that takes all the river's flow during dry periods but which is overtopped during higher discharge allowing a somewhat depleted surface river to continue. The sink itself is impenetrable but an entrance in a nearby crag gives access to over 365 m of cave that descends over

40 m below the river. After another 1.5 km there is a complex of sinks collectively known as Redhurst Swallet, where a further 300 m of flood-prone cave has been explored. Continuing downstream there are several other sinks, some of which lead to cave passages. As each sink is able to take more water, the river's flow decreases downstream until eventually the surface channel is dry. As the flow arriving onto the limestone decreases, the areas of dry channel migrate upstream so the Manifold behaves in the opposite way to most watercourses, which dry from the top down. The water from the underground Manifold emerges from three risings below Ilam Hall. A major Manifold tributary, the River Hamps, shows a similar pattern. As flows decrease, this river also sinks, flows underground to Ilam and emerges at a different point to the Manifold water.

Although there are few streams, the surface of the White Peak is dissected by a complex network of sinuous mostly dry valleys (Fig. 29). In the past some were thought to be collapsed caverns but it is now accepted that they are part of the drainage system that developed on the Millstone Grit cover superimposed onto the limestone as erosion lowered the surface. They then became dry as drainage migrated underground; a process that has, in many locations, been greatly accelerated by extensive lead mining activities. The dry valleys carried surface streams again during the Pleistocene cold periods when permafrost inhibited underground drainage. The longest dry valley system extends over 9 km from Perryfoot to Miller's Dale but is only totally dry during drought conditions. At other times a small stream fed by perched springs flows down Monk's Dale's lower stretch. Other long dry valleys include Long Dale from Sparklow to Hartington and another Long Dale which extends from north of Friden to Gratton Dale, a tributary of the River Bradford.

The very steep (1-in-5) Winnats Pass, west of Castleton, is a dramatically different dry valley (Fig. 35). Barely 1 km long but over 130 m deep, it is one of the most imposing dry gorges in Britain made famous reputedly for a particularly violent double murder in the 17th Century. The gorge probably began in the Visean as a shallow channel in the reefs and this became part-filled by sediments during the Namurian. The sediments are thought to have been largely excavated by meltwater running off a snow-and-ice field near Windy Knoll in the later Pleistocene.

The other characteristic karst landforms are closed depressions (dolines) which pit the plateau surface. Solution dolines and suffusion dolines both occur, although analysis of their form and pattern is complicated by the past activities of lead miners, by recent reworking of their waste hillocks for fluorspar, and by farmers who infill depressions to create level fields. Collapsed dolines are less common although there are a few good examples in the Castleton area.

FIG 35. The Winnats Pass, 1-in-5 steep road at the bottom of a deep cleft in the limestone (note the size of the car in the bottom). Steep north- and south-facing slopes on either side with rock outcrops.

Downstream of the sinking streams and beneath the dry valleys, dolines and soil-covered fields, there is a complex conduit network transporting water to springs. Most conduits are too small for entry but there are over 280 accessible caves with a total length exceeding 67 km. In some of the White Peak a very early phase of cave development occurred during the Carboniferous when the limestones were exposed at the surface and there was groundwater circulation. Later, when the limestone was buried by younger rocks, some of the voids were utilised by hot fluids that were rising from depth. Locally these fluids were sufficiently aggressive to dissolve the limestone and form linked cavity systems that were later mineralised and are locally called pipes. Larger vein cavities were also formed. There was then a very long period of time when the limestones were under a great depth of younger rocks and groundwater circulation was very slow. Over time these rocks were gradually removed and during the Neogene (Table 1) the limestones were first exposed allowing direct entry of water and initiation of a modern phase of cave development. Ford & Gunn (2010) consider that the early cave systems developed slowly, with some conduits intersecting ancient vein cavities. Bottomless Pit is one easily visited example in Speedwell Cavern, discovered by lead miners driving a canal now used for tourist boats. They tipped several thousand tonnes of rock into the Pit and as the water level did not rise it was assumed to be bottomless. We now know that this was because the water enters conduits that function in a similar manner to a sink overflow.

Early water circulation would have been slow, largely beneath the rock cover, but as surface rivers incised into the limestone, they intersected conduits allowing much more rapid flow. Conduit development was focussed on particular limestone levels that were especially susceptible to dissolution, called inception horizons. Early conduits would have been water-filled so surfaces dissolved at similar rates producing roughly circular passages called phreatic tubes. Where conduits developed on inception horizons that were quite close together and valley incision was rapid. They drained quickly leaving largely unmodified relict phreatic tubes. With slower drainage, erosion was focussed on the floor producing keyhole-shaped passages. In some cases the cave streams cut down deeply and these are known as vadose trenches. Cave passages that are no longer

FIG 36. Excellent examples of ceiling and floor flowstone and a curtain with small stalactites above the figure in Convenience Cave, part of an ancient system that was intersected by Eldon Hill Quarry near Castleton. (John Gunn)

being eroded by flowing water are called relict, whereas a fossil cave is one that has been completely filled by sediment.

When a cave moves from being water-filled (phreatic) to having some air (vadose), percolating water enters from the roof. The air in the cave contains much less carbon dioxide than soils so dissolved carbon dioxide is lost from the percolation water that is then saturated with calcium carbonate which is deposited as speleothem. Common forms are stalactites (hanging from the ceiling), stalagmites (growing up from the ground) and flowstone sheets that coat cave walls. Where a stalactite and a stalagmite meet they form a column and where water runs down a sloping surface it can form a curtain. In the Peak District most speleothem are white to yellowish in colour due to iron or soil derivatives, although some are greenish from copper traces or blue-grey coloured by manganese (Ford, 2002) (Fig. 36). Speleothem can be aged by measuring the uranium, lead and thorium isotopes they contain or by examining their magnetism.

Other than the ancient vein cavities, the oldest known cave in the Peak District is Elderbush Cave in the Manifold Valley which contains stalagmites tentatively dated to over 1.7 million years. The cave itself must be much older as stalagmites can only grow after a cave has been drained of water. Water Icicle Close Cavern near Monyash is another very old cave system as the oldest stalagmite formed almost 648,000 years ago showing that it had already been drained by then. Masson Cavern, part of the Heights of Abraham tourist complex, contains sediments that are magnetically reversed and were deposited over 700,000 years ago.

Most of the caves away from the limestone margins are relict and, with the exception of Lathkill Head Cave (Fig. 39), all of the main active cave systems are fed by sinking streams. The longest at 18 km, the ninth longest in Britain, is the Peak–Speedwell system in Castleton comprising natural and mined passages. It also has the third largest vertical range of any British cave, 238 m from the highest point in Titan to the lowest point so far explored by cave divers, 74 m below the water surface in Main Rising. Titan is Britain's largest natural shaft, with a vertical range of about 170 m, including a free-hanging drop of 142 m. The majority of the Peak–Speedwell passage length consists of long, largely relict, phreatic tubes developed primarily on three horizons. The most unusual features of the cave system are the number and vertical extent of its mineral vein related cavities. There are over 50 more vein cavities with a cumulative vertical extent of over 2,600 m and a volume of about 185,000 m³. The Speedwell Cavern part of the system is fed by the Perryfoot Swallets described above and the water eventually emerges at Russet Well and Slop Moll on opposite sides of Peakshole Water in Castleton. Peak Cavern is fed by percolation water that emerges at a rising below

Peveril Castle at the head of the Peakshole Water. However, at times of flood the Speedwell conduit cannot accommodate the volume of water, which backs up and enters Peak Cavern.

Relict caves are useful 'museums' preserving evidence of past environments. Isotopes in speleothems provide information on climate and there are 45 caves that contain archaeological remains and a further 15 that contain ancient animal bones (Ford & Gunn, 2010). A famous example is the recovery in 1823 of a near complete skeleton of a woolly rhinoceros (*Coelodonta antiquitatis*) from Dream Cave near Wirksworth. A Bos/Bison bone from the same deposit was recently dated at 45,130–41,530 rcy BP (McFarlane *et al.*, 2016). A dozen caves are known to have been visited by humans in the Upper Palaeolithic, roughly 40,000 to 10,000 years ago and caves were used as places of burial from the New Stone Age to the Bronze Age, roughly 6,000 to 3,000 years ago (see Chapter 5 and Fig. 41).

The main cave systems in the White Peak are all recognised for their scientific interest and there are eight cave SSSIs: Castleton, Bradwell Dale, Hamps & Manifold valleys, Lathkill Dale, Masson Hill, Poole's Cavern, Stoney Middleton Dale and Upper Lathkill Dale. The caves are also an important recreational resource used by many thousands of cavers each year. Barker and Beck (2010) provide descriptions of those open to suitably equipped explorers and Chapman (1993) describes some of the historical and current explorations possible in these systems, with highly evocative names like 'Crabwalk', 'Great Relief Passage', 'Poached Egg Passage' and 'Chambers of Horrors' that reflect the difficulties or delights experienced. There are many caving clubs most of which are members of the Derbyshire Caving Association, the regional body which co-ordinates access and undertakes conservation work including monitoring SSSIs for Natural England. There is also a national website, https://newtocaving.com, for those interested in exploring caves.

Ford & Gunn (2010) and Gannon (2010) give routes (walking and cycling) and commentaries for visiting caves, swallets and other limestone landmarks. Seven caverns are also open to the public, four in Castleton (Blue John, Peak, Speedwell and Treak Cliff Caverns), one in Buxton (Poole's Cavern, Fig. 37) and two in the Heights of Abraham complex at Matlock Bath (Masson and Rutland Caverns). Each provides the visitor with a different underground experience. At Poole's Cavern there is commonly a large underground river which is one of the sources of the River Wye, quite deafening at times but dry during drought periods as the water follows a lower pathway. Poole's Cavern is also famous for remarkable rapidly growing speleothem, including the 'poached egg stalagmites' which are influenced by the quicklime (calcium oxide) waste heaps above the cave from old lime kilns.

FIG 37. The 'long view' in Poole's Cavern, Buxton, an SSSI managed as a tourist cave by the Buxton Civic Association. The cave is considered to be several hundred thousand years old and is largely relict as the stream that flows through it only runs when lower elevation conduits are unable to accommodate water that has its origins on Stanley Moor and Axe Edge to the west. (John Gunn)

Defoe unkindly dismissed this as 'another of the wonderless wonders of the Peak' (although it is a splendid cave) and the name derives from an outlaw, Poole, who reputedly used the cave from which to rob travellers in the 15th century!

Blue John Cavern is a relict cave that has excellent exposures of the eponymous mineral and a magnificent canyon thought to have been formed by rising, not descending, water. Treak Cliff Cavern is also relict and has two sections. An old miners' tunnel gives access to an ancient phreatic cave dissolved out of a mid-Carboniferous Boulder Bed and containing Blue John mineral (see Chapter 6 and Fig. 60). You take a boat ride along the tourist section of Speedwell Cavern into a passage excavated by miners in the 1770s to the Bottomless Pit.

Peak Cavern has the largest natural cave entrance in Britain, the entrance chamber of which was used until the mid-20th century for making rope – with a demonstration part of your visit. It also houses atmospheric musical events, so watch out for these. It is marketed under its old name, the Devil's Arse, derived from the cave being regarded as a 'gateway to hell' (one of several in the area!) and through a natural siphon near the entrance that emits a gurgling noise representing the Devil breaking wind. At the Heights of Abraham a cable car takes visitors to two caverns that have been extensively modified by lead miners. Masson Cavern began life in Carboniferous times and the early voids were then partly filled with minerals by hydrothermal fluids. There was then a very long period of relative quiescence before phreatic dissolution re-commenced in the early Quaternary followed by partial infilling with glacial meltwater outwash sediments that commenced over 700,000 years ago. The main interest in Rutland Cavern is the lead mine heritage, but there is a fine exposure of Carboniferous volcanic ash deposit in the entrance passage.

Some dales carry ephemeral or seasonal streams when water tables rise in winter or after heavy rainfall events. Some of these are controlled by basalt outcrops in the dale bottoms, as in Monk's Dale and the top of Cressbrook

FIG 38. The upper end of Cressbrook Dale flooded after wet winter weather. Note the abundance of ant hill mounds on the right-hand side above the water.

FIG 39. Water flowing from Lathkill Cave (just beyond the wall on the left) after a wet spring upstream of Holme Grove Risings.

Dale (Fig. 38), for example, which is regularly flooded in winter as indicated by its wetland plants. For part of the year Deep Dale (Topley Pike) has a fast-flowing stream that rises as springs below two relict caves with entrances high up the valley-side, Thirst House Cave and Deepdale Cave. Lathkill Dale has a complex hydrology that has been modified by soughs (lead mine drainage levels). The river's perennial source is the large Bubble Springs below Over Haddon. Upstream is characterised by prolonged low flows and some sections dry up completely most years necessitating fish rescue. There have also been losses of sensitive aquatic species and an increase of terrestrial vegetation in the channel. In winter the river flows from Lathkill Head Cave (Fig. 39), various dale-side springs and occasionally from higher up the dale. As underground flows decrease, the cave ceases to flow but water continues to emerge from Holme Grove Risings, about 330 m downstream extending to the footbridge at the bottom of Cales Dale. Before the construction of Magpie Sough, which discharges into the River Wye upstream of Ashford-in-the-Water, the Holme Grove Risings, and probably also Lathkill Head Cave, flowed throughout the year. Now the cave flows for about a third of the year, the uppermost Holme Grove rising flows for about six months and the lower Holme Grove rising flows for only about 85 per cent of the year.

In addition to these springs there are many other sites where water discharges. Some water has been in the limestone for months to years (which can keep the rivers cool, Chapter 14), while others discharge water that has been underground for only hours. Notable springs along the River Wye include Otter Hole, Wye Head, Rockhead (a mineral water source being developed for bottling), Ashwood Dale, Wormhill and the Lees Bottom spring-group. Three main springs discharge into the Dove upstream of Hartington (Crowdewell, Ludwell and Sprink) and downstream of Milldale there is an east-bank spring group that extends to Nabs Spring. The springs at Ilam, discussed above in relation to the River Manifold, are amongst the largest in the Peak District. Lead mine soughs together discharge large volumes of water that would once have emerged from the natural springs.

Finally, mention must be made of the mineralised thermal springs associated with anticlines or fault zones that stimulated spa town developments drawing visitors to our area. St Ann's Well in Buxton emerges at a constant 28 °C and is believed to have taken some 5,000 years to reach it after slowly filtering through a mile of limestone. The Buxton's health spa properties date back to Roman times around 80 CE when the first baths were built, although the Celts possibly considered the springs as a sacred shrine much earlier. Buxton's heyday as a spa town after the style of Bath in the late 18th and 19th centuries was developed by the Dukes of Devonshire. Similarly, in Matlock Bath, a thermal spring emerging at a steady 20 °C was first discovered in 1698, with two more found nearby subsequently providing the foundation for the fashionable spa resort and Hydropathy centre. Famous visitors included Lord Byron and Princess Victoria.

Where highly saturated water emerges from the ground, carbon dioxide is given off and the calcite precipitates out to form pale brownish tufa deposits. Particularly substantial deposits are associated with thermal springs in Matlock where tufa can encase objects within a few years. There is a spring in the nearby Via Gellia, where there was enough tufa deposited to be quarried and used to build the quaint Tufa Cottage around 1830, originally as a gamekeeper's cottage and thought to be the only tufa-built house in Britain.

The main period of tufa deposition occurred at a landscape scale when the climate was warmer and more humid, some 7,000 to 4,500 years ago in the Holocene's Atlantic and Sub-Boreal periods, when river flows would have been greater and depositional conditions ideal, but there are also older ones dating to various periods in the Pleistocene (Kitcher, 2014, Banks et al., 2012). Tufas can develop barrage-like structures across streams or rivers, typically forming chains of elongated pools, each one separated by an arcuate, thin, well-cemented tufa barrier from the next (Taylor *et al.*, 1994). There are visible examples in Lathkill Dale, where there are steps up to some 5 m in the valley floor below Raper Lodge (which was growing at least in the Late Devensian), marking several tufa deposits, believed to be the best natural tufa barrage in England (Andrews *et al.*, 1994). Tufa generation persisted to about 4,200 to 3,800 years ago, possibly affected then by the effects of forest clearance, as we will see in Chapter 5 (Kitcher, 2014). Banks *et al.* (2012) postulate that tufa barrage deposits within the Wye catchment are largely related to dominant inception horizons in the Monsal Dale and Eyam Limestone foundations and located in zones of river capture associated with faults and steps in the bedrock. There were two main phases of tufa development related to the hydrological evolution of the Lower Lathkill and zones of river incision on the eastern side of the White Peak. Small-scale tufa deposits are still accumulating where springs emerge or streams flow (as in Monk's Dale and at Pudding Springs in Lathkill Dale), and are found especially around the limestone boundary south of Bakewell and around Youlgreave (Gunn, 1985).

FIG 40. Limited patch of limestone pavement on High Edge, south of Buxton, showing classic clints and grykes on a small scale.

Compared with other limestone areas, there are no significant areas of limestone pavement in the Peak District because they are considered to be derived from glacial stripping of the bedrock followed by post-glacial chemical weathering and biological processes and the last Devensian ice advance did not cover our area. More of the limestone is covered too with other deposits as described above, so masking the rock surface. We do have minor fragments of limestone pavement where exposed blocks of rocks (the clints) are separated by fissures (grykes) 30–60 cm deep or more, all sculptured in swirls, sharp edges and hollows as if attacked by an oversized ice-cream scoop. The best examples are on High Edge and Upper Edge south of Buxton (Fig. 40), with smaller fragments west of Hay Dale on Middle Hill, north of Earl Sterndale near Hindlow Quarry, to the east above Dowel Dale, on Carder Low north of Hartington, and at the top of Biggin Dale. Although minor in extent (no more than 10 ha in total), our limestone pavement has not been fully evaluated in terms of its history and evolution. If the last ice advance did not cover the areas where the pavements lie, then it has to be asked if they are a product of an earlier ice advance, or whether the interpretation of the latest ice advance needs to be revised.

How this is eventually resolved does not detract from limestone pavement's rarity within Europe. England holds some 78 per cent of the European limestone pavement resource, the significance of which is recognised by its inclusion in the EC Habitats and Species Directive as a priority habitat. That makes our little patches of high value, more from a geological and geomorphological perspective though since there are only hints of the wonderful floristic features typical of more extensive pavements elsewhere.

RECENT SEDIMENT DEPOSITION

More sediments accumulated along the river valleys below the upland plateaux. Except when frozen, the rivers would have continued to deposit sediments throughout the Pleistocene and onto the modern period (Table 4). Vegetation cover and, therefore, the level of human activity as well as climate played a role in determining the sediment loads the rivers captured and then deposited. Most of the wider river valleys have a layer of alluvium deposited across their flood plains, although there are many instances now where historic deepening and straightening of river channels has left the flood plain disconnected from the water course, as in the Manifold River's middle section. Nevertheless, numerous river terraces can be identified, particularly in the Dark and South

West Peaks. Some of the hill-slope gullies are believed to have been cut in recent times, marked by fan-shaped alluvial deposits and debris cones where they hit flatter ground.

That these processes are still operating today became clear in a major storm that hit Longdendale and Glossop in 2002. A significant cascade of sediment and stone in Crowden Great Brook resulted in tonnes of rock tumbling down from Rakes Moss. Careful observations of this debris and sediment cone showed that an old landslip pre-dated the remains of an enclosure wall, it being clearly built over the slippage, but another had squashed some of the wall, indicating a date later than mid-19th century, with the 2002 debris on top. The stream was littered with large boulders and stone, much of which was dumped in the adjacent Torside Reservoir, forcing the Water Company to clear the debris. In nearby Heyden Brook, huge car-sized blocks of peat and rushes had partly or wholly dislodged from the hillside, some landing in the stream, others slipped but balanced precariously further up the slopes.

Smaller scale slips, slope gullies and sediment deposition can be seen in the Dark Peak in particular, often associated with the softer shales where heavy storm events lubricate the rock layers. With increased storm events predicted with climate change in the future, more such gullying, sediment deposition and land slips might be expected.

THE GEOLOGICAL RESOURCE

This chapter has focused on the natural development of the landscape, but the Peak District scenery is also characterised and sometimes scarred by human exploitation of its geology – a subject that is worthy of its own chapter where it is integrated with our local wildlife and human history.

CHAPTER 5

Plants, Animals and People: From Ice Age to Iron Age

W E HAVE ALREADY SEEN HOW the Peak District's Carboniferous flora and fauna were often related to but different from those present now. We have also hinted at the onset of human activities and their relationship with the changing nature of our wildlife in more recent times. However, the story merits more detail to appreciate the extent to which human activity has shaped our natural history and landscape. We need to start towards the end of the last ice age and examine the subsequent changes in climate, vegetation and animals and weave in the influence of human activity. This chapter will take us up to 500 BCE when humans were influencing the landscape on a large scale. All that has happened since then deserves a chapter of its own (Chapter 6).

Evidence for the early part of our story comes largely from a synthesis of sources like pollen analysis (see Glossary) and of macro-fossils preserved in peat or organic-rich materials; cave and limestone fissure deposits; archaeological excavations and, for more recent times, the written record. John Tallis (Manchester University), plus Sheila Hicks and Vera Conway (see Bibliography), and some more recent investigators have been the key researchers applying pollen and peat analyses to reveal vegetation history and how it tied in with human impacts on the environment. These analyses tell us much about the moorland environment, but much less about the White Peak's vegetation history, although there are limestone heaths (Chapter 14), and their drier peaty soils have been analysed by Dave Shimwell. Analysing the plant and animal remains like aquatic Ostracods (small Crustaceans called seed shrimps), algae, mosses, freshwater snails and higher plants trapped in tufa (Chapter 4) over time is also possible. The Ostracods have a symmetrically hinged outer

FIG 41. One of many caves in Dovedale, this one near Mill Dale. Archaeological remains providing evidence of human and animal activities have been found in several caves in Dovedale and the nearby Manifold Valley.

shell made of magnesium calcite that survives as fossilised material in these cores, which can, like peat remains, be radiocarbon dated.

The Peak District is one of the few places in the country where evidence of animals and human activities are found in limestone caves (Fig. 41) and fissures (Chapter 4). Some animal remains plus other environmental evidence like plant fragments, seeds and invertebrate sequences such as beetle cases (beetles are good indicators of climate change) give insights into the contemporary environment. Some of the earliest systematic explorers of cave deposits in the Peak District were Sir William Boyd Dawkins and his protégée, Dr J. W. Jackson in the late 19th and 20th Centuries. Buxton Museum and Art Gallery was the recipient of Boyd Dawkins' library and papers and also the Jackson collection after their deaths. Boyd Dawkins was fascinated with fossils, especially ancient human and mammal remains recovered from caves (Cliffe, 2010). Dr Jackson was an expert in fossil bones from caves in the area, working at Manchester

Museum. Dr Don Bramwell, excavating in the 20th century, advanced the exploration and interpretation of cave deposits, particularly in Elderbush (Manifold Valley) and Fox Hole Cave (High Wheeldon), both of which are now Scheduled Ancient Monuments. Buxton Museum and Art Gallery displays include a timeline incorporating many of the fossil bones of various animals and human artefacts found in these caves.

Relevant and accessible written records are available mostly from the 18th century onwards. Early accounts gives us travellers' views on their impressions of our area (some glowing, some not). These have been collated into *A Peak District Anthology* (Smith, 2012) and are widely quoted by Bull (2012). Particularly relevant though is the *History of the Agriculture and Minerals of Derbyshire*, by Farey, commissioned by the Government and in three volumes from 1811 to 1817. C. E. Moss produced the seminal *Vegetation of the Peak District* in 1913, which not only described and gave lists of plants for different habitats, but also included vegetation maps of much of our area from which we can detect change over time.

THE APPEARANCE OF THE MODERN FLORA AND FAUNA

The earliest evidence for a more modern fauna compared with Carboniferous animals comes from cave deposits found in a quarry now buried under a refuse dump in Dove Holes, dated to the Pliocene (some 5 million years ago). This was at a time when modern groups of mammals, including marsupials and placenta-bearing groups had replaced the earlier Mesozoic ones and some of the modern orders had evolved (Yalden, 1999). The finds had an exotic feel, with remains of Sabre-toothed cat (*Homotherium latidens*), hyena (*Pachycrocuta* spp.), Mastodons (*Anancus arvernensis*), Southern Mammoth (*Archidiskodon meridionalis*) and a type of horse (*Equus* sp.). The habitat is likely to have been a forest cover largely of conifers with hemlock (*Tsuga* spp.) – now only native in North America, Pine (*Pinus* spp.) and spruce (*Picea* spp.) prominent. This is the last time spruce would be native to the country (Musk, 1985). The fauna and flora would change with climate though with birch (*Betula* spp.), alder (*Alnus* spp.), elm (*Ulmus* spp.), oak (*Quercus* spp.), and Hornbeam (*Carpinus* spp.) featuring in warmer phases. In colder periods a more heath-like vegetation took over.

There is a dearth of further cave deposits until bone fragments dated to 45,800 years ago from Peak Cavern, suggested either Bison or Aurochs (*Bos primigenius*) in the area. The bones presumably entered the system in sediments after death, but some had signs of gnawing by possible predators (Derbyshire

Caving Association website: https://thedca.org.uk). The Middle and Late Pleistocene (Table 1) interglacial period prior to the last Devensian ice advance is considered to be when Brown Bear (*Ursus arctos*) and Wolf (*Canis lupus*) first appeared. Many other more familiar species were widespread at this time like Pigmy and Water Shrews (*Sorex minutus* and *Neomys* spp.). Mole (*Talpa europaea*), Beaver (*Castor fiber*), and the three commonest rodents now; Wood Mouse (*Apodemus sylvaticus*), Bank Vole (*Myodes glareolus*) and Field Vole (*Microtus agrestis*), along with Red Squirrel (*Sciurus vulgaris*), Wild Boar (*Sus scrofa*), Red and Roe Deer (*Cervus elaphus* and *Capreolus capreolus*) would mostly be recognisable too. There were also some more exotic animals in the British fauna, but no direct evidence from our caves for their local presence.

The closest cave deposit revealing the nature of the bird fauna prior to the last Devensian ice advance is at Creswell Crags, only 50 miles east of the Peak District, in which 98 species probably representative of the whole region have been identified from a warmer period within the Devensian Glaciation. They include geese, various ducks, birds of prey such as Golden Eagle (*Aquila chrysaetos*) and Osprey (*Pandion haliaetus*). Red Grouse (*Lagopus lagopus scotica*), Black Grouse (*Tetrao tetrix*), Ptarmigan (*Lagopus muta*), a variety of waders and owls (some of which could have been predating lemmings in colder periods). There were also many smaller birds typical of woodland such as finches and tits, upland species like Ring Ouzel (*Turdus torquatus*), and those of open habitats like Skylark (*Alauda arvensis*) (Yalden & Albarella, 2009). The dating of hyena bones to between 42,000 and 23,000 years ago indicate a time period for these remains.

Late Glacial communities, 17,500–8000 BCE

Although many of our modern species were present before the last ice advance, their abundance would have changed with climatic variations. Few of the current mammal fauna would have tolerated the Arctic conditions and would have migrated south or perished (Yalden, 1999). The periods of rapid erosion of bare ground described in Chapter 4 are not suggestive of a vegetated cover required as a prerequisite for most wildlife. These periglacial conditions are thought to have continued until some 10,000 years BCE.

Table 4 summarises what happened next, much of which has been reconstructed from pollen and peat analysis. There were alternating colder and warmer periods termed the Older Dryas, Allerød (warmer) and Younger Dryas. *Dryas octopetala*, Mountain Avens, is a beautiful, white-flowered arctic–alpine plant abundant now only in Scotland in the UK. It shared the vegetation community with miniature trees like Artic Willows (*Salix* sp.) and Dwarf Birch (*Betula nana*). Prostrate Juniper (*Juniperus communis*), Crowberry (*Empetrum nigrum*),

TABLE 4 A synopsis of events in the Holocene Epoch since the last ice advance in the Peak District

Geological divisions	Climatic period	Approx. date for climatic period	Climate	Culture	Habitat	Human effects
Glacial (Pleistocene)		Pre 17,500 BCE	Arctic	Upper Palaeolithic	Dwarf willow, arctic herbs, freeze-thaw cycles dominate	Flaked stone & worked bone tools found in caves
Late Glacial	I Older Dryas	17,500–11,500 BCE	Tundra		Arctic-alpine herbs e.g. Mountain Avens (Dryas)	
	II Allerod	11,500–8800 BCE	Increased warmth		Open birch woodland, pine, herbs	
	III Younger Dryas	8800–8000 BCE	Return to cold arctic		As Older Dryas	
	IV Pre-Boreal	8000–7600 BCE	Sub-arctic, increased warmth		Open birch wood, development of stable soils	
	V/VI Boreal	7600–5500 BCE	Increased warmth, dry	Mesolithic	Pine & Hazel replace Birch, Elm & Oak increase towards end, small amounts of Alder. Sea levels rising	Burning for improved grazing to attract wild animals. Flint scatters found eg March Haigh, Holmfirth
Post Glacial (Holocene)	VII Atlantic	5,500–3000 BCE	Increased temperature & rainfall, oceanic sea level rose	Mesolithic	English Channel forms, increased Alder. Oak, Elm, Hazel, Lime deciduous woodland. Peat formation begins	Some land clearance, especially White Peak, Lismore Settlement, Buxton
	VIII Sub-Boreal	3,000–500 BCE	Drier, more continental	Neolithic then Bronze Age c. 1,400 – 500 BCE	Elm declines, appearance of first weeds. Expansion of Ash	Deforestation, pastoralism, cultivation, hill-top settlements eg Mam Tor
	IX Sub-Atlantic	500 BCE to present	Colder and wetter	Iron Age, Romans, Saxons, Normans,	Increased podsolisation & spread of Heather. Rapid growth of peat in wetter periods. Increased Birch, decline in Lime.	Continuous changes to natural vegetation, deforestation etc

Note: Different authors give varying dates for the different periods of climate change and for archaeological periods, which will depend on geographical location.

FIG 42. Crowberry in flower on Nether Moor below Blackden Edge, Kinder. This plant was recorded in post-glacial times and is still a prominent plant on blanket bogs and dwarf-shrub heath, especially in the Dark Peak.

Sea Buckthorn (*Hippophae rhamnoides*) and Rowan or Mountain Ash (*Sorbus aucuparia*) were characteristic of this tundra vegetation, which would have formed a type of heathland. The proportions and locations of the taller trees would probably have been determined by variation in shelter and aspect. Crowberry (Fig. 42) notably is still a prominent component of our blanket bog and heathland plant communities (see Chapters 8 and 9) and Rowan is widespread, but the others have gone. Open Birch woodland with Aspen (*Populus tremula*) followed the heathland phase as the climate warmed before reverting to the tundra-like conditions favouring Mountain Avens again (the Younger Dryas, Table 4).

Remains in Ossom's Cave (Manifold Valley) dated to the Younger Dryas show a local bird fauna not dissimilar to that preceding the Devensian ice advance, comprising Ptarmigan, Black Grouse, Golden Plover (*Pluvialis apricaria*), Eagle Owl (*Bubo bubo*), Jackdaw (*Coloeus monedula*) and Raven (*Corvus corax*): very similar to those recorded for the Older Dryas indicative of open-ground habitats, but probably with more species (Yalden & Albarella, 2009). The assemblage resembles a northern one of open moorland or tundra, perhaps with loosely wooded valleys. This type of assemblage has also been found further south in other caves, so the Peak District was not particularly distinctive at this time.

Mammals, being warm-blooded, are more tolerant of a range of climatic conditions and therefore provide less precise evidence compared with invertebrates, and the beetle fauna generally indicates conditions suitable for tree growth and productive vegetation that could support large numbers of herbivores in the warmer Allerød period (Yalden, 1999). The mammal fauna in the early Post-glacial period was represented by a mixture of Scandinavian and Central Steppe species together with some now extinct species. Caves in the Peak District show this was the last period when Woolly Mammoths (*Mammuthus primigenius*) roamed the area (Yalden, 1999), remains being found outside Peak Cavern in 1878 (Derbyshire Caving Association).

Ossom's Cave has also yielded remains of Horse, Reindeer (*Rangifer tarandus*), Red Deer and Steppe Bison (*Bison priscus*), along with flint artefacts associated with Later Palaeolithic, Mesolithic and Neolithic people (Table 4) showing human occupation as the ice receded and conditions ameliorated. Palaeolithic people roamed the Peak District in small groups as hunter-gatherers, and archaeological evidence has been found in over ten caves in the Peak District, including Ossom's, where Reindeer meat was stored in a stone box-like structure (Barnatt & Smith, 2004). Red Deer and Aurochs are essentially woodland animals, so were probably more prevalent in the better climatic conditions in a woodier environment. There is no evidence that the now extinct predators like Hyena (*Crocuta crocuta*), and Lion (*Panthera leo*) returned after the last glacial maximum (Yalden, 1999).

Amongst the largest deer of the Late Glacial was the Irish Elk (*Megaloceros giganteus*), a large relative of the Red Deer lineage with palmate antlers larger than those of the true Elk, remains of which have been found in Hoe Grange Cave. One skull from another site has been dated to 10,850 BCE, placing it in the period of open Birch woodland in the Allerød (Table 4, Yalden, 1999). Predators of this Late Glacial period nationally included Lynx (*Lynx* sp.), Brown Bear, Wolf, Red and, uncommonly, Arctic Foxes (*Vulpes vulpes and V. lagopus*) and Wolverine (*Gulo gulo* possibly). Of these, Arctic Fox and Brown Bear have both been found in local cave deposits (Yalden, 1999). The Wolverine (a large mustelid of the stoat and weasel family now native to North America and Northern boreal forests and tundra) was a rare mammal in the Late Glacial period. One, found in Wetton Cave (Manifold Valley), was associated with Arctic Fox and Reindeer remains and a mixture of Mesolithic implements. They are thought to have been present during the Allerød or Younger Dryas (Yalden, 1999, Table 4).

Small mammals were represented in this period by two species of Lemmings, a Vole, the Steppe Pica (*Ochotona pusilla*) and Mountain Hares (*Lepus timidus*), the latter commonly, all of which have been found in local cave deposits. Of these, only the Mountain Hare lives now in the Peak District, but it had to be reintroduced after becoming extinct (Chapter 8). The loss of the other now extinct mammals is surmised to have been largely due to climate and associated habitat changes. Some of the species seem not to have been abundant, but even those that were, such as Reindeer, disappeared as forests returned and spread northwards (Yalden, 1999).

POST-GLACIAL FLORA AND FAUNA

The Pre-Boreal and Boreal Periods, 8000–5500 BCE

A prolonged period of increasing tree cover marked this period, until altered by human activity. Soils had stabilised sufficiently to support considerable forest cover in the Boreal period when the climate was essentially warm and dry. Studies in the Dark Peak have shown a very different scene to that we currently see – much more wooded, no peat, open canopy forest with a range of woodland and open ground species on raw, developing soils, where leaching would not yet have removed the dissolvable minerals or reduced the pH (Tallis & Switsur, 1990).

As the climate ameliorated, species like Hazel (*Corylus avellana*) became abundant and probably formed a scrub habitat in the more sheltered cloughs and valley heads in the Dark Peak by around 6800 BCE. This was not necessarily uniform; Johnson *et al.* (1990) suggested that it developed in Middle Seal Clough, north of Kinder Scout, before it reached further north and higher on Robinson's Moss where Tallis and Switsur were exploring the site's history. Pollen analysis from Middle Seal Clough also reveals Juniper being mixed with hazel scrub or as an understory along with broadleaved herbs like Meadowsweet (*Filipendula ulmaria*) and some of the daisy family associated with a hazel and willow cover. *Dryopteris* ferns were also abundant. If modern comparisons hold true, these are likely to have been Male and/or Broad Buckler Ferns (*Dryopteris filix-mas* and *D. dilatata-mas* Fig. 43).

More light-loving plants like Heather (*Calluna vulgaris*), Crowberry and various grasses occupied open areas at higher altitudes. Pollen of other species found at the bottom of Robinson's Moss' peat profiles are indicative of more base-rich soils found now more in lowland meadows, such as Devil's-bit Scabious (*Succisa pratensis*), Sorrels or docks (*Rumex* spp.), sedges, and *Geum* and *Potentilla* species (related to Avens, Tormentil (*Potentilla erecta*) or Cinquefoil)

FIG 43. *Dryopteris* ferns were abundant in the first scrub and woodland, which may have looked like this: mostly Broad Buckler Fern.

(Tallis & Switsur, 1990). As temperatures continued to increase in the Boreal period, some of the warmth-loving trees such as oak, alder, elm and lime (*Tilia* spp.) began to appear in the pollen profiles and birches were displaced to higher altitudes. The higher hillslopes around Robinson's Moss were covered with a birch/willow scrub with open grassy moorland at higher altitudes (Tallis & Switsur, 1990). Pine forest had been confined to lower altitudes at around 7000 BCE. Over the next 2,000 years, however, the scrub spread rapidly upwards, reaching the higher moorlands, and probably extending as far up as Shelf Moss, Wain Stones and Kinder (all between 595 and 610 m in the Dark Peak) in response to a rapid rise in temperatures. At the same time, upland pine forests developed, invading and compressing the scrub zone by about 5500 BCE when the next wetter and warmer Atlantic period commenced.

On the limestone, the tufa analyses from the Rivers Wye and Lathkill suggest a spread of birch woodland as temperatures increase to a maximum in the 6565–5885 BCE period (Taylor *et al.* 1994), which is a fairly late date compared with those for the Dark Peak. Changes would have taken time to develop at different altitudes, with warmth-loving species appearing on lower ground first and possibly taking several hundred years to spread to higher altitudes. The higher altitude scrub communities also seemed to be quite sensitive to climatic or other changes during this period (Tallis & Switsur, 1990), suggesting alternating scrub and more open conditions.

Cave remains again tell us about animal life during this period of developing forest. Dowel Cave, near Earl Sterndale and Wetton Mill rockshelter in the Manifold Valley, have revealed a typical Mesolithic bird fauna (Yalden & Albarella, 2009): a large one of 22 species in the latter site, which included Water Rail (*Rallus aquaticus*), Red (Fig. 104) and Black Grouse, Capercaillie (*Tetrao urogallus*) and Grey Partridge (*Perdix perdix*). Stock Dove (*Columba oenas*) and Great Tit (*Parus major*) come from Dowel Dale, together with some undeterminable small birds, whilst Wetton Mill reveals Common Buzzard (*Buteo buteo*) and woodland species like Chaffinch (*Fringilla coelebs*), Spotted Flycatcher (*Muscicapa striata*), Blackbird (*Turdus merula*), Song Thrush (*Turdus philomelos*), Redstart (*Phoenicurus phoenicurus*) (Fig. 205), Robin (*Erithacus rubecula*) and Tawny Owl (*Strix aluco*). Remains of Greylag Goose (*Anser anser*) and Mallard (*Anas platyrhynchos*) were probably associated with the river below.

The bird fauna uncovered in Demon's Cave, near Ashford-in-the-Water, shows a larger number of wetland birds, possibly attributable to Beaver (also found in the cave deposits) creating pools on the River Wye nearby. The mixture of species in all these deposits indicate wooded conditions along with wetlands and open-ground. Ptarmigan and other tundra birds have disappeared, as might be expected, with the change of habitat. It is suggested that Mesolithic people were catching a range of birds for food, which would contribute to the remains found (Yalden & Albarella, 2009).

One particularly interesting bird species found from the Devensian ice age up to the Mesolithic variously at Demon's, Langwith (Derbyshire) and Ossom's Caves is the Eagle Owl. It was probably never common as a large predator, but Yalden & Albarella (2009) consider that there are a trickle of records, not just from the Peak District. Speculatively, small mammal and amphibian remains found associated with Neolithic/Beaker barrows on Longstone Edge, have been interpreted as from Short-eared (*Asio flammeus*) and Eagle Owl pellets, although no owl bones have been recorded alongside them. Could it be a native species the authors ask?

As far as the mammals are concerned, there is no evidence of recolonization after the colder periods by Bison, but Aurochs were widely distributed and would be large enough, plus browsing deer, to create clearings in the forest for other species to colonise along with plants favouring more open conditions. Wild Boar would have been present, along with Elk, Wild Cat (*Felis silvestris*), Lynx and Wolves, plus Mesolithic people who would have been hunting some of these animals (especially deer). Beavers and Otters (*Lutra lutra*) would have been associated with the rivers, with good numbers of Stoats (*Mustela erminea*), Weasels (*M. nivalis*) and Polecats (*M. putorius*). Remains of most of these, along with Badger (*Meles meles*), Brown Bear, Red Fox and Mountain Hare have been found in different caves in Staffordshire and in Derbyshire, demonstrating that they were widespread in the Peak District at this time (Yalden, 1999).

It is not clear when species began to decline or disappear from our fauna. Lynx and Elk were lost, although when is unknown as dating evidence is limited. Lynx might have been in Scotland as late as the Roman Period, but the extent to which its distribution was wider than this is not known. Similarly, dates for Elk bones include Mesolithic ones in Yorkshire and Berkshire, and its loss could be related to increased woodland cover and climate change. Reindeer met the same fate, with last dates in Scotland probably in Atlantic times (Table 4, Yalden, 1999).

In the Boreal period, starting at around 6700 BCE, the forested landscape attracted tribes of Mesolithic people living as hunter-gatherers. Evidence for human occupation is in the form of microliths (small stone implements of flint and chert), which are widespread and frequent in the Dark Peak in particular, possibly owing to the persistence of conditions unaltered by agricultural disturbance and to peat erosion uncovering them. Hey (2014) describes local efforts that have discovered evidence for considerable Mesolithic activity comprising several thousand flints, workshop sites and worked tools on the moors on the northern edges of our area at Warcock, Cupwith, Flint, March, Pule and White Hills within the Meltham and Marsden Moors. The suggestion is that the sites were part of summer hunting party routes within a largely wooded landscape full of potential game. Mesolithic people may have burnt some of the vegetation regularly, more particularly close to the tree-line, where there would have been more scrub and a relatively open forest. This was possibly to attract higher densities of herbivores like deer, which would benefit from the flush of more nutritional growth that develops after fire. Such burning could have impacted the extent to which this upland scrub regenerated and lead to the development of heathland and peat (Jacobi *et al.*, 1976, Tallis & Switsur, 1990).

The Atlantic period, 5500–3000 BCE

This period was marked by a significant increase in rainfall, warmer temperatures and more of an oceanic climate; probably warmer and wetter than it is now. The increased melting of the ice sheets and glaciers to the north, plus adjustment of the land as the weight of ice was removed (termed isostatic rebound) were responsible for changes in the relationship between land and sea that first cut off Ireland from the rest of Britain and then severed Britain from the continent by 5000 BCE. This prevented new land plants and animals with poor powers of dispersal reaching Britain and accounts for our relatively impoverished flora and fauna compared with that of continental Europe.

This warmer and wetter time was the period of maximum broadleaved woodland development, but also marked peat formation on the higher plateaux. We need to remember that soils take time to develop. After the last ice retreat, soils would have been thin, skeletal and low in organic matter. It takes time to develop the standard divisions of soils into the upper A horizon with higher organic matter, and a lower B horizon over the weathered bedrock (C horizon), as explained in Chapter 7. Thus we need to envisage climate changes affecting soils as well as wildlife and people.

As the climate improved, competition would have increased and species like birch, a relatively short-lived early coloniser, would have reduced as Hazel expanded. Pine too would have become marginalised as a more northerly species. Analysis of tree stumps (Fig. 44) remaining under or in the lower layers of peat from a wide range of sites in the Dark Peak shows that Pine was the main species

FIG 44. An ancient Pine tree stump at the bottom of the peat above Hollingworth Clough, near Hayfield.

in 29 sites and Willow in 14. Alder was only found below 425 m altitude, whilst Oak extended only locally above 500 m. Birch and Willow both occurred at higher altitudes. This represents the key components of the forest at the higher elevations at this time (Tallis & Switsur, 1983).

At lower altitudes below about 425 m, Oak, Elm, Lime and Alder became the dominant trees. Ash (*Fraxinus excelsior*) and a *Sorbus* (possibly Rowan) were only a small component of these forests. Hazel was probably an undershrub, as it is now, as well as a scrub component. This major expansion of lowland forest, at least around Longdendale, took place between 5600 and 4800 BCE. The higher altitude forest spread upwards until about 3500 BCE, although there was some fluctuation in conditions and therefore in forest-edge locations, possibly exacerbated or produced by burning (Tallis & Switzur, 1990; Tallis, 1991). Remarkably, pollen analysis from the landslip at Coombs above Charlesworth, shows that Beech (*Fagus sylvatica*) and Hornbeam were both present in the area in the later Post-glacial, although in small quantity compared with other main trees (Franks & Johnson, 1964). These species are not considered to be native in Peak District woods now.

On the Eastern Moors, pollen analysis shows a dense mixed oak forest by 4000 BCE, thinning with altitude with a more herbaceous or heathland flora on the plateaux, Birch–Alder woodland where conditions were suitable and reed swamp on the eastern part of Totley Moss (Hicks, 1971). As the climate became wetter, the lower altitude mixed oak woodland was replaced by Pine and Hazel with Alder/Birch swamp and reed beds in wetter areas. This Birch may have been Hairy Birch (*Betula pubescens*), which is better adapted to wetter conditions compared with Silver Birch (*B. pendula*) that prefers well-drained soils.

So far the forest descriptions are for the Dark Peak and its environs. We must assume similar forests and scrub habitats at this time in the South West Peak – there being little evidence to show otherwise. In the White Peak, however, evidence for woodland cover comes from one of the most important Mesolithic/Neolithic sites to be excavated at Lismore Fields in Buxton. Pollen analysis from its organic-rich deposits suggests increases in Alder and Birch between 5130 and 4090 BCE, consistently high Oak values, and constant amounts of Lime and Elm in a forest that had a maximum development by about 5000 BCE. However, glades and gaps in the canopy were implied by plants needing higher light levels, including Sorrels/docks and Bracken (*Pteridium aquilinum*) (Wiltshire & Edwards, 1993). Such glades could be related to human activities as well as grazing animals such as deer. Analysis of tufa deposits in Lathkill Dale supports the general picture painted by the Lismore Fields analysis, but without waterlogged woodland in which Alder flourished. Mixed deciduous forest of Oak,

Hazel, Lime and Elm, but with no or very little Ash (now the dominant tree in the Limestone Dales) was revealed (Taylor *et al.*, 1994).

It is pertinent that Vera (2000), supported by detailed examination of the beetle fauna (Whitehouse & Smith, 2010), considered the original forest of this time in Europe, including Britain, to be discontinuous, exhibiting a cyclical pattern of glades and forest over time interacting with grazing animals prior to significant human impacts. This would also account for pollen of open ground species being present and provide refugia for these species which then spread after forest clearance.

Forest development was not the only story from this time. This is the beginning of peat accumulation that is draped over so much of our moorlands. The wetter weather stimulated greater leaching, resulting in soluble minerals like calcium, manganese, iron and aluminium along with nutrients like phosphorus, moving out of the upper soil layers into lower layers, or being lost in soil water. The most susceptible areas would be more freely draining sandy soils, largely overlying the gritstones rather than the shales and in the loess deposits on the limestones. This leaching results in acidification of the top soil and reductions in fertility. Earthworms are much reduced over time and the soil fauna changes. Decay of dead material is more limited and an organic layer builds up. This is the environment that would suit Heather and other heathland species rather than broadleaved woodland. Hazel, being a species of soils with a moderately high pH, would be particularly susceptible in the higher altitude scrub in what is now moorland.

Woodland cover tends to protect soils from leaching as trees recycle minerals from greater depths, depositing them in the leaf litter to decay later. However, combined with the evidence for burning and opening out of the Boreal woodland by Mesolithic people, this could have had a significant effect on the nature of the forest. Charcoal layers have been found around the bottom of the peat, along with tree stump remains already mentioned, but some of these show evidence of being burnt in what are interpreted as major summer hunting grounds during the Mesolithic period (Tallis & Switsur, 1990). This is possibly the first time that human activity may have exacerbated the effects of climate change resulting in quite different vegetation.

Synthesis of information from several sites reveals peat formation over a 5,000-year timespan in the Peak District (Tallis, 1991). It seems to have accumulated first in basins which trapped water, commencing prior to the Atlantic period at around 7000 BCE, particularly on the shale grits such as Featherbed Moss (south of the Snake Pass) and Ringinglow Bog, where unstructured clays within the shale grits form a more impermeable base

(Tallis, 1991). Peat formation began at this early stage too on Robinson's Moss, but later on Alport Moor, with Willow scrub associated with the early peat formation under a vegetation dominated by sedges (Cyperaceae) and Meadowsweet type pollen. Hicks (1971) notes Alder–Birch woodland associated with waterlogging on lower-lying basins on the Eastern Moors where peat development also began early. It is noteworthy that these descriptions represent a totally different vegetation compared with that found currently on these areas.

Peat formation accelerated from 5000 to 4500 BCE developing on Kinder, Bleaklow, the cols on Featherbed Moss (Tallis, 1991), Holme Moss (Garton, 2017) and on Ringinglow, Leash Fen and Totley Moss (Hicks, 1971) on the east side of the Peak District. A third phase of peat expansion occurred from about 3500 BCE on Tintwistle High Moor (north of Longdendale) and at Leash Fen (Eastern Moors). Garton found peat development starting later, at the turn of the third to second millennia BCE, at the peat margins of lower sites at Arnfield and Ogden Cloughs below Tintwistle High Moor, where there seemed to be widespread evidence of tree growth below 480 m prior to peat development, but with more open areas of heathland at the same time. This shows the sequential development of peat over time progressing downslope well into the 2nd and 3rd millennia BCE.

Tallis considered there to be enough evidence linking the first and third phases of peat development to clearance of scrub and burning by Mesolithic people, but the second phase to be more firmly rooted in the increasing wetness of the Atlantic period, although burning could have played a part. However, Garton's (2017) investigations suggest that Mesolithic activity could have been much more variable than previously appreciated. He notes that peat development began after mostly small-scale erosion and burning episodes on Holme Moss at 500 m, but at the lower sites to the west there was a much longer time gap with peat inception starting later, well after early dates of burning evidence, but coinciding with increasing climate wetness. He postulates that there may have been repeated burning, possibly on a small-scale, into the second millennium BCE during the Neolithic based on radiocarbon dating plus the number of Neolithic artefacts found within the moorland environment, but that this was concurrent with change to wetter conditions.

The third, more rapid phase of peat formation witnessed a massive spread of peat in the whole of the South Pennines, spreading out from the former hollows, developing further on lower-lying ground such as Leash Fen and the moorlands to north and south of here, covering the high plateau tops and extending well down slopes of up to about 20 degrees, but stopping abruptly where there are significant changes of slope, as around the hill tops north of Longdendale

FIG 45. Peat covering the flatter tops of hills stops abruptly at the change of slope, visible here above Oyster Clough in Woodlands Valley where the pale colour of Mat-grass-covered mineral soils on the steep slopes contrast with the dirty-pinkish vegetation on the peat on the plateaux. Bleaklow and Howden Edge in the distance.

(Fig. 45). The peat depth is mostly 2–3 m, or more, on plateau tops, but 0.5 to 1 m or so on slopes. There are though pockets 6 m, or more, deep in places (on Ringinglow Bog and Leash Fen for example).

It is difficult to re-construct the nature of the vegetation overlying the developing peat. Pollen analysis and plant remains in the peat profiles suggest an abundance of sedges (which would include cottongrasses), Ericaceae, such as Heather, Cross-leaved Heath (*Erica tetralix*), Bilberry (*Vaccinium myrtillus*) and other closely related species, and *Sphagnum* bogmosses. Pollen of some individual species cannot be differentiated so are represented at the genus level. Crowberry (Fig. 42) was also a regular component, as it is now. The main cottongrasses, like now, would be Common Cottongrass (*Eriophorum angustifolium*) and Hare's-tail Cottongrass (Fig. 6, *E. vaginatum*). There are likely to have been some grasses as well, possibly Wavy Hair-grass (*Avenella flexuosa*) and Purple Moor-grass (*Molinia purpurea*), which are also familiar components of current vegetation. It is the proportions and patterns of this vegetation that are hard to visualise.

Sphagnum bogmosses are the most important group of species in a bog owing to their unique structure. They are distinctive in having a 'head' or capitulum of densely packed branches covered with leaves (Fig. 46), which contain large, dead hyaline cells with pores that can absorb and hold water. As a result, *Sphagnum* can hold between 16 and 26 times as much water as its dry weight, the amount varying between species. What is more remarkable, is that the phenols (a class of organic compounds) in the cell walls can also resist decay by inhibiting bacterial action. These special properties of *Sphagnum* have been harnessed for centuries. During the First World War for example, it was collected en masse, including from the Peak District, as a cheaper alternative to cotton-wool and used for wound dressings owing to its high absorbance capacity and anti-bacterial properties. By 1918, one million dressings per month were being sent from Britain to hospitals in continental Europe and elsewhere. Additionally, the thin film of water on the *Sphagnum* plant is more acidic than the soil water and this also resists decay. However, not all *Sphagnum* species are the same. Those with higher resistance tend to contribute more to the development of peat since

FIG 46. *Sphagnum* Bog-mosses, purple coloured Magellanic Bog-moss and several other species growing with Common Cottongrass (with wine-red ends of leaves) and Common Haircap moss in a blanket bog community. Bog-mosses were abundant on our peatlands in the Atlantic period and are a major component of the peat.

their dead material decays least. It is calculated that 99.5 per cent of the material forming peat consists of *Sphagnum* remains and that when the bog is dominated by Heather, cottongrasses and Deergrass (*Trichophorum germanicum*), decay is greater and peat accumulation less as a result. In drier interludes, *Sphagnum* cover would have declined or disappeared, locally or on a wider scale, depending on the detail of topography and precipitation, with plants like Heather and Crowberry increasing in cover. Thus, peat accumulation is likely to be greater in wetter conditions in which *Sphagnum* grows better.

This helps explain why peat development has not been a steady continuous process, but one which responded to climatic variations influenced by altitude, topography and human activities, particularly burning. Natural grazing animals like deer would also have had an effect. Both concentrated trampling and fire can damage *Sphagnum*. Tallis (1964) shows that there are *Sphagnum*-rich layers in peat profiles, sometimes with little in-between that are interpreted as drier interludes. Interestingly, one of the main *Sphagnum* species of the deeper, central peats is *S. imbricatum* (Imbricate Bog-moss, now divided into two species), which does not occur in the Peak District today but which was indicative of the wettest conditions on the bog surface, distant from any drying from gullies or streams. Other *Sphagnum* also feature in the peat profiles, more particularly *S. acutifolium* (now renamed *S. capillifolium* Acute-leaved Bog-moss), *S. magellanicum* (Magellanic Bog-moss now divided into two species of which one, *S. medium* occurs here), *S. cuspidatum* Feathery Bog-moss and *S. papillosum* Papillose Bog-moss. All these species still occur on the moorlands, although some more frequently than others.

Little information is available for our area during this period, but we can assume that a woodland and woodland edge bird assemblage was present. The mammal fauna of the Atlantic period would have largely been established by the end of the Boreal, but the spread of forests would have favoured some over others. Aurochs and Red Deer were hunted by Mesolithic people. Roe Deer would have benefited from the extensive woodlands, hunted by Lynx. Brown Bears and Wolves were already present, along with Red Fox, Otter, Beaver, Wild Cat, Badger and Pine Marten (*Martes martes*).

Sub-Boreal, 3000–500 BCE

The climate became drier and slightly cooler, although still warmer than today, in this period and the vegetation and animal life responded accordingly. At the same time, the practice of agriculture spread from the Middle East across Europe and into Britain. This marked a significant change from a generally hunting/gathering culture to farming. Domesticated animals and cereal crops were introduced, as people enclosed land and herded stock for the first time on a significant scale. In the Peak District, evidence of Neolithic occupation comes from widespread scatters of stone tools, commonly including knives, scrapers, borers and arrowheads made of flint (which would be imported into the region) and chert. Evidence for Neolithic presence is equally distributed across the limestone plateau, the Wye and Derwent valleys and the Eastern Moors. The most notable settlement site, of national importance, is at Lismore Fields in Buxton where Neolithic people followed the earlier Mesolithic occupation (Barnatt & Smith, 2004). Investigations have revealed emmer wheat, flax seeds, hazelnuts and other fruits, and seeds of crab apples, reflecting some of the plants that were cultivated or available in the vicinity. Barnatt & Smith (2004) postulate that the main valleys like the Derwent and Wye, although heavily wooded, would have provided important sources of fish, fruit and fowl as well as shelter in winter, whilst the higher limestone plateau and Eastern Moors supported summer pasturing where the canopy was probably lighter. Both the limestone and gritstone areas support lighter soils that would have been easier to till with hand implements, whilst the higher moors provided both hunting and grazing opportunities, all of which place Neolithic people throughout the Peak District landscapes. The introduction of polished axes, often made from hard igneous rock types from as far away as Cumbria and North Wales, provided more scope for woodland clearance and management on a significant scale. Timber was needed for buildings, fuel, fences and for various artefacts, and was cleared for agriculture.

This is the time of the first major monuments – the henges and chambered barrows. There are eight known chambered cairns built by the Neolithic people between 4500 and 2000 BCE, the earliest being small mounds with one or two chambers, the later ones being more massive, all on the limestone plateau. A good example is Minginglow, accessible via a concessionary footpath from the High Peak Trail, which evolved from an earlier small mound with a chamber to a massive, near-circular mound (Barnatt & Smith, 2004). Significant stone circles were erected within the two Peak District henges – Arbor Low (extant) south of Youlgreave (Fig. 47) and Bull Ring (removed) next to the community ground in Dove Holes. Defined by banks with internal ditches, the henges and stone circles were centres of territories divided by the River Wye and were used as communal monuments. There are also many later small stone circles on both the limestone and gritstone areas, sometimes with a bank, some with large stones, others smaller and some believed to have astronomical links (Barnatt & Smith, 2004). Look on Ramsley Moor (and elsewhere on the Eastern Moors), Beeley Moor, Eyam, Abney and Offerton Moors or the Nine Stone Close near Robin Hood's Stride on Harthill Moor.

FIG 47. Arbor Low stone circle, near Monyash. This henge was probably built in the Later Neolithic between 3000 and 2000 BCE. It is defined by a bank with internal ditch and is large enough to hold many people. Most of the stones have fallen or been pulled down. (John Barnatt)

You are more likely to see circular burial barrows in the Peak District – there are over 500, probably mostly constructed in the late Neolithic and early Bronze Age. They are generally about 10-30m across, grass-covered mounds that might have stood out in the landscape as bright white limestone or heaps of gritstone stones when freshly constructed on hill tops or ridges (Barnatt & Smith, 2004). Many of these are named 'low' as at Merryton Low and Cock Low, which may be derived from the Old English word *hlaw*, meaning a mound, as in Warslow (Hey, 2014).

There is evidence elsewhere that Neolithic people managed woodlands to supply timber for different functions. Barnatt & Smith (2004) believe that coppicing (cutting stems down near the base from which they re-grow) would have been possible with the available tools. Clearance for agriculture would initially have selected the more sheltered and south-facing locations with easily workable soils. Thus Neolithic people began to change the appearance of the landscape – a process that has continued through the ensuing Bronze and Iron Ages and carried on to this day. From this time onwards, the habitats and the plants and animals that occupy them will be intimately tied to the fortunes of the people and cultures that lived in the Peak District.

Only shortly into this period there was a catastrophic, Northern Europe-wide decline in elm pollen in all the profiles, and the Peak District suffered equally. The causative factors have been much debated, but Parker *et al.* (2002) suggest disease as the one cause capable of the large, abrupt and widespread drop in elm pollen (the major Dutch elm disease episode in Britain in the 1970-80s had a similar effect), but in combination with disturbance from the new Neolithic agriculture and possibly facilitated by climate change. It has been established that the Scolytidae beetles which carry Dutch elm disease, were present in mid-Holocene England and these are typical of edge habitats which would have increased with widespread harvesting and the opening out of woodland. Elm is highly palatable (containing high phosphorus and calcium levels). Moreover, they are in UK at the northern edge of their range and sensitive to winter cold, which would have been more prevalent with the more continental climate developing in this period. These factors in various combinations resulted in a major loss of Elm in the woodlands that had developed in the previous period (Fig. 48).

FIG 48. Wych Elm in Monk's Dale. The elm decline in the Sub-Boreal period saw a massive decline in elm pollen across the whole of Northern Europe.

Not just Elms were lost though (and we do not know which Elm species as their pollen grains are not readily distinguishable), pollen analysis confirms major forest clearance during this period by Neolithic people, although this was not always continuous nor consistent across the Peak District, with forest recovery at times. On the Eastern Moors, Hicks (1971) suggests only small groups of Neolithic herdsmen in the area at the beginning of this period, with minor clearances, possibly undertaken by semi-nomadic pastoralists with herds of cattle and pigs, growing a little grain in temporary forest clearings. More small-scale forest clearance took place in subsequent centuries using polished stone axes, but it was not until around 1500 BCE that forest clearance became widespread, attributed to Bronze Age activity. Hicks (1971) provides indications of regional forest cover being replaced by heathland on the Eastern Moors in general, with pasture more important than arable cropping and Hazel still growing locally. Long *et al.* (1998) noted a sequence of change between the second and mid-first millennium BCE in a detailed exploration of Stoke Flat. High woodland pollen, representing a diverse assemblage including Elm, Hazel and Beech in an open canopy with ferns and herbs below, showed increasing disturbance from the mid-second millennium BCE and it was concluded that a prehistoric arable field system on Stoke Flat was associated with local woodland management and pastoral activity. When this was abandoned, possibly related to soil loss, tree pollen declined further, which could result from upland grazing and climate change as peaty soils developed.

The Eastern Moors are immensely important for their Bronze Age settlements. There is widespread survival of field systems, cairnfields and enclosures on the high moorland shoulders east of and overlooking the River Derwent, where the altitude is low enough for mixed farming but which has fortunately escaped later loss through agricultural changes. These moors, now mostly heather-clad, would have supported many more families and farms than they do today (Barnatt & Smith, 2004).

There seems to have been considerable variation in the composition of woodland in the Sub-Boreal period across our area. Shimwell (1977) undertook pollen analyses in podsolised soils and from cave deposits in various sites across the White Peak and, despite interpretation issues of poorly preserved deposits, concluded that the forest in the period 1200 to 500 BCE comprised Oak, Alder and Hazel with low amounts of Birch. This is not dissimilar to that in the Dark Peak except that Lime and Ash occur in small quantities and in the Dark Peak, Alder was more abundant. However, even by this time, Shimwell envisages extensive open ground with the forests removed on the limestone plateau. This concurs with tufa analysis which suggests deforestation of the landscape adjacent to Lathkill Dale at about 2100 BCE, with episodic mud events detected in the tufa deposits possibly resulting from storms delivering sediment into the river (Taylor *et al.*, 1994).

The mammal fauna had evolved and spread with forest development into the open tundra and grasslands but changes now focus on losses rather than gains and in changing fortunes for woodland and open landscape animals. The major change in this period is the introduction and spread of domestic livestock associated with early farming. Sheep, goats, cattle and pigs were initially involved, with horses being domesticated and introduced rather later (Yalden, 1999). This could have reduced the effect of hunting on wild animals, at least at first. Although wild game was still on the menu for Neolithic people, especially Red Deer and Wild Boar as evidenced in cave remains, their importance was reduced. Aurochs seem to disappear at the beginning of this period in the Peak District, although they were still in southern England in the Bronze Age (Yalden, 1999). The last dated remains in the Peak District are 5–6,000 years ago in Carsington Pasture Cave (Cliffe, 2010).

Increasing dependency on domesticated animals will have changed attitudes to predators, with persecution adopted probably at times to protect livestock. Yalden considered that Brown Bears were still present, with remains identified from Fox Hole Cave (on High Wheeldon), which are thought to have been killed and eaten by Beaker people at the beginning of the Bronze Age, along with Red Deer, Roe Deer and various birds. Brown Bear teeth pierced to use in a Bronze-age necklace have also been located in Harborough Cave (near Brassington). Yalden (1999) suggests that wild Brown Bears became extinct around 2000 years ago in the whole country, even though they continued to be regarded highly (or not as you may determine) for bear baiting, dancing bears or for their skins. This interpretation is challenged by a recent re-assessment of Brown Bear remains in Peak District caves, which suggests a Late Glacial disappearance date, although material has been found mostly dating from the Mesolithic and Neolithic and even into the Anglo-Saxon period in the Yorkshire Dales, not far away (O'Regan, 2018). We know very little about the Wolf from this period. There are no later cave finds. Wolves were still in Britain according to later written evidence, but with domesticated dogs being present too, separating out dog from Wolf bones is not always conclusive.

With the reduction in woodland cover and spread of grasslands for grazing livestock, small mammals more typical of this habitat would be expected. Bronze Age dated finds in Dowel Dale and Fox Hole caves and Wigber Low (near Ballidon) reveal Water Voles (*Arvicola amphibius*) and Field Voles, plus smaller numbers of Common Shrew (*Sorex araneus*), Bank Vole and Wood Mouse, suggesting widespread grasslands. A couple of other mammals are of interest. Yellow-necked Field Mouse (*Apodemus flavicollis*) remains were found in the Neolithic layer in Dowel Dale but it no longer occurs in the Peak District. This is a deciduous woodland species which Yalden (1999) considers to have had a wider range in the past. The other is Hazel Dormouse (*Muscardinus avellanarius*) which is seldom found in archaeological deposits. However, again Dowel Dale harboured Neolithic remains and Ossom's Eyrie Cave (Manifold Valley) later Romano-British examples (Yalden, 1999). These finds hint at a once wider distribution, although Hazel Dormouse has been re-introduced into the Manifold Valley in recent years (Chapter 13).

Another species with a much contracted range now only in south and west Britain is Lesser Horseshoe Bat (*Rhinolophus hipposideros*) – a woodland animal that drops down on passing insects from woodland perches. Neolithic dated remains were found in Dowel Dale Cave and Wetton Mill Rock Shelter and the species persisted in the Peak District to Romano-British times at least around Ossom's Eyrie Cave nearby. This suggests a previously greater abundance and wider distribution, the loss of which could be linked to post-Neolithic woodland clearances (Yalden, 1999).

Bird data from caves is much less frequent than for mammals, but the Peak District has the distinction of having the best English examples of Neolithic bird remains from Dowel Dale and Fox Hole Caves. Both provide evidence of woodland and open environments at this time. Woodland birds were part of a large fauna found in Dowel Dale Cave. Great Tit was most numerous, but others which are still familiar included Robin, Common Redstart, Bullfinch (*Pyrrhula pyrrula*), Greenfinch (*Chloris chloris*), Goldfinch (*Carduelis carduelis*) and Blackbird. Hawfinch (*Coccothraustes coccothraustes*) was an interesting find, as it is now quite scarce. Predators in the deposit included Tawny Owl and Goshawk (*Accipter gentilis*), both woodland birds. But there were open ground species too like Skylark, Grey Partridge, Kestrel, Barn Owl and Stock Dove, although some need tree holes for nesting (Yalden, 1999). Interestingly, the remains of what is considered to be a Red-backed Shrike (*Lanius collurio*) were also found in this cave – now a very rare vagrant in Derbyshire, although once breeding close to the Peak District fringes in the 19th century (Frost & Shaw, 2013).

Fox Hole Cave also yielded Capercaillie – suggesting not only woodland but a coniferous component, whilst Black Grouse are more typical of scrubby edges. Sadly, Capercaillie is not now a Peak District species. Golden Eagle remains in the same cave suggest possible transport of the larger birds as prey from some distance on higher ground since most of the other remains were represented by woodland species, including Greater Spotted Woodpecker (*Dendrocopos major*), Jay (*Garrulus glandarius*) and Corvids.

CONCLUSION

During the 3,000 years since the Neolithic something not unlike the present-day countryside of farmland and small woodlands was established, in contrast to the previous woodland with clearings that had predominated. This is the time when our present-day mammal and bird fauna were established with more of those typical of open grasslands than in the earlier forested period. Other animal groups are likely to follow the same general pattern, but there is no evidence for this, nor is it sufficient to be able to establish the relative balance between species. This is the period when human activity began to impact significantly on the Peak District's wildlife, an effect that was about to grow and expand over the next 2,000 or more years, as portrayed in the next chapter.

CHAPTER 6

How the Last 2,500 Years Have Shaped the Peak District

THE LAST CHAPTER SHOWED HOW key elements of the Peak District's flora and fauna evolved, appeared, and sometimes disappeared in response to climate change or human activities. Some climate-related changes continue but there are now additional influences on our wildlife related to increased utilisation of our resources: soils for agriculture, Royal Forests for game, peat for fuel, mineral exploitation, industrial development, water use and forestry; together with some repercussions from these, sometimes from outside the Peak District such as air pollution. Together, these have played a major role in shaping the current landscape and its wildlife, but they cannot be disconnected from concomitant developments related to the success or otherwise of these ventures, which all add to the palimpsest of cultural and landscape heritage layers. These include introduction of new species with new incomers, development of the villages, transport routes and stately houses with their major estates, which are all major players in the Peak District now. This chapter continues at first with the time-line established in Chapter 5, but then focuses on the key land-use effects of resource use individually for clarity so you need to knit some of the strands together through the ages.

Further climate change marks the start of our story as the Sub-Atlantic period (Table 4) that followed the warmer and drier Sub-Boreal period commenced about 600–500 BCE and continues to the present. It is markedly colder and wetter than the preceding period, although it is suggested that the switch to this new climate was gradual. There have been some significant variations since too, including the Little Ice Age which followed a Medieval Warm Period.

THE IRON AGE STARTS

By the time Iron Age peoples were building hillforts and homes at the beginning of the Sub-Atlantic period, the temperatures were 0.5 to 1°C lower than present day, with more surface water. Living in a Peak District hillfort does not sound too comfortable, although they may not have been continuously occupied. There are eight Iron Age hillforts in the Peak District bestowing striking landmarks in key spots, mostly with a single rampart and ditch, although there is a double rampart at Castle Naze above Chapel-en-le-Frith. The largest hillforts are on Mam Tor, Burr Tor above Great Hucklow and Fin Cop above Ashford-in-the-Water, all on prominent hilltops, and possibly Canes Fort near Youlgreave. Carl Wark (on a conspicuous rocky hill above Hathersage) may be an Iron Age fort. A substantial single rampart, outer ditch and counterscarp bank crown the famous landslide on Mam Tor (Fig. 49). Evidence of circular buildings, small platform terraces to overcome the internal slope and a timber palisade have been found, some of

FIG 49. Mam Tor Iron Age fort built probably after the landslip scars were in place, although subsequent erosion has truncated some of the earthworks. A single rampart, outer ditch and counterscarp bank probably derived from regular cleaning of the ditch are visible along with many small platforms cut into the slopes. There are two older hilltop barrows within the fort. (National Trust)

which could pre-date the hillfort, along with two Bronze Age mounds (Barnatt & Smith, 2004 & Bevan, 2005). Owned by the National Trust, access is from the car park below and surrounding footpaths.

Some areas of earlier cultivation were probably abandoned as growing cereals and other crops would be more challenging in the wetter, colder climate and pastoral use became more prominent on today's middle level moorland, as shown by the persistent loss of trees in the pollen rain. Increased levels of soil erosion have been identified possibly related to the development of packhorse routes on some of the Eastern Moors (Heath, 2003).

INCREASED PEAT COVER AND DEPTH

At the same time, with increased rainfall and wetness, peat accumulation was re-invigorated for some 1000 years, as has been shown on the Eastern Moors and in the Dark Peak, although at different rates and times in different areas. This re-invigoration is attributed partly to tree clearance and to upland grazing (Long *et al.*, 1998) and partly to climate change. Water levels in the peat rose enough in places to include pools, as evidenced by fruits of Bogbean (*Menyanthes trifoliata*) and Bog Sedge (*Carex limosa*), the latter no longer occurring in the Peak District, found preserved in the peat. Rapid phases of peat growth were replaced by drier periods when cottongrasses and the moss Woolly Fringe-moss *Racomitrium lanuginosum* replaced the abundance of *Sphagnum* as the peat surface dried.

The pollen and plant remains suggest a peat bog with abundant *Sphagnum* or cottongrasses depending on wetness, but with other species not dissimilar to today like Heather, Crowberry, other *Erica* species presumed to be Cross-leaved Heath (*Erica tetralix*), and Cloudberry (*Rubus chamaemorus*) found in the top 130 cm of peat. The net result of changes in rates of peat accumulation are recognisable in the layers in the peat profile of dark-coloured, well humified peat (decomposed in more aerobic conditions) and lighter brown unhumified layers, which is when plants decay less. This layering occurs several times in most peat deposits in Northern Europe as a whole, and are known as recurrence surfaces. Tallis describes four for the Peak District, dated to 1000–2000 BCE, 600 BCE, 400 CE and 1300 CE, the second of which is suggested to be the start of the Sub-Atlantic period when climate deterioration started throughout Scandinavia and Germany as well as here. Between 1000 and 1200 CE there was a warm climatic standstill resulting from prolonged anticyclonic conditions over Europe.

Subsequent climatic deterioration greatly affected agriculture and land use in the whole country, culminating in the Little Ice Age between 1550 and 1700 when depressed crop yields led to widespread famine. From 1700 to about 1820 there was a significant increase in wetness, humidity, cloudiness and coldness of springs and summers, resulting again in renewed peat accumulation and wetter bog surfaces favouring *Sphagnum*. Between 40 and 300 cm was added to peat layers, the largest proportion on sites forming depressions as on Totley Moss and Ringinglow (where some of the deepest peats lie). This was the period when peat cover spread out over the flatter moors, joining up smaller patches and forming the interconnected blanket we see today (Fig. 45), although its condition has changed markedly (see Chapter 8).

THE FIRST HEATHER MOORLANDS

During the early Sub-Atlantic period, pollen analysis shows that woodland continued to reduce dramatically and permanently as the population increased and used more efficient tools. Simultaneously, there was an upsurge in heaths and grasses, including in the White Peak between 500 BCE and 1000 CE (Shimwell, 1977). This is probably the first time we see expanses of the now familiar Heather

FIG 50. Heather moorland on Gun Moor on the southwest edge of the National Park looking towards Back Forest covered by more Heather moorland. Shutlingsloe is the conical hill in the left-hand distance.

moorland (Fig. 50). The continual deforestation on the well-drained soils on the gritstones and loess on the limestone in particular, plus higher rainfall and cooler temperatures, would have led to more rapid leaching and soil impoverishment. The forest failed to regenerate probably owing to continuing grazing pressure (wild deer plus domestic stock). Since this time, a variety of factors has determined periodic expansion and contraction of Heather moorland up to the present (see Chapter 9). Heathland on the limestone might seem to be a contradiction since it favours acid soils, but as much of the limestone is overlain by acidic, windblown deposits (Chapter 4), leaching at a time of high rainfall could easily lead to heathland development if grazing prevented tree regeneration (Fig. 224). We know that there were extensive limestone heathlands in and prior to the 18th century (Farey, 1815 & Barnatt, 2019), although very few now survive (Chapter 14).

FURTHER LOSS OF FOREST

The evidence for continuing woodland clearance from the Iron Age onwards comes mostly from the pollen and peat analysis on the Dark Peak and Eastern Moors (particularly Tallis & Switsur, 1973), and limited pollen analysis from White Peak heathlands (Shimwell, 1977). There was still forest on the lower slopes in the moorland areas featuring Alder, Birch, Hazel, Pine, Elm, Lime, Ash and Beech. Woodland herbs were present, but there were also docks and various composites (Dandelion etc), which suggest cultivation or grazing disturbances (Long *et.al.*, 1998). At the same time as the White Peak plateau heathlands expanded with Ericaceous pollen dominating, Lime and Oak disappear from the profiles, although Birch increases. This does better on acidic soils and often expands after disturbance and abandonment. Lime is thought to have been one of the commonest trees in lowland Britain before the Sub-Atlantic period. Its loss may be climate related or exacerbated by woodland clearance. Lime was a tree of a thousand uses – forage for livestock, timber, good for furniture and carving, fibre for rope and sandals and producing a food sweetener, so it would have been an important resource. However, Small-leaved Lime (*Tilia cordata*) – one of the two native species in the Peak District – is known to be infertile in conditions that are too cold (Fig. 183; see also Chapter 7).

In general, the proportion of tree to non-tree pollen and the species involved suggest further intense clearance to have been initiated by the 10th century. There are hints in the pollen profiles of a temporary reduction in woodland clearance possibly explained by the extent of Norman Forest Law which was applied to large areas of the Peak District (as described below) but also to the impact of the Black Death in 1348–9 CE, which killed over a third of the British population. However, after a rapid expansion of population post 1550, there was renewed evidence of pollen associated with the spread of agricultural activities that has continued to the present.

NEW SPECIES ARE INTRODUCED

Until the beginning of the Sub-Atlantic, most colonising species had arrived naturally. With more invading peoples and new cultures and customs, there were opportunities to introduce species purposefully or accidentally. There is evidence in the pollen analyses of the first Walnuts (*Juglans regia*) as agriculture expands around the time of the Roman occupation and the Romans are credited with spreading the tree throughout much of Europe. But Britain is too cold with short growing seasons to produce a good nut crop. There are very few records now in the Peak District, although one Walnut tree with an amazing girth of 2.75 m has been found in Castleton (Wilmott & Moyes, 2015).

FIG 51. Sweet Cicely, a pot herb introduced by the Romans, flowering along a roadside verge in the White Peak.

More abundant plants that the Romans reputedly introduced include Goutweed (*Aegopodium podagraria*), variously also called Bishop's Weed and Ground Elder (as its leaves are similar to those of Elder) and Sweet Cicely (*Myrrhis odorata*, Fig. 51). Both are members of the Cow Parsley family (Umbellifers) and were introduced as pot herbs. Sweet Cicely has a sweetly aniseed flavour (rub the leaves between your fingers), whilst Ground Elder makes a more stringy, tangy dish. Ground Elder was also used to treat gout. Both plants are now regular features of road-side verges and waste ground (Chapter 12).

Newly introduced animals included the Black Rat (*Rattus rattus*) from Asia, possibly brought here by the Romans (Yalden, 1999). Its importance lies in its role as a vector of diseases like the Black Death, although it is now extinct in Derbyshire (Mallon *et al.*, 2012), being replaced by the Brown Rat (*Rattus norvegicus*), another introduction of Asian origin, but much later in the early 1700s, spreading throughout the country and fast becoming a pest everywhere by 1754 (Yalden, 1999). Brown Hare (*Lepus europaeus*) is also believed to have been introduced and is not uncommon across the lower parts of the Peak District (Fig. 175). Neolithic and later farming provided the ideal habitat for it.

FIG 52. Fallow Deer stag in Lyme Park. Fallow Deer were introduced by the Romans, but were kept in Deer Parks later by the Normans.

Fallow Deer (*Dama dama*) originated from Central Europe and are known to have been introduced into at least one site by the Romans (Fishbourne Roman Palace, Sussex) in the 1st century CE, although they were thought to be confined in 'vivaria' rather than running wild. The dearth of Fallow Deer bones in archaeological excavations suggests very limited numbers at this time and they did not become widespread until the Normans adopted them for venison in their deer parks. As then, Fallow Deer are mostly in old deer parks now in the Peak District, as at Lyme Park, Disley (Fig. 52) and Chatsworth, and in one free-living population originating from Stanton-in-the-Peak deer park.

MEDIEVAL AND LATER INFLUENCES

There are snapshots and hints of what the Peak District might have looked like at various times in its history, with many relict features still represented in the landscape and place names today. An interpretation of the Domesday Book (1086) in the Peak District by Barnatt & Smith (2004), shows a preponderance of pasture located in the White Peak and arable concentrated around the lower southern and eastern sides of the District in the valleys on more productive soils. Woodland was more widespread in the Dark Peak (probably along the valleys rather than on the higher moors) and southeast corner of the Peak District. Waste settlements are conspicuously dense in the northwest corner, around Glossop and Longdendale, but also along the Dove Valley and South West Peak. Bearing in mind that the Domesday survey was a record of the taxable wealth of the country, waste represented untaxable areas; small settlements and scattered farms below the threshold of viable subsistence agriculture and prone to abandonment. Domesday Book also shows that the majority of Peak District villages were established by this time, although many could be much older. Many still retain their medieval layout pattern, despite most of the buildings being replaced over time. Limestone plateau village layouts are either typically linear along a single street, with the farm yards and barns backing onto surrounding strip fields, as in Wardlow, Flagg and Chelmorton (Fig. 56), or built round a market square, as in Castleton, Tideswell, Longnor (not on the limestone, Fig. 53), Hartington and Alstonefield. Other village patterns can be seen in places like Ashford-in-the-Water and Warslow.

An important development, although one where the evidence on the ground is now rather thin, was the turning of large tracts into forests by the Normans for hunting deer and other game such as wild boar and hares. Peak Forest was one of the largest in the country, although not as extensive as Sherwood Forest or that

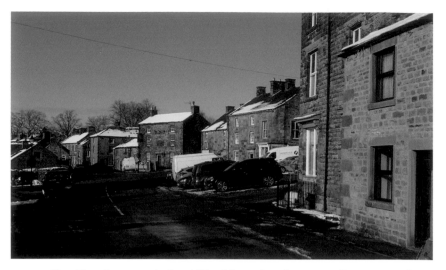

FIG 53. The old market square in the middle of the nucleated Longnor Village, now used as a car park.

focused on Pickering in North Yorkshire (Simmons, 2003). Hey (2014) suggests that the Forest officers had plenty of opportunity to illegally remove deer, graze their own stock and fell the woods as it was seldom visited by the kings. Macclesfield Forest was much smaller but abutted Peak Forest on its northwest flank and continued south to embrace much of the Staffordshire Moorlands in the South West Peak (Fig. 54). It seems to have been a pre-Conquest hunting ground for the Earls of Mercia, but during the 13th century began to change to pasture for horse studs and cattle-rearing farms. Two iron forges had been built by the Earls by the 14th century, the tenants of which had the right to cut oaks and clear land, thus contributing to the loss of further tree cover (Hey, 2014).

'Forest' comes from *foris*, meaning outside – in this case outside the common law, but inside the forest law (Simmons, 2003). It does not mean that trees were a major part of the landscape – and we have already seen how deforestation continued apace from the Iron Age and how the blanket bog spread out to drape the higher flatter areas. These Royal Forests were more likely therefore to be a mixture of heathland, blanket bog, woodland, scrub and other uncultivated land.

The Royal Forest of the High Peak comprised three districts: Longdendale, Hopedale and Campana and Edale Cross still stands where these three districts met on the route across the southern edge of Kinder Scout. Peak Forest had

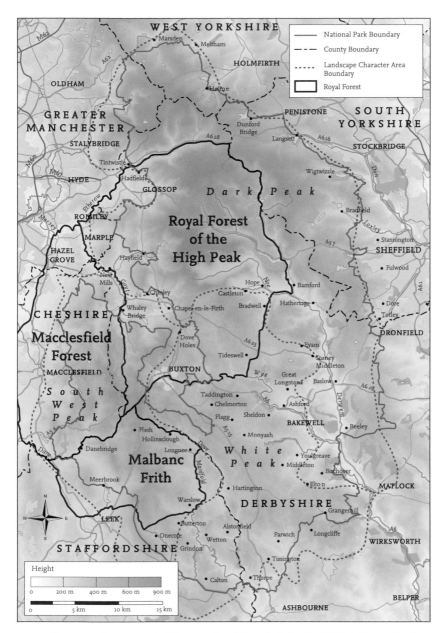

FIG 54. The approximate extent of the main Royal Forests in the Peak District in medieval times. (Contains OS data © Crown copyright and database right, 2020)

its own laws and forest officials and Peveril Castle above Castleton was an administrative centre and prison for those apprehended breaking forest law. The forest officers based near Bowden first built their own chapel in the forest, Chapel-en-le-Frith, in about 1225 (Hey, 2014). The forest laws were relaxed in the mid-13th century and gradually fell out of use. Activities that affect our story include the control of predators. For example, one set of officers set traps for wolves in spring and autumn and hunted them with trained mastiff dogs. Although wolves remained in the area until Tudor times, this 'control' would have placed them on the road to extinction.

The Eastern Moors were incorporated into a third, short-lived Royal Forest east of the River Derwent, while some of the eastern valleys like the Rivelin and Loxley were chases or friths in Hallamshire, an ancient district stretching across the southwest Yorkshire moors to Stanage (Stanage Pole marks its western boundary). Hey (2014) gives much more detail on the history of land ownership throughout the Peak District moors during the medieval period and beyond and makes a particularly striking comment relating to the condition of the vegetation from a review of complaints about overstocking on the Campana ward in 1516. This revealed that the grass was so bare from grazing nearly 1,000 cattle, four times that number of sheep plus 130 horses that the 360 deer would not last through the winter. Such heavy grazing would have fashioned the nature of the moorland vegetation communities as we will see in Chapter 9.

One of the earlier forms of resource use was peat extraction for fuel, which was quite widespread in the Peak District in centuries past, not only on the peat-covered moors but also on the limestone plateau, and evidence can still be detected on the ground. Medieval documents indicate rights of turbary (a legal right to cut turf or peat for fuel) existed in several areas on the limestone plateau, as on Sheldon Moor, and there is evidence of medieval peat extraction on a significant scale in the Edale Valley (Bevan, 2005) and in the Upper Derwent valleys (Barnatt & Smith, 2004), which continued to the 19th century. Smaller extraction sites would have been accessed for local farms and cottages. Shimwell (1977) concluded that the heaths on the White Peak plateaux had been largely denuded by the late 18th and early 19th centuries when the rights of turbary seemed to disappear.

Ardron's (1999) detailed investigations found that there has been near complete removal of peat down to the mineral soil around the fringes of the upland plateau, mostly now obscured by colonising vegetation. Some of the turbary rights specify removing the sods before extraction and subsequent replacement to re-provide grazing, so removal may not have resulted in bare, eroding peat on any significant scale. Look carefully and it is clear that there

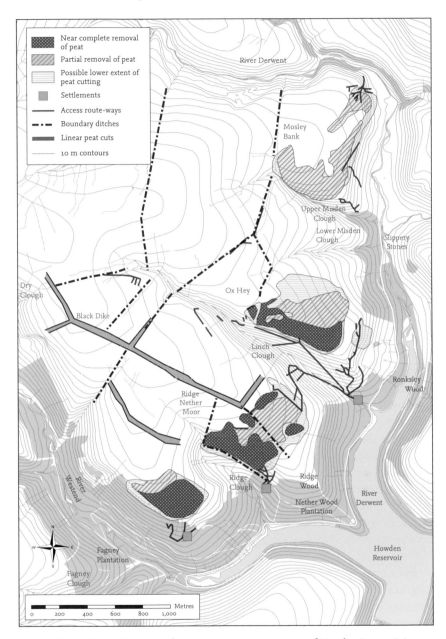

FIG 55. Peat cutting and associated access routes on moors west of Howden Reservoir in Upper Derwent Dale. (Paul Ardron)

is often an abrupt and linear change in vegetation from grasses like Purple Moor-grass (on wetter slopes) or Mat-grass (*Nardus stricta*, on drier soils) to cottongrasses with Heather still occupying deeper, more intact adjacent peat. The often straight, cut peat edge is frequently visible crossing low hills and running across the contours as on Span, just west of Crowden Brook, on Lockerbrook Heights and at Fagney Clough, both west of the Upper Derwent Valley, and rectilinear cells where peat has been cut are visible on Axe Edge Moor south of Buxton. The evidence is everywhere on the moorland fringes. Some of that on Axe Edge was much more recent and supplied the thermal baths at the Crescent in Buxton, which included a mineral or mud bath, also called a Moor Bath, involving immersion in a bath of peat mixed into the thermal water (Porter, 1984).

Ardron has calculated that something in the order of 34 million m³ of peat could have been removed from land above 370 m, with more from the thin peats on lower ground. The huge significance and scale of this is put into perspective if compared with the 30 million m³ calculated as extracted from the Norfolk Broads. Associated with the extraction sites are broad drains or linear cuts up to 50 m wide and 2 km long, often interlinked by drains or earthen boundaries to enclose a land holding. Scanning aerial photographs and the 1:25,000 OS map of the Dark Peak, reveals such long drains on the blanket bog west of Linch Clough (Black Dike) and further west on Ridge Nether Moor, in Upper Derwent Dale (Fig. 55). These cuts do not all function as drains, but their impact on the hydrology of the adjacent blanket peat is likely to be significant, with localised drying and reduction of *Sphagnum* Bog-moss cover and diversity. The cuts link to settlements or farms below the moor by hollow-ways down which peat was extracted on sledges. Figure 55 shows some of Ardron's interpretation of the evidence for peat extraction in part of the western section of Upper Derwent Dale and Barnatt (2019) provides another for Edale with peat being cut well into the 19th century, only being replaced by coal with the advent of the railway in 1894. There are numbers of remnant or well-developed hollow-ways, some of which are now footpaths, all around the moorland edges.

The evolution of land ownership and its use for agriculture gives us one of the most conspicuous features of the Peak District now, largely post-dating the Royal Forests: drystone walls. How these evolved helps explain landscape patterns. The medieval villages had large open fields incorporating access lanes between the long thin strips into which each was divided to share the richer and poorer land. These strips and the open fields were not necessarily bounded by hedges or walls at this time and most may have been important as arable land. However, where used primarily for grazing on the poorer land, some kind

FIG 56. The linear village of Chelmorton with medieval layout including back lanes and fossilised narrow, linear medieval open fields stretching out on both sides. The large rectangular enclosures to the far left are former commons enclosed in the late 18th and early 19th centuries.

of enclosures may have been present. These open fields were surrounded by unenclosed common pasture incorporating heathland and rough grassland. The dales could have provided timber and brushwood for domestic fires and some may have been more thickly wooded throughout the medieval period. An increase in Hazel, Oak and Birch woodland is evidenced in the pollen on the limestone plateau heaths at this time, for example (Shimwell, 1977).

The strips in the open fields began to be enclosed at least from the 14th century onwards, groups of which are fossilised in the landscape still today, often with their reverse S-shape still visible. There are examples near Taddington, Flagg, Monyash, and Chelmorton (Fig. 56) (Barnatt & Smith, 2004; Barnatt, 2019). Generally, it is thought that large pastures and arable areas not related to villages were defined by boundary banks or walls but not subdivided by walls or hedges. This pattern is visible on an abandoned medieval farm you can explore on Lawrence Field (at the northern edge of the National Trust's Longshaw

Estate), where there is an encircling bank and ditch and regular northwest to southeast aligned linear clearances with associated cairns of boulders and stones. House footings are visible closer to Padley Gorge wood. These remains have subsequently been reclaimed by heathland.

By the early 17th century, many areas had been enclosed into mixtures of larger rectangular fields or smaller irregularly-shaped ones. Enclosure has been piecemeal over a long period, with some open strips surviving until the Enclosure Awards of the 18th and 19th Centuries. Much of the land that was enclosed at this time was common land, and remarkably, there is very little common left in the Peak District today compared with other upland areas (like the Lake District). The Enclosure movement resulted in geometric field patterns over the old wastes and commons, with ruler-straight boundaries imposed on the landscape, often regardless of the topography. There are clues, therefore, for the broad date of wall patterns, although the walls themselves could be any age, depending on when they were last repaired or rebuilt. The consequence of this history is hundreds of walls in the Peak District, built using the closest source of stone; so limestone walls in the White Peak (Fig. 13) and gritstone walls in the Dark and South West Peak, with a little intermixing on the geological boundaries.

As part of the agricultural exploitation of the Peak District, a large number of granges were established, often on new sites, but also replacing existing farms in the 12th and 13th centuries during a warm, dry period. Over 50 granges are known to have existed, many of which are still farms today. These were important in exploiting often quite remote areas on the high moors, such as in the Upper Derwent Valley, as well as the limestone plateau. These granges were owned by monasteries outside our area. Most were used for cattle and sheep grazing and sometimes horse studs. About half were owned by the Cistercians, who were highly involved in the wool trade and the Peak District became one of the great sheep runs of 14th-century England. This has had a lasting effect as will be described in Chapter 9. There are still place names that reflect this part of the Peak District's history, such as Grange Hill.

Associated with the increasing wealth generated in or near the Peak District, many halls and manor houses were established, but only a few are of particular importance now. After the dissolution of the monasteries, courtesy of Henry VIII, Sir William Cavendish, the third husband of Bess of Hardwick (Hardwick Hall lies near Chesterfield), bought the Chatsworth property and began to build a new house in 1553 using the Ashover Grit. The Earl of Shrewsbury bought the Basingwerk Abbey manor of Glossop and Welbeck Abbey's land holdings in the Upper Derwent Valley but sold them to Sir William Cavendish by 1554, thus consolidating the early Chatsworth Estate.

Said to include a deer-park when Bess of Hardwick purchased Chatsworth, the present one surrounds the house and formal gardens. The South Park, south of the gardens, contains many veteran oak trees, which are of prime ecological importance (see Chapter 10). The park itself has been enlarged and has changed over time, more particularly when Lancelot 'Capability' Brown was commissioned to transform the garden and park into a then fashionable naturalistic landscape between 1750 and 1765, the main features of which still prevail (the park and garden are registered Grade 1 by Historic England). Many trees were planted, singly and in belts or clumps, including a variety of non-native species, some of which were specially imported in 1759 from America. Carriage drives were routed to capture carefully planned views; smooth, rolling grassland and a natural-looking lake were created; the ground contours altered and the river straightened and dammed in order to provide more attractive views of water from the house.

In 1826, Joseph Paxton, who trained at Kew Gardens, was appointed Chatsworth's head gardener and was to become the most innovative garden designer of his time. His influence on the Chatsworth gardens can still be seen, where he sometimes altered Brown's designs. A number of features still survive, including the Arboretum, a set of greenhouses and the Emperor Fountain. The latter was installed as the world's highest fountain to impress Tsar Nicolas I of Russia, who unfortunately died before he could visit. It is served from a lake high on the moors to the east to provide the head and has attained a splendid 90 m height when on full power. The high lake was later used to generate electricity for the house before mains were available and has now been re-serviced with a new turbine. This is the Joseph Paxton who worked with Edward Milner to design the Paxton Suite and associated buildings in Buxton. The Dukes of Devonshire were instrumental in supporting the development of Buxton as a spa town in the later 19th century, having built the Crescent and contributed to other developments. We will see more of the influence of the Dukes of Devonshire in relation to mineral extraction.

The second great historic house in our area, Haddon Hall (a Grade I listed building and Grade I Registered Park and Garden), has a different history. Its origins date back to the 11th century. It was held by William Peverel (giving his name to Peveril Castle above Castleton) in 1087 at the time of Domesday, but was acquired by the Vernons in 1170. Dorothy Vernon married the first Earl of Rutland, Thomas Manners in 1563, and the house remains in this family. Much of the hall, with additions in the 16th century, remains (Fig. 57). The beautiful terraced gardens overlook the Wye southeast of Bakewell, and the Park (closed to the public) consists mostly of open pasture with numerous scattered trees, including Oak and Limes dating from 1879.

FIG 57. Haddon Hall, situated above the River Wye east of Bakewell. It is a magnificent Grade I listed building and Registered Park and Garden, originating from the 11th century.

Lyme Park, Disley, is the third important grand house. Familiar to anyone who enjoyed the BBC's adaptation of *Pride and Prejudice* as Darcy's country estate, the mansion (Grade 1 listed by Historic England) is surrounded by formal gardens and set in a medieval deer-park, which originally formed part of Macclesfield Forest. It has sweeping moorland from which extensive views towards Manchester and the Cheshire Plain can be enjoyed. The Legh family acquired the estate in 1388 and it remained in the family until 1946, when it passed to the National Trust. The house, the largest in Cheshire, dates from the 16th century, although there were later modifications.

EXTRACTION OF GEOLOGICAL RESOURCES

Our rocks and associated minerals are the key natural resource that has been extensively exploited in the Peak District over many centuries. The landscape visible today bears the signature of limestone, mineral ore, coal and some unique stone extraction. Some of the products are of national economic importance, whilst others are more historic or decorative.

The Orefields

The veins containing minerals are labelled according to their structure as rakes, scrins, flats and pipes. These are mostly local lead miner's terms for the types of mineral deposits. A rake marks minerals found at geological faults, sometimes several metres wide and up to hundreds of metres deep, which run in lines across the landscape (Fig. 58) – although sometimes the term 'rake' is used for all ore features within an orefield. A scrin is smaller than a rake, shorter and often only a few centimetres wide. A flat is a deposit that lies roughly horizontally, following the bedding planes in the surrounding rock and often over a volcanic layer, usually followed underground. Finally, a pipe, represents ore deposits that are long and narrow, but lie horizontally. They may be ancient caves filled with mineralised material (Barnatt & Penny, 2004).

Lead, zinc and copper ores as well as gangue minerals (those not of prime importance associated with the others); fluorite, baryte and calcite have been important since before the Roman occupation. Lead, probably exploited in

FIG 58. Two parallel lead rakes running along the slopes above Tansley Dale, a tributary valley of Cressbrook Dale. The characteristic hummocks, trenches and waste heaps are all visible.

prehistory, was the most important ore mined from Roman times, as evidenced by several Roman-period pigs of lead (the unrefined lead was cast in blocks called pigs) that have been found. Mining continued subsequently, with mines around Wirksworth mentioned in the Repton Abbey records in the 9th century. The Danes later destroyed Repton and took over the mineral rights of the High Peak limestone areas which subsequently became known as the 'King's Field'. Lead mines were mentioned in Domesday Book and continued to expand in Norman times. Their heyday was in the 16th and 17th centuries when the greatest volume of ore was extracted, along with new veins opening such as on Eyam Edge. Gunpowder, introduced in the 1660s to blast adjacent limestone and access the thin ore deposits, revolutionised mining across the orefield from the early 18th century enabling faster extraction, with further innovations in the 18th and 19th centuries including the use of compressed air for drilling shotholes, smelting hearths and cupola furnaces (Barnatt, 2019).

Britain has been Europe's largest lead producer to which the Peak District made a significant contribution (Barnatt & Penny, 2004). Calibrating lead levels in alpine ice cores, peat cores, speleothem and human remains has shown peaks of activity from the Peak District in the 12th century, but with a dip related to the Black Death in 1349. As the toxic effects of lead in the body compromise immune systems, miners were more susceptible to disease and rats spreading the Black Death would have been plentiful in mines. Lead was exported widely by ship with many uses such as construction projects like palaces and abbeys (Loveluck *et al.*, 2020). One use was white paint and Vermeer's captivating painting, *Girl with a Pearl Earring* (c. 1665), used white paint made from Peak District lead to highlight the earring. From about 1850 lead mining declined as the price dropped, the best veins had been extracted and drainage became more of an issue, although extraction continued into the 20th century in Millclose Mine near Darley Bridge and 'miner-farmers' continued to extract lead to supplement incomes (Barnatt, 2019). The mining laws of the Derbyshire Orefield were codified in 1288 and the Barmote Courts established, one of which still operates today to resolve disputes.

There are some 2,000 named mineral veins that have been worked, criss-crossing the White Peak, which are highly visible and form distinctive features in the landscape. The remains of the lead mine at Magpie Mine, a Scheduled Monument cared for by the Peak District Mines Historical Society (PDMHS), are the best preserved in the Peak District (Fig. 59). There are heaps, hummocks, shafts and remains of buildings on the site, including an engine house for the large Cornish pumping engine (1869), deep mine shafts (one reaches 224 m), a circular chimney built in 1840 for an earlier engine, a square chimney to serve the winding engine (1840), a winding drum on the outside of the engine house, and

FIG 59. Magpie Mine near Sheldon. From right to left: an 1840 chimney first built for an earlier engine, the 1868 pumping engine house in front of which is the main shaft and 1950s headframe, the winding house, and the 1840 square chimney with one end of its boiler house.

a circular powder house. Two gin circles are visible. Heritage days are run by the Society (check the PDHMS website for more information), but public footpaths cross the site, which is also in open access land.

As mine excavations went deeper, water tables were encountered requiring more drainage, particularly from the 17th century onwards. Many drains, called soughs, were driven deep into mines and are still visible, for example in Lathkill Dale. Over 100 soughs have been recorded according to Ford and Gunn, some of which are over 1 km long and draining catchments of up to 50 km². Such extensive drainage has lowered the water table in many dales, revealing caves and sometimes reducing flows. Some soughs have captured water that would otherwise have flowed into the dales and deported it elsewhere (see Chapter 4).

From the floristic perspective, the heavy metals are very toxic to plants such that only well adapted species occur on those wastes with the highest residual levels. These form a distinctive and internationally recognised and important plant community termed Calaminarian grasslands, more of which in Chapter 14. Ford and Gunn (2010) give estimates of the total number of ore-field shafts at 50,000 to 100,000. Although many are filled or partially obstructed, a good number are still open and care needs to be taken if visiting them.

Other minerals were also exploited early in mining history. One of the most important was copper from Ecton Mines in the lower Manifold Valley, along with some lead and zinc. There is archaeological evidence that copper was extracted in the Early Bronze Age: it being only the second such site in England (Barnatt, 2013). Deep Ecton Mine, the richest of the mines, was the deepest in the country in the 18th century (over 400 m deep – a staggering 100 m below sea level), and very profitable for some 30 years, but less so for much of the 19th century. The near vertical ore deposits were difficult to work, facilitated by investment in new technology by the Duke of Devonshire, who owned the mineral rights. The Boulton and Watt engine house perched on the hilltop above the mines built in 1788 is believed to be the earliest surviving example in the world used for winding out ore. Now restored by the National Trust, it is accessible on guided tours.

The mining relics are numerous and important archaeologically and ecologically, with surface hillocks and hollows, old waste heaps with bare coarse scree (Fig. 182), the remains of buildings and mining structures. Some of the extensive underground workings are still accessible, apart from those below the River Manifold which are flooded. There are extensive passages, underground engine chambers and major levels for extraction and drainage. The Ecton Mines Project (Barnatt, 2013), provides a wealth of information.

Fluorspar and barytes

The main minerals being exploited now are fluorspar and barytes. Fluorspar is the only significant source of fluorine used in the production of hydrofluoric acid and small amounts are used as flux in steel manufacture and in other industrial applications. Production in Britain is confined to the Peak District orefields after the North Pennines' ones closed in 1999. Although extracted from a National Park in a landscape of high value, it is the balance between a nationally important resource and an equally important landscape that has to be considered. Barytes is used in drilling fluids in oil and gas exploration and as a filler in paint and rubber. It occurs with fluorspar and is recovered as an important by-product.

Extraction of these ores is concentrated along fissure veins several kilometres long and up to 10 m wide, mostly in the past as open-cast workings. The main extraction sites lie within the Monsal Dale limestones (the highest strata). The Hucklow Vein on Eyam Moor, the main workings for fluorspar, has permission for mined extraction until 2029. There is a processing plant at Cavendish Mill near Stoney Middleton, which includes current and former waste lagoons. This is the sole supplier of fluorspar and barytes in the UK.

Unique Blue John stone

One of our most famous minerals is Blue John Stone; a unique material found worldwide only under Treak Cliff at the edge of Castleton. It is a semi-precious form of fluorspar discovered by miners looking for lead, which incorporates coloured bands of purple, blue, white or yellow. Apart from being part of the mineralisation process, its origin is obscure. Historically there have been only 14 distinct veins of Blue John found in Treak Cliff and Blue John Caverns, with names like 'Miller's Vein', 'Old Tor Vein', or just '5 Vein'. You can see some on these show caves' tours. Blue John was fashionable in the Regency period in the early 1800s as tableware, jewellery and other decorative pieces. It has graced famous tables including in Buckingham Palace and Chatsworth House. A new, 15th vein was found in Treak Cliff Cavern in 2015 more by accident than design, the first since the mid-1800s, and was called 'Ridley Vein' after its founder (Fig. 60). Only small amounts are extracted each year, largely still by hand with

FIG 60. Ridley Vein, Treak Cliff Cavern, the first new vein of Blue John Stone found since the mid-1800s, named after its founder. (Gary Ridley)

FIG 61. An example of the fine decorative pieces made from Blue John Stone, here a bowl with beautiful markings. (Treak Cliff Cavern)

picks and shovels, from which beautiful, mostly quite small, decorative objects are made (Fig.61). You can see these and impressive rock samples in several Castleton gift shops, those associated with the caverns and in some other specialist jewellery shops.

Coal mining

Preservation of the remnant Coal Measures within the western synclines and the proximity to minor coal seams in the eastern side of the Peak District have been critical for the development of a small, but important, coal industry. Mines extended from Mosley south nearly to the Roaches and along the eastern moors with concentrations around Ringinglow, Owler Bar, Robin Hood and Beeley Moor (Barnatt, 2019). Mining began in medieval times and peaked in the 18th century, but increasingly mines were abandoned through the 19th century as better coal could be imported from elsewhere. Chatsworth Estate extracted more of its coal in the First World War when fuel was in short supply to support the house and greenhouses. Locals in Longnor also describe how coal was extracted on the west side of Axe Edge for domestic use in the Second World War when fuel was scarce. The moorland location of many mines has ensured survival of visible evidence, which is much rarer in lowland Britain (Barnatt, 2019).

There are seven separate coal seams, most of which were less than 600 mm thick, although one, called the Yard Seam, was up to 1.5 m thick, making its extraction more viable. This was the main coal seam exploited in the western fringes. Faulting is more significant in the south of the seam, effectively breaking it into pockets which were separately extracted in local mines. Some of this could be obtained by digging tunnels into the hillsides, particularly at the top of the Goyt Valley where faulting brought the seam closer to the surface. Elsewhere, extraction needed shafts and tunnels as shown by Barnatt (2019), who provides plans and details of several mines in the western Peak District. The Match Marsh and Goyt's Moss collieries are of exceptional archaeological importance lying high on Axe Edge Moors extending south to Orchard Common. These illustrate site complexity and details of evidence still visible, including over 300 shaft sites, often with associated waste heaps, hollow ways, tracks, causeways, tramway beds and ruined buildings. Dranfield (2008) tells the story of the coal mines of the Goyt Valley, especially those associated with Errwood Hall – the remains of which are still embedded in the hillside above the Reservoirs. The coal seams include one delightfully named 'Big Smut Seam', only 225 mm thick, extracted in Whaley Bridge, and the 'Simmondley' Seam, a thicker 375 mm exploited on the Errwood Hall Estate. This latter seam, although difficult to extract, produced good quality 'candle' coal because it caked on the fire and needed poking to produce a candle-like flame. This coal was in high demand for forge work and the Buxton Lime Co., subsequently ICI, purchased a regular supply. Much was also used for the local lime-burning industry (more of which later).

In the 1790s, coal mining would have been a major local source of employment, with possibly half of Burbage's working residents employed in the South West Peak industry. The Buxton Mines, not being large enough for significant investment, largely remained dependant on physical labour rather than machines, with only one stationary steam engine for haulage at Cistern Clough and horses for coal winding and transport (Roberts & Leach, 1985). Despite these limitations, it has been estimated that Buxton coal mines produced about 1.8 million tons (1.77 million tonnes) of coal at competitive prices, which supported the local economy at this time.

It could have been the first of the area's turnpikes that gave better accessibility facilitating the use of horses and carts rather than packhorses that stimulated more coal extraction. Further improvements for coal haulage came with the development of the canal system – the Peak Forest canal and the Cromford canal south of Matlock are the closest, but were linked (due to the impossibility of constructing a canal economically across our hills) by the Cromford and High Peak Railway, now mostly a National Park Trail.

On the eastern side of the Peak District, mining relics comprise clusters of close-spaced shaft hollows and mounds with an occasional linking causeway, sites of adit entrances and gin circles. Barnatt (2019) provides layout maps of two of the more important sites at Stanage Collieries and Baslow Colliery at Robin Hood, where the Ringinglow Seam was worked from medieval times. It is notable how many of the names for the mines are still on OS maps and used locally, although sometimes with altered spellings, and many of the footpaths and old tracks we now use are believed to have been either for access for moving coal or for workers to reach the mines.

Bearing in mind that there was very little coal compared with the thickness of overlying clay layers in the Coal Measures, enterprising exploiters used these where suitable to make bricks. Dranfield (2008) describes brick works in Furness Clough and at Spons Colliery, both of which had brick kilns adjacent to the pit head, thus minimising travel for the extracted coal to fire the kilns.

Bunting (2006) mentions accidents from inhaling methane and falling into a bell-pit from ropes in the 13th and 14th centuries. One indirect repercussion of the local coal mining described by Roberts and Leach was the response to the unacceptable death toll from mining accidents in the country as a whole in the 1880s, of which several were associated with Peak District mines. On seeking a centrally located, safe location for undertaking mining safety research in the early 1920s, a site at Harpur Hill, Buxton was identified close to existing noisy quarries where generating more noise would not be considered a nuisance. Thus the Safety in Mines Research Establishment was founded, now transformed into the national Health and Safety Laboratory and an important local employer.

Rock quarrying

Chert, the flinty silica found mostly in veins in the uppermost limestone beds, was worked into tools in prehistoric times, being chipped like flints to produce flakes and sharp edges. Both scrapers and awls were made from local sources by Late Mesolithic people (Barnatt & Smith, 2004), with examples found across the Peak District. The main historic chert quarries and mines are in the Bakewell area, at the edge of the limestone landscape. Holme Bank and Pretoria Mines, were accessed through adits at Bank Top. The chief demand for chert was developed by Josiah Wedgewood for grinding calcined flint used to whiten porcelain. Initially moved by cart or packhorse to Staffordshire and Yorkshire, the advent of the canals allowed transport to Stoke-on-Trent via the Cromford Canal from 1793. One of the more important uses was for the new creamware developed by Wedgewood in the Potteries near Stoke-on-Trent. The chert was ground in flint mills such as that at Cheddleton, just south of Leek, a grade II listed complex dating back to 1253, but used for flint grinding from the late 18th century. Other chert pits were west of Great Longstone, in Ashford-in-the-Water, Over Haddon pastures and Bonsall. All the mines had closed by the 1960s (Bunting, 2006).

Just west of the chert industry was another of significance: Ashford black marble. This is not true marble, but an impure form of limestone impregnated with bitumen which turns black when polished and has been much in demand for decorative purposes. Although a piece of Ashford marble was found capping a Bronze Age cist burial on Fin Cop (Archaeological Research service), its heyday was mostly from the late 17th to the early 20th centuries. The main sources were Arrock Quarry and another in Rookery Plantation where it occurred in narrow beds. Large slabs were taken to a mill nearby for cutting, grinding and polishing. Etchings and engravings were used for the first designs but this was revolutionised by the application of inlaying after 1835 with striking floral and geometric designs patronised by royalty and local grand houses like Hardwick Hall. There were other locally extracted 'marbles' like the Duke's Red Marble, actually a heavily hematised limestone capable of being polished to produce a blood-red colour, believed to have been taken from Alport, Youlgreave or Newhaven quarries (Thomas & Cooper, 2008). Small fragments are visible in the beautifully made chessboard table (Fig. 62). There is a lovely inlaid table in Ashford-in-the-Water's church, but you will find the best local collection in Buxton Museum and Art Gallery, several items of which were exhibited by Derbyshire inlayers at the 1851 and 1862 Great Exhibitions.

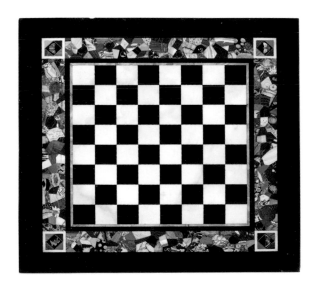

FIG 62. Ashford black marble chessboard table top, made probably in the second half of the 19th century, with inlaid materials including Duke's Red Marble, shells and stalactitic barytes from near Youlgreave. Purchased with assistance of the V&A. (Buxton Museum and Art Gallery)

Limestone has been used for hundreds of years in field walls, buildings and field barns, but it was also used extensively for spreading on the fields on the acidic loess soils and on land throughout the Dark and South West Peaks. Lime is a good ameliorant on shale soils as it flocculates clay particles and helps produce a better tilth. Lime from the Peak District was recognised by farmers for its high quality and was much sought from miles around. Farey (1813) describes the 'avidity with which Derbyshire farmers search after limestone ... and how they toil with it over the hilly roads of an uneven country, to the distance of eight or ten miles'. Moreover, he expresses some surprise that the 'farmers of Cheshire and Yorkshire come near twice these distances, to fetch the Peak Lime, in carts; and that by means of the canals, it is distributed around from Crich and Peak Forest to the distance of 30, 40 or more miles for Agricultural purposes!' He notes too that the dark-coloured limestone makes a very white and flowery lime when slaked, which had the greatest repute with farmers.

This combined demand for agricultural liming and the quality of the Peak District limestone led to a major industry, mostly on a small scale. Lime was burnt to produce quicklime (calcium oxide) and this and hydrated lime (quicklime mixed with water to make it safer and less caustic) were extensively used for mortar and cements and were in demand as a flux for lead and iron smelting, as well as being applied to the land. There is even some evidence of its use in Buxton by the Romans. Lime kilns were scattered all round the White Peak in particular, but also in other areas, with remains still visible in fields or waysides. A walk along the Monsal Trail east of Miller's Dale Station will take you to some commercial kilns where buildings and the chute system are all preserved. A large, well-preserved kiln remains by the footpath to Bostern Grange in Tissington. Smaller domestic scale kiln remains and the trenches for them can be seen widely across the limestone fields (Fig. 63).

According to Farey there were also concentrations of lime-burning associated with the canal basins in Buxworth (called Bugsworth then), Whaley Bridge and along the canal in Disley, all using limestone from Barmoor and Dove Holes' quarries in Peak Forest. In addition, there were private kilns on farms where the farmer could sell any surplus made. The spread and extent of these works accounts for many of the hillocks, hollows, odd stone enclosing walls and other structures, especially in the White Peak today.

The biggest effect on current landscapes and wildlife are where lime-burning was concentrated – particularly on Grin Hill, Peak Forest, Dove Holes and around Ashover, close to local sources of coal. Farey describes the enormous lime ash waste material from the kilns that had accumulated (and expresses some surprise that they too had not been used on fields), covering large areas. The tops set and

FIG 63. Old limestone kiln in upper part of a Dovedale tributary with small quarry to provision it behind.

hardened by degrees, and after being slaked by rains, could be excavated into permanent structures. Many Buxton inhabitants who worked the kilns lived in these huts on Grin Hill, sometimes with thatched roofs of grasses, on which, incongruously, cattle might be grazed (Bunting, 2006). Tragic collapse of a hut roof resulted in the death of two women and two children in Dove Holes in 1863. Farey considered the lime waste heaps and conditions on Grin Hill to be 'disgusting desolation' and suggested tree planting to hide it, which is how Grin Wood came into being – planted 200 years ago for the owner, the 5th Duke of Devonshire. The remains of three huts still survive on Grin Hill and a display depicting the scene is exhibited on site. Peak lime-burning occurred in the 18th and 19th centuries.

Not only was the intensive kiln burning having a direct effect on the landscape, but air quality would have suffered as well. Bunting (2006) quotes from contemporary observers. Stoney Middleton was considered a particularly unhealthy place, smothered by billowing acrid black smoke; the kilns lighting up the night sky with flames like miniature volcanoes. The kilns in Small Dale were in a dell which would have led to the dale resembling a 'cauldron of steaming smoke', with flashes from the fires below, 'curling into mid-air … rolling over our heads in murky volumes'. This choking smoke and acrid smell, with polluting particles and a high sulphur content (from coal burning), would have had a deleterious effect on many plants, particularly lichens and mosses, that we can only guess at. When you next walk around the White Peak, imagine the conditions and sights of these times and be thankful for the cleaner air we breathe today.

Limestone extraction is still important but the scale has changed radically. By far the greater part of the extraction industry now is concentrated in large quarries worked by equally large machines leaving much more strident scars. The demand is for limestone rock and the White Peak is one of the key sources in the country. Estimates for the Derbyshire part of the Peak District put the total at about a quarter of the country's output in 2002–11 (Derbyshire County Council *et al.* 2013). The consistency in quality (strong, durable and of low porosity), ease of extraction from thick, fairly level beds and proximity to key markets all help make it a desirable product. The very high chemical purity derived from the lack of mineralisation of ore veins (98.5 per cent calcium carbonate) lead to a high demand for industrial uses such as lime production, glassmaking and as a filler in paint, rubber and plastics. It is the Bee Low Limestones that are of the highest consistent purity and chemistry and which therefore provide a high proportion of the limestones quarried in the area particularly around Buxton and Wirksworth. Indeed, Tunstead Quarry in Buxton is reputed to have the longest, purest limestone face in Europe. The younger Monsal Dale and Eyam limestones are more variable and less pure, but are crushed at several sites for rock aggregate. At Hope, the Monsal Dale limestones are extracted from Pindale Quarry (mostly well hidden from general view) for the Hope Valley Cement Works. There are now fewer quarries in the White Peak than previously as planning consents run out or small quarries become uneconomic. Eldon Hill Quarry, west of Castleton, for example, closed in 2005 after permission for an extension was refused on account of its increasing prominence and concerns about its impact on the geological Castleton Caves SSSI.

Compared with limestone, there are now few gritstone quarries in the Peak District. Some extraction still takes place near Glossop (Shire Hill) and Hayfield as well as near Kerridge (a pink gritstone), Eyam, Grangemill and Stanton-in-the-Peak. The buff-coloured stone from Shire Hill furnished the steps to London's Victoria and Albert Museum. There are also many small quarries that would have provided stone for building, quoins and lintels, gate posts and troughs as well as for the many miles of enclosure walls. But it is the gritstone millstones that have made the Peak District famous far beyond its boundaries. This industry was focused in the edges above Hathersage and Baslow.

FIG 64. Gritstone flat-faced millstones ready for sale but then abandoned at Bole Hill Quarry lying along a straight, embanked grassy trackway. Over 375 stones have been counted here. This quarry later supplied the stone for Derwent and Howden dams.

Millstones were produced in the Peak District from at least the 13th century generally for milling corn. Each stone measured about 2 m across, flat on one surface and domed above. By the 16th century Peak District millstones were being shipped to the Thames Estuary and being used in eastern and southeastern England. Hey (2014) calculates that a hewer could make a millstone in two weeks, and many were part time, being farmers as well. This trade collapsed in the 18th century with the availability of better French stone, rising again during the Napoleonic wars when this became unavailable. Our gritstone was unsuitable for milling white flour, when white bread became fashionable, as the stone tended to colour the flour grey. In the 19th century, many of the millstone quarries had a new lease of life responding to demand for grindstones used in the cutlery and tool-making industries based in Sheffield, as well as being used for grinding other materials, until composite stones of superior quality that produced less dust replaced them (Barnatt & Smith, 2004, Hey, 2014).

Any visit to the area along the edges and old quarries from Millstone Edge near Surprise View car park south to Gardom's Edge will reveal grinding stones in all stages of extraction and fashioning, mostly dating from the 19th and 20th centuries. Some were awaiting transport to their destinations, whilst others are broken pieces or half exposed still embedded in their gritstone blocks (Fig. 64). Now clothed in lichens and mosses and buried in Bracken, the old stones are a testament to our past industries. Regarded as so emblematic of the Peak District, the millstone is adopted as the National Park's symbol and appears as the Park boundary marker on our major roads.

EXPLOITING OUR WATER

As an upland region, with a high rainfall compared with surrounding areas, the Peak District's water has been harnessed for many years. Water mills have existed since at least late Saxon times, with Domesday Book recording mills at Bakewell, Ashford-in-the-Water and Youlgreave. Early mills were probably for local use, including corn mills, saw mills and, later, paper mills exploiting the clean water supply. The water would have been diverted to power the wheels, affecting river flows and habitat.

The first industrial mills were built for making textiles, especially from cotton, but also wool, flax and silk, in the late 18th century. The beginning of the industrial revolution in terms of the first organised factory system is considered to have begun in Cromford where Richard Arkwright built his first mill in 1771. He is credited with inventing the spinning frame, which, once converted to water power, was renamed the water frame. His mill is now part of the Derwent Valley Mills World Heritage Site and well worth visiting. Other mills developed soon after, including at Cressbrook, Litton, Ashbourne, Bakewell, Calver, Bamford, Hayfield and Edale. Most have now been converted to other uses, but you can capture something of the times by reading Margaret Dickinson's interpretation of life in Cressbrook Mill and the Wye Valley in the 19th century in her novel *Pauper's Gold* or Mrs Trollope's *Michael Armstrong* and Mrs Banks' *Bond Slaves*. The atrocious conditions for orphan apprentices sent into bondage, suffering from malnutrition, overwork and factory accidents are common themes being brought to the country's attention (Shimwell, 1981).

By the 19th century, the focus of textile manufacture moved to the northwest towards the major Lancashire cotton-production area. Glossop, New Mills, Stalybridge and Marple were transformed into industrial towns. It is said that there were as many mills in this part of Derbyshire as in the rest of the county.

The White Peak, with inadequate seasonal flows or lack of water, was spared most of these changes, leaving the Wye and rivers in the gritstone country more vulnerable to development pressures. Rivers were diverted to form races and provide the head for the water wheels. Weirs were built, some of which still are in place as on the River Dane and River Derwent. Mill ponds were developed – some still in place, as in Watford Lodge near New Mills, a Derbyshire Wildlife Trust nature reserve, or on the Wye at Bakewell which supplied Lumford Mill. Many rivers still bear traces of this industrial past.

Reservoirs are a key feature of the moorlands (Fig. 5). As demand outstripped local resources and industrialisation increased the need for water resources downstream in cities to the west and east, water providers looked to the hills where annual rainfall was sufficient to supply urban life. A series of Acts of Parliament authorised flooding of valleys and displacement of hamlets and farms. Reservoirs at Crookesmoor and later at Redmires on Stanage Moor and then Loxley Valley were constructed to provide water for Sheffield. In 1800 the three Wessenden reservoirs south of Marsden were completed by mill owners. Three more followed in the Holme Valley in 1837, but one, at Bilberry, was too weak and burst in 1852 after heavy rain, reaching Holmfirth three miles away in about 20 minutes, with disastrous loss of life – 81 people died and many buildings were destroyed to the extent that 7,000 people were put out of work. This was not the only disaster – the embankment at Dale Dyke reservoir north of Strines collapsed in 1864 and the ensuing flooding in the Don Valley in Sheffield killed 240 people and 693 animals: the worst 'natural' disaster in Britain at that time. The dam was rebuilt, along with Agden and Strines in 1888.

Construction of six reservoirs in Longdendale and above Arnfield commenced in 1948 and gave Manchester the distinction of having the longest chain of reservoirs in the world. Unfortunately, the scars this produced in the landscape were augmented by the giant electricity pylons that stride up the valley, although now undergrounded for a short section through the former Woodhead railway tunnel. The dramatic sequence of reservoirs in the Upper Derwent Valley were constructed between 1901 and 1916, with Ladybower at the bottom being added during the Second World War. The story of their building is told in interpretation panels and maps along the road above the reservoirs, now operated by Severn Trent Water. The huge building project employed 2,753 navvies, over 1.2 million tons of stone came from Bole Hill Quarry, near Grindleford, a rail line was developed to bring the stone to site and a village called Tin Town was built to house many workers and their families (Hey, 2014). There are other smaller reservoirs, in sequence or single, such as in the Goyt Valley (completed in 1967), and on the eastern flanks at Midhope, Broomhead and Underbank and below Kinder at Hayfield.

FIG 65. Derwent reservoir dam with impressive towers between which the water overflows (just visible at right-hand bottom). The structure was used by 617 Squadron to practice low level Lancaster bomber flights and target for bouncing bombs, known as the Dam Buster raids. (Kev Dunnington)

In order to ensure sufficient water reached the reservoirs, a number of drains and diversions were created. On the high moors, you can follow large drainage ditches constructed to catch flows and divert them into the reservoirs, or into pipes feeding these. Some of these drains are visible on the Wessenden and Bobus moors southwest of Marsden and are similar to peat cuts except that they follow the contours. Large volumes of water are also diverted from streams to reservoirs, or extracted from rivers as well. For example, water from the Noe in the Edale valley is taken through the hills between and enters the Upper Derwent reservoirs near the dam just above Fairholmes' car park. More famous are the Upper Derwent dam's massive stone structures which were used by 617 Squadron to practice low level Lancaster bomber flights and targeting for bouncing bombs, known as the Dam Buster raids aimed at destroying German dams in the Second World War (Fig. 65). The RAF still uses the Upper Derwent Valley for low-level flying practice.

TRANSPORT ROUTES

Developing the Peak District's resources needed transport between villages, to towns and beyond the region. There are groups of ancient hollow-ways indicating old routes in several locations, some of which are now footpaths. Packhorse trains carried goods of all sorts between the market towns and growing cities outside our area. Lead, stone, millstones, grindstones and wool went out of the Peak District for example, while salt and cheese were moved from Cheshire eastwards and Sheffield knives (noted for their production as early as the 14th century) went westwards and elsewhere.

The hollow-ways used for trade generally occurred in the cultivated valleys and avoided the highest moorlands covered in blanket peat, which would have been difficult to cross. A few routes were flagged, termed causeys, on particularly difficult terrain. Thus we have Doctor's Gate, known in 1627 as Doctor Talbot's Gate, which is partly flagged and believed to be a medieval and later hollow-way running up the Woodlands Valley and over Snake Pass, partly close to the current A57. There is no evidence for it overlaying a suggested Roman route between Melandra (Glossop) and Brough. Another example is the Long Causeway that climbs up beside Redmires Reservoir from Sheffield to Stanage Pole and Stanage Edge. The pole also marks the Derbyshire and South Yorkshire border, and possibly that of the ancient kingdoms of Mercia and Northumberland. A pole has been in place since at least 1723, although there are records of older graffiti dates on the surrounding stones (McGuire, 2016). The pole, though has been replaced intermittently, the latest being in 2016 with a public ceremony attended by over 400 people.

The packhorse routes would need overnight grazing for the ponies and wayside inns and other facilities for their masters, locally called jaggers. We have a Jaggers Clough at the southeast end of Kinder and there are Jaggers Lanes in Hathersage and on Darley Moor. There is also the strange story of John Turner, a jagger on the Derby to Chester road who lived in Saltersford. There is a stone on a lane south of Rainow inscribed: 'Here John Turner was cast away in a heavy snow storm in the night about the year 1755'. The reverse side reads, 'A woman's single footprint was found by his side in the snow'.

It is suggested that the date should read 1735 and that his train of ponies survived, but the riddle of the woman's footprint will probably forever remain a mystery. Try reading Alan Garner's novel *Thursbitch* for one interpretation, interplayed with a more modern story.

Many of the current networks of roads would have started as hollow-ways, many braided, some very deep, especially on the shales, which are more easily

FIG 66. The old stone-built packhorse bridge, a Grade II listed structure, at picturesque Three Shires Head over the River Dane, where Cheshire, Derbyshire and Staffordshire Counties meet.

eroded by running water than their limestone equivalents. It was not until turnpike roads were established in the 18th century with better surfacing and drainage that travel was transformed. The earliest turnpike routes were the Derby to Hurdlow (1738) and Buxton to Hurdlow (1749), joining up with the Manchester to Buxton turnpike in 1724 to complete the route. By the later 18th and early 19th centuries, much of the present road network was established. Stone, or iron, milestones from the 18th or 19th century still survive along many roads, identifiable by unique styles and carving typical of each Turnpike Trust. Stone packhorse bridges replaced earlier wooden ones or fords after the 1697 Act of Parliament and many still stand. For example, one was rescued before flooding the Goyt Valley and lies to the south crossed by a footpath; another graces Three Shires Head (Fig. 66); one crosses Burbage Brook at Padley and there is a splendid one in Ashford-in-the-Water over the Wye with an old sheep dip adjacent. Sections of modern road routes have bypassed some of the old alignments, leaving abandoned sections of 18th century turnpikes now used as footpaths or bridleways, such as the tracks across Houndkirk Moor to Ringinglow and the old road to Buxton from the top of the Goyt Valley. Apart from providing opportunities for easier travel, the widespread use of limestone as hardcore on the road surfaces has left its mark still on some roadside flora (Chapter 12).

The coming of the railways had a major impact on some of the extraction industries. Although there is only one public cross-route now (and a glorious ride it is through the Hope Valley) between Sheffield and Manchester, another line to Buxton also from Manchester and several mineral routes in the Buxton area, there were additional railway lines once, most of which are now public trails owned and managed by the NPA. The first railway opening was the Cromford and High Peak line in 1831, which later included a connection enabling the London and North Western Railway to open a through route across the Peak District. The elevated routes of what are now the High Peak and Tissington Trails give splendid views over the surrounding countryside. These linked the Cromford Canal and the Peak Forest Canal (at Bugsworth) and Ashbourne in the south, moving limestone and other goods. What is now the Longdendale Trail to Woodhead was the first Manchester – Sheffield link, with a long tunnel (4,886 m) at Woodhead that took six years to build and in which at least 32 men were killed and many more injured in the dangerous working conditions (Hey, 2014).

1863 saw the line from Manchester via Chinley to Matlock opened to provide a direct link from Manchester to London for the Midland Railway. Some of this spectacular route is now the Monsal Trail, with dramatic viaducts, dripping tunnels and high dale-side glimpses of the Wye Valley's glories from Chee Dale to Monsal Head. Although we now benefit greatly from the grandeur of the route through the Wye Valley, when planned, John Ruskin, the eminent 19th-century literary and social critic, opposed this and other Peak District railways, arguing that the Dales were sacrosanct, although his pleas ultimately failed (Shimwell, 1981).

The current Hope Valley railway was opened in 1894 after excavating even longer tunnels between Chinley and Edale and from Grindleford to Dore. In contrast the Manifold Trail was first established as a line from Hulme End to Waterhouses in the Staffordshire part of our area, mostly through the dramatic dales of the Manifold and Hamps valleys. It had a short life, from 1904 to 1934, designed primarily to carry milk from local dairy farms, but was also used by passengers. It was unusual in being a narrow-gauge line, thus saving on construction costs.

AIR POLLUTION

We have seen how different industries have developed and sometimes declined in the Peak District over time and their links to industrialisation all around us, but we now have to deal with some of the fallout of these sometimes more distant developments in terms of air pollution and its effects, which have been significant and wide-reaching. Local air pollution has already been described, but it was not until the industrial revolution and mass development of factories in the surrounding regions that larger scale pollution became all pervasive. The culprits were principally soot and sulphur dioxide, primarily derived from coal burning, and these have had a greater effect in the Peak District and South Pennines than anywhere else in the country, especially in the upland environment. The main effects have been on lichens everywhere and on blanket bog, particularly on bog-mosses, where there is less natural buffering in acidic soils. Blanket bog species live in an ombrotrophic environment – with nourishment supplied by rainfall and deposition rather than from rocks or soils – rendering them more vulnerable to acidifying air pollution.

Evidence for the effects of air pollution came from the large-scale disappearance of bog-moss remains in the peat profiles detected by Tallis between about 1750–1800 CE, coincident with layers of soot. Experimental work showed that the losses were largely due to increasing sulphur dioxide. This combines with rain to form dilute sulphuric acid deposited both in the mist that envelopes our hills regularly (termed 'occult deposition') and in rain. This acidified the peat and its water, with pH levels as low as 2.8 recorded. Bearing in mind that clean rain should have a pH of 5.5, and blanket peat elsewhere might have a pH of around 3.5–4.0 (and that the pH scale is logarithmic), then gross acidification is clearly evident. The impact of this has been researched more on blanket peat than in other habitats.

Sulphur is not the only pollutant. Heavy metals including lead, zinc and copper have also been deposited in the upper layers, mostly within the top 5–30 cm – but with less within the most recent top few centimetres – although not in a spatially consistent pattern, but related to proximity to, and severity of, the pollutant source. Compared with other sites, the Peak District peats bear the distinction of being the most contaminated in the world for lead (Rothwell & Evans, 2004). Sulphur dioxide levels are now well below damaging thresholds and soot levels are also low. However, their legacies are still evident in the high acidity of the peat and upland streams. You can also still see the effect of soot on the blackened stones. There is more concern now over nitrogen as nitrous oxides from industrial and traffic exhausts and ammonia from agriculture have not decreased to the same extent. The total nitrogen loading (the amount deposited on the vegetation) is still high and one of the highest in an upland area in the country; exceeding the critical threshold for healthy heathland vegetation. Another pollutant that could be an issue, but on which there is little evidence yet, is low level ozone. This is a product of photochemical reactions between nitrous oxides and volatile organic compounds. Ozone is increasing and normally higher in rural than urban environments (Caporn & Emmett, 2009).

SCENE SETTING

This brief but wide-ranging overview of resource exploitation and its relationship with the Peak District's cultural and industrial history sets the scene for the more-detailed examination of the habitats and species found in our area. It provides some explanations for the landscape's patterns, shapes, bumps and hollows and shows not only how its history is critical in determining where many species might be found, but also helps you read and interpret the scenes you witness on your travels. The next chapter adds to the determining factors of soils, climate and the geographical location of our area within the country, all of which also contribute to the character of this special place.

CHAPTER 7

The Accident of Geography

BEFORE DELVING INTO HABITATS AND species, we need to explore some more factors that determine their presence or absence and their distribution. These are largely tied to local climate and soils along with their interaction with the geology and landscape. First, being at the southern end of the Pennine Chain, we are the first significant hilly country travelling north from lowland England. Consequently, our generally higher altitude compared with surrounding lowlands shapes our climate. This has a significant effect on many Peak District plants and animals. Moreover, climate acting on the varying character of the underlying rocks and structure of the landscape in terms of slopes and aspect results in different soils, even within the same dale or clough. These are reflected in the plant communities and their associated fauna with subtle or sometimes more dramatic differences over often very short distances, especially in some limestone dales.

The Peak District is renowned for being the crossroads where northern and southern species meet owing to the combination of these topographical, climatic and soil factors. Additionally, a few species are more generally western or eastern in their national occurrence and are at the margins of their normal geographical spread here. Cloudberry (on blanket peat), and Northern Marsh-orchid (*Dactylorhiza purpurella*), found in damp grassland, are good examples of northern plants, while Stemless Thistle (*Cirsium acaule*) and Dogwood (*Cornus sanguinea*), both on the limestone, are largely southern in their distribution. Bilberry Bumblebee (*Bombus monticola*) is another northern and upland species, whilst Red Grouse and Golden Plover (Fig. 88) are essentially upland birds. But looking at any species group will reveal a wide range of mosses, liverworts, higher plants, lichens, beetles, flies and birds, amongst many that have quite distinctive distributions with the Peak District on their range edges.

The meeting and overlapping of species with different geographical distributions produces vegetation communities and animal assemblages that, in detail, are not found anywhere else, conferring a real sense of place to the Peak District. Habitats further north, in the Yorkshire Dales for example, will be missing the southern species that peter out here, whilst those to the west such as in Wales, might be missing the northern species, or have more abundant western species. As a result, plants like Ivy-leaved Bellflower (*Wahlenbergia hederacea*) or Bog Pimpernel (*Lysimachia tenella*) might be much more common in western hills, or Round-leaved Sundew (*Drosera rotundifolia*) visible at every step on some Lake District Hills, but all are quite rare and worthy of a special search here. Understanding more about the Peak District climate and soils and how they affect species distributions, helps interpret our habitats and species assemblages.

PEAK DISTRICT CLIMATE

We experience weather on a daily basis, but climate represents the average weather patterns for an extended period composed of all these daily observations. A neat definition is that climate is what you expect but weather is what you get! Detailed information on the Peak District's local climate is not readily available. The Meteorological Office website shows how the main features of the Peak District climate compare with the rest of the Midlands region, with additional information provided by the National Park Authority's website. The most comprehensive set of records available for the weather in our area is from Buxton (at 310 m altitude) and daily maximum and minimum temperatures and rainfall from 1976 to 2019 have been obtained and analysed. Where data are missing, other weather stations' data have been used but interpolated to account for altitudinal differences. Some information on sunshine and other aspects of the local weather is available on the Buxton live weather internet site (https://www.buxtonweather.co.uk; also a source of information on local weather and road conditions throughout the year).

The mean annual temperature over the Midlands as a whole is 8–10 °C, with the Peak District's at the lower end of this range. For the period 2011 to 2019, for example, the annual average (for the whole day not differentiated into the maximum or minimum) was only 8.58 °C. This compares with over 11 °C in Cornwall and 7 °C in the Shetlands. This immediately underlines our continental geographical location and higher altitude. The annual temperature range tends to be quite high as we are far from marine moderating effects.

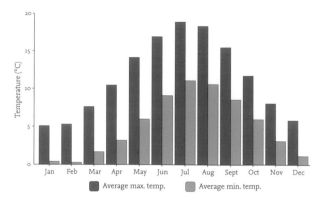

FIG 67. The average monthly maximum and minimum temperatures 1976–2019 for Buxton.

January and February are usually the coldest months, accentuated more in the many frost hollows, and snow cover is higher compared with the rest of the Midlands. The average minimum temperature for January and February is around 0.2 to 0.4 °C but December 2010 was an exceptional month with an average temperature of -1.32 °C, coinciding with heavy snowfall and a prolonged cold period. The average number of days with air frost is over 60 here compared with only about 40 in the lower Severn Estuary for example. Only August is frost free in Buxton, although July frosts are very rare (only one 1976–2019). July is usually the warmest month, with the mean daily maximum for the Peak District at 18.9 °C (1976 to 2019 data) compared with 22 °C in the south and east Midlands, 23.5 °C in the London area and 15 °C in the Shetlands. August is similar to July (Fig. 67).

Analysis of the Buxton 1976–2019 data shows that seven out of ten of the hottest days have been within the last 20 years and all but two of the ten lowest maximum temperatures all occur in the 1979–1987 period. Moreover, comparing monthly average maximum temperatures shows increasing trends for some months (Table 5). Minimum temperatures show a similar trend in winter but not in summer (Table 6). Note that it is daily average temperatures that are used in countrywide predictions rather than the maximum and minimum numbers used here. The trends shown on the tables though are in line with those found more generally in the UK and are likely to continue. They are matched, for example, by records on Holme Moss (520 m) from 1992 to 2006 where an upward trend in annual mean temperature was recorded (Caporn & Emmett, 2009).

Buxton does not fare well for sunshine totals. Average sunshine duration is 1,227 hours per year compared with 1,600 hours in the south Midlands, 1,750 hours along the English southern coast and 1,900 hours in the Channel Islands. The Shetlands are worse off though with only 1,100 hours. Our lower totals are partly due to the higher altitude, with more cloud, mist and hill fog. Buxton is

TABLE 5 Buxton: average maximum monthly temperature (°C)

	1976–85	*1986–95*	*1996–2005*	*2005–19*	*Difference 1976–2019*
January	4.07	5.29	5.37	5.47	**1.4**
June	16.3	16.3	17.1	18.0	**1.7**
July	18.5	19.0	18.4	19.7	**1.2**

TABLE 6 Buxton: average minimum monthly temperature (°C)

	1976–85	*1986–95*	*1996–2005*	*2006–19*	*Difference 1976–2019*
January	−0.77	0.41	0.9	0.89	**1.66**
February	−0.77	0.11	0.94	0.63	**1.4**
June	9.45	8.55	9.11	9.34	**−0.11**
July	11.47	10.89	10.73	11.3	**−0.17**

prone to fog as well, as it sits in a hollow, so will not be representative of the whole area. It also has the distinction within the Midlands of recording the dullest winter month with only 4.1 sunshine hours in January 1996. However, totals are much higher in May and July (168 hours) on average.

Rainfall increases with altitude, so Buxton and the Peak District are generally wetter than the surrounding lowlands, with over 1,000 mm at higher elevations and around 800 mm in lower areas. The annual average for Buxton is 1,334mm (1976–2019 data) and the total each month is higher than the national average, although it is very variable. Analysis of rainfall data between 1976 and 2019 shows an interesting picture (Fig. 68). The blue line, known as a smoother, is one way of

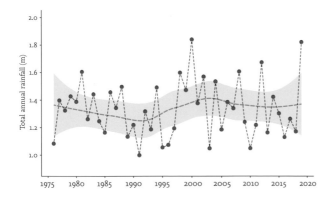

FIG 68. The variation and trend in total annual rainfall from 1976 to 2019 for Buxton.

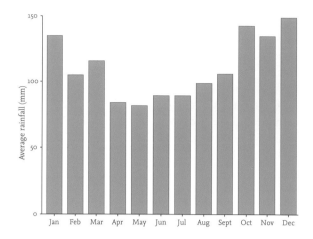

FIG 69. Average monthly rainfall 1976 to 2019 for Buxton.

representing a trend, with the grey band effectively the standard error. The graph shows distinct periods of relative dryness in the mid-1980s, around 1990 and in the mid-1990s, followed by a wetter period around 2000. For example, 1991, at 998 mm, was the driest year and 2010, a year with several droughted months, saw a total of only 1,015 mm. Compared with 2012, which was renowned for being particularly wet across the country, we received 1,591 mm. However, 2000 and 2019 were wetter, both witnessing totals over 1,820 mm. The overall average compares with annual amounts of around 500 mm in drier parts of East Anglia to over 4,000 mm in the Scottish Highlands, which clearly shows where the Peak District sits within the country's rainfall gradient.

Rainfall generally is well distributed throughout the year (Fig. 69), although February and the summer months are typically drier and January, October and December tend to be the wettest months, when fronts associated with more vigorous Atlantic depressions are prevalent. The variability is demonstrated by a high of 230 mm and a mere 11.6 mm in June 2012 and 2018 respectively. There is evidence that the pattern is changing too with over 50 per cent of months with more than 250 mm rainfall occurring in the last 20 years, and 67 per cent of those with less than 25 mm of rain prior to 1990. This suggests greater storm activity throughout the year despite otherwise droughted periods. The number of days with rainfall over 1 mm (termed 'wet days') tend to match the monthly rainfall pattern. 40 to 45 such wet days per winter and 30 in summer (June to August) are more typical in our area compared with only 30–35 in winter and 20–25 in summer for the lowland Midlands.

Snowfall is clearly linked with temperature, which has to be lower than about 4 °C for snow to lie. Snow usually falls between November and April, although falls have been known to occur in October and May. An exceptional snowfall on 2 June 1975 disrupted a county cricket match in Buxton – the only place in the UK where snow stopped play (and much else besides) which has now entered local legend. More usually, snow lie is restricted to December to March. In the Peak District, snow falls on 35 days on average per year, compared with only 20 days in lower lying areas, although this total is declining with climate change. There is, on average, an increase of about five days of snow falling per year per 100 m increase in altitude. This means that the higher hills will hold more snow for longer, particularly in the gullies, cloughs and north-facing slopes, than the lower ground. Average snow lie total is about 20 days a year in the Peak District, compared with over 60 days in the Scottish Highlands and three in South West England. The total number of days, however, is highly variable from year to year. As an area with important cross-peak transport routes between Manchester and Sheffield and other conurbations, it is the closure of the Snake Pass (A57) and Cat and Fiddle Road, the well-known public house at the watershed on the A537 between Buxton and Macclesfield, that grab the national news when snow stops traffic.

The windiest parts of the country are usually those closest to the Atlantic facing stronger winds associated with depressions, especially in winter from December to February. Thus the Peak District is less vulnerable to these. On the other hand, this is moderated by being at higher altitude. The Peak District experiences about five gales per year compared with only two per year over the Midlands in general. The wind strength and speed will, however, be modified by topography, with steep slopes, deep valleys, irregular shapes and high hills all contributing to increased gustiness and local conditions. There are certainly occasions when it is hard to stay upright in the wind, particularly on high, west-facing edges.

Any Peak District inhabitant knows that there is considerable variation in weather and climate across the area, although there is a dearth of supporting data. In general, there is a reduction in temperature of 1 °C for every 100 m ascended. Bearing in mind the additional wind speed and exposure, this accounts for the different climate and local weather conditions on hilltops and the need to take extra clothing when walking there. Such a difference is more critical in winter conditions of course or when the cloud level descends engulfing the unwary. People can and do die of exposure and hypothermia on our highest hilltops if they are unprepared.

In anticyclonic conditions with light winds and high air pressure, temperature differences can be inverted, with colder conditions in valley bottoms as cold air gravitates down the slopes. Thus the Edale and Hope valleys and the Glossop and Buxton basins for example can develop severe frosts, much colder

FIG 70. A temperature inversion resulting in mist filling a deep valley, Edale.

than on their valley sides. They are also often shrouded in mist or fog when the hills above are clear and sunny producing beautiful, ethereal views across the landscape (Fig. 70). Slope and aspect will also determine how quickly frosts melt, with north-facing slopes and hollows being colder than south-facing ones. With dales and cloughs often deep with steep slopes, this can be an important determinant for some plants and animals. For example, Bracken fronds are frost sensitive so may be killed or damaged where exposed. This in turn exerts control on Bracken density or presence within a complex of slopes.

The greatest precipitation of the area occurs on top of the highest hills: Kinder Scout and Bleaklow, with upwards of 1,650 mm, whilst lower ground on the west of the Peak District would be closer to 1,000–1,250 mm. In the south and east of our region, however, annual rainfall is nearer 875 mm. This also means that sunshine totals are generally higher where rainfall is lower. It can take a day for fronts reaching the western hills to bring their rain to the eastern hills, with usually far less rain too.

This summary of the main features of the weather and climate in the Peak District shows how the upland environment differs from the surrounding lowlands, helps explain why species might be at their range edges here and shows the variation you might experience across the area. The climate also determines to some extent the nature of the local agriculture and accounts for the predominance of pastoral stock and dairy farming along with the general dearth of arable cropping. The short growing season and low average summer temperatures, together with high rainfall compared with adjacent lowland environments would militate against growing crops. The same climatic features also control the abundance and occurrence of many plants and animals, which must be adapted to, or tolerant of, the highs and lows of different climatic parameters.

SOILS

They are mostly hidden and too often forgotten, but soils are fundamental – what food can be produced depends on them, how carbon is stored and accumulated is determined by them, and how water is trapped and moves through the landscape is controlled by them, mostly modified by our treatment of them. In the Peak District, with such contrastive rock formations and topography, soils vary significantly in depth, acidity, organic material content, texture, water holding capacity, nutrient totals and metal content, such as lead.

The main soil types are linked to the rocks on which they sit, except for peat, which is probably the most extensive (Chapters 4 and 5). Deep peat is normally defined as being over 0.5 m deep (although there are variations in this). Peaty soils are shallower. Soils within the moorland environment will be more sandy or gritty on the gritstones and more clay-based on the shales. These provide quite different soil-forming properties. On the limestone plateau, soils tend to be deeper, developed over the various materials on the surface and less commonly directly on the limestone. In the limestone dales, in contrast, the rocks are often close to or exposed at the surface, with a typical, shallow, lime-rich soil on the steeper slopes, especially those facing south. On north-facing slopes, leaching is greater than evaporation and capillary action, leaving more acidic, usually deeper soils.

Soils develop distinctive profiles over time related to the predominant action of water and capillary activity and the build-up of organic material. Below the vegetation and incorporated into the upper organic 'O' horizon is an organic layer of dead and decomposing material (Fig. 71). The rate of decay is dependent on the degree of waterlogging (which reduces decay) and the acidity in particular. Acidic soils (usually in the 3–4 pH range) support more fungi, moulds and anaerobic bacteria, whilst more neutral and calcareous soils favour earthworms. Decayed organic matter gives a dark stain to the upper layer of the soil, which can permeate down the upper A horizon (Fig. 71) through leaching. The rock or parent material (sediments of various kinds) form the bulk of the soil profile, with much less

FIG 71. Soil profile showing a degree of podsolization developed over loess on the limestone plateau south of Buxton.

Thick organic layer: the litter and humus layer showing plant roots

Upper A horizon: dark humic stained, lower eluviated horizon leached and paler

Iron pan: showing undulating surface

B horizon: where iron and other minerals accumulate, low in humus. Merges below into C horizon of parent material

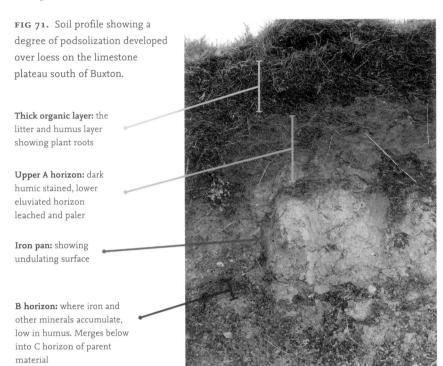

mixing with organic material in the lower B horizon (Fig. 71). These differences are fundamental to the vegetation and associated animals and an understanding of their key components is essential to appreciate where to see different species and habitats. Cranfield University (2018) provides Soil Survey of England and Wales maps and descriptions. The following provides a non-technical summary.

Starting with the peat that overlays the highest hills on the gritstone moors, this is the extreme version of a soil dominated by organic material derived from dead plants and animals. Peat consists essentially of organic material but is not uniform throughout its depth. Rather, there should be an upper looser, less compacted layer which water can permeate called the acrotelm containing the plant roots. In an intact peat mass, this layer may be waterlogged with a high water table in winter, but a lower one in drier summer months. The acrotelm is usually about 5 to 10 cm deep but will vary. Below this is a layer of compacted, compressed older peat called the catotelm. This should be consistently saturated, but is so dense that water is held tightly and does not flow readily. Where peat has been eroded, gullied and the water table lowered, as over much of the Dark

Peak blanket bogs, these horizons are less clear or absent. Water tables tend to be more than a metre down these profiles and peat decay is accelerated in the oxygenated environment. We will return to this theme later.

The character of peat as a growing medium for vegetation varies. Our peat is naturally acidic, but the effects of past sulphur dioxide pollution in particular, but also current nitrogen deposition, both reduce the pH further (Chapter 6). The more acidic a soil, the less available nutrients are to plants. Thus, the blanket bog habitat is a nutrient-poor one, with low available nitrogen, phosphorus and potassium – the three key nutrients on which growth depends. However, the acidic environment also results in reduced forms of iron and aluminium and it is the latter in particular that can limit some plant species in boggy conditions where water flow and oxygenation are minimal.

Peaty podsols are the main soil type on the slopes below blanket peat, mostly around the higher Dark Peak moors and under the heathland to the west and east of the region. They are associated with the sandstone or grits, often on steep or moderately sloping valley sides, along with frequent terracettes, gullies and rock exposures on gritstone escarpments. There is a variable depth of organic material in the upper layer. Podsol, a Russian word, refers to the pale ash colour of the lower A horizon, below the organic-rich layer, that results from leaching (Fig. 71). The clay, iron and other soluble materials are washed out and accumulate lower in the profile in a distinct orangey/rusty coloured layer termed an iron pan (Fig. 71). This is rarely horizontally or vertically even and can fluctuate over very short distances, thus giving different depths of leached soil above, which is reflected in plant responses. If sufficiently consolidated, this iron pan can become impermeable, resulting in wet soil profiles above it in what is otherwise a dry and well-drained gritty soil. Wet heath favours these soils. Soil depth varies with slope, being generally thinner on steeper slopes, and podsols are characteristically low in nutrients and acidic, with the majority being less than pH3.5 and most less than 4.0 (Lloyd et al., 1971). Variants of this soil type with deeper profiles and less developed iron pans can be found on the head deposits in the uplands, such as south of Combs Reservoir and in the Hayfield area. The peaty podsols also occur on the loess over the limestones (Fig. 71).

Wet, acid, loamy and clayey soils are found on the shales and shale grits surrounding the gritstones and on the Coal Measures, so are more prevalent in the northeast from Holmfirth south to Curbar below the main moorland areas, in the valley below Rushup Edge on the limestone/gritstone junction and in the South West Peak. These are slowly permeable, generally waterlogged soils owing to their fine loamy texture over a clay base. They often have a peaty top layer on flatter ground and are acidic, cold, sour soils which are heavily gleyed – that is waterlogging where oxygen depletion results in iron changing colour from brown/orange to grey in its reduced form, producing grey mottling in the lower horizon, especially in wetter conditions in winter. These soils are also acidic, but the median pH is closer to 3.4–4.0 compared with the lower levels in the podsols (Lloyd *et al.*, 1971).

Many of these soils have been extensively drained and improved agriculturally with the addition of lime and fertilisers, particularly on lower-lying and less steep land, but the widespread appearance of rushy, wet, ill-drained fields in many of the shale-based areas shows how difficult it is to subdue their natural nature completely. Such rushy fields are the backbone of some of our more important breeding wader populations as well as marshy habitats.

The brown earths typical of the Dark Peak and South West Peak exhibit a dark greyish brown, slightly stony sandy loam in the upper A horizon, but a yellowish brown, moderately stony lower B layer. This is a free-draining soil with no gleying above 40 cm depth, tending to occur at a lower altitude than the podsols already described. These brown earths have been targeted for agricultural improvement and underlie many of the walled pastures below the moorland blocks.

In general, the soil types are spatially more consistent over the gritstones and shales, giving rise to large expanses of some habitats like dwarf-shrub heath or acid grassland, changing as shales give way to grits or vice versa. The same degree of uniformity applies to much of the limestone plateau, but the limestone dales are quite different, showing interestingly subtle differences over very small areas. Understanding this helps account for what you might find there. The sequences of soils found on the Carboniferous Limestone were investigated by Balme (1953), Pigott (1970) and Lloyd *et al.* (1971). Each of the distinct limestone deposits (described in Chapters 3 and 4) displays varying degrees of resistance to weathering owing to differences in porosity and erodibility. This variability is inflated by the appearance of silica residues from chert layers and concentrations of magnesium derived from the dolomitised limestones. The resulting soils have different amounts of coarse scree material at the base of their profiles (associated with reef limestones) or of fine sediments (from basin limestones), both rich in calcium. Weathering of the more resistant dolomite and chert layers is slower and magnesium-rich debris features below some of the dark-grey tors associated with these limestones, as between Friden and Elton and around Brassington. The fine aeolian loess deposits, containing high levels of more acidic silica (Chapter 4) occupy much of the limestone plateau and some along with plateau chert residues have been washed out in the past in variable quantities down dales-sides, thus sometimes contributing an acidic silty material to these slope soils.

Add to these various parent materials the local effects of climate and aspect, and we have a veritable soup of different conditions in which soils form. The principle processes are leaching, insolation and the interaction with plants. Given that the rainfall differs across the White Peak, from the higher north down to the lower southern areas, leaching is greater and insolation less in the northern damper, more cloudy environment. This is accentuated on north-facing slopes and reduced on south-facing slopes, with gradual gradients between all other aspects as dale slopes twist and turn. Slope angle also plays an important role. Typically, slopes are 30–35° at their steepest (discounting rock outcrops), supporting the shallowest soils. North-facing steep slopes though could be in shade for significant periods, thus adding to the leaching, low temperatures and lack of insolation, resulting in increased organic matter and soil wetness.

All these factors result in a range of soil types at the local scale as demonstrated in Figure 72, which shows an idealised cross section of north and south-facing dale sides, with different materials and altitudes on either side and the soil types shown below. The variation in soils along this cross-section is demonstrated by the pH range found from 183 soil samples from 14 locations on the Carboniferous Limestone with the majority in the 7.0–7.5 alkaline class, but a

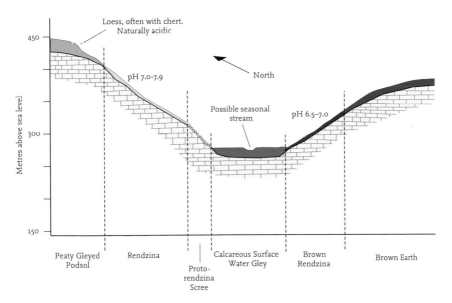

FIG 72. A diagrammatic cross-section across a hypothetical limestone dale to show the variation possible in soils on different slopes and aspects.

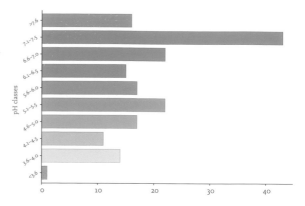

FIG 73. The range and relative frequency of soil pH across 183 soil samples from 14 locations on the limestone. (Data from Lloyd et. al., 1971)

wide range as shown in Figure 73. In general, pH measures were similar down the profile, as might be expected where leaching is the predominant process.

Starting on the plateau tops of the cross section, there are brown earths derived from the silty loess, particularly in the northern half of the White Peak above Bradwell, Little Hucklow and Great Longstone. These have long been the focus of agricultural activity and few now exhibit natural profiles. Naturally, these slightly acidic soils (pH around 5 to 6.4) contain more quartz than weathered limestone residues. In other profiles the loess is mixed with deeply weathered limestone residues, which results in a higher pH where these predominate (Pigott, 1962). In the southern area of the limestone plateau, the soils contain highly plastic yellow or red clays, possibly inheriting their iron-rich colour from the limited Triassic deposits (Chapter 3). These brown earths are generally well-drained, although some gleying can occur, and are not droughty owing to their moderate depth. Again, there are variants that are more loamy, or more stony and gravelly, particularly where there are chert layers. Most have been modified for agricultural use: limed, fertilised and often ploughed, but many are also disturbed through mineral workings (see Chapter 6).

Figure 72 also shows peaty gleyed podsols on the plateau tops, which in places can extend to the dale edges, but rarely onto the steeper slopes. The chert and loess insulates the soil from the underlying limestone's influence and rainfall is high enough to wash out calcium. Podsols on the limestone show considerable variation in terms of edaphic maturation (bearing in mind that a full iron pan formation takes time to develop). In a fully developed podsol under limestone heathland, the soil typically has up to 15 cm of acid peaty humus over a podsolised profile, usually with a thin iron pan where iron deposition is concentrated, visible in Figure 71. Beneath this is an orange-brown loam which might contain chert fragments and where pH and calcium content will increase down to the underlying limestone. A less well developed podsol such as on Wardlow Hay Cop above Cressbrook Dale has a thinner layer of humus and a less well developed leached layer, with no iron pan. The pH is between 4.2 and 4.8. Intermediate forms also occur such as on the edge of Coomb's Dale near Stoney Middleton.

There is evidence that some heathland plants like Heather, Bilberry, Wavy Hair-grass and Mat-grass can modify soils through the accumulation of acidic, black, dead material producing a 'mor' humus. Iron oxides are more mobile in this, thus contributing to the podsolization process. Under Wavy Hair-grass, this black humus has been shown to be denser with less aeration than the more moderate 'mull' humus, with fewer earthworms and therefore also less burrowing and aeration. These changes result in reductions in growth and establishment and eventual exclusion of many calcareous or neutral grassland species as Wavy Hair-grass spreads, thus emphasising the acidic nature of the soils. Experiments showed that the competitive exclusion by Wavy Hair-grass resulted more from excess manganese as a result of the wetter humus in which manganese becomes more soluble, although this was not toxic to plants typical of acidic conditions such as Heath Bedstraw (*Galium saxatile*). This demonstrates how complex the chemistry and interactions of plants with the soils can be (Pigott, 1970). These podsolised soils on the limestone are now quite restricted in occurrence owing to agricultural improvements, which improve drainage by breaking up the iron pan, add lime to counteract the acidity and fertilisers for more vigorous plant growth.

On the steepest limestone slopes especially with a southerly aspect, the typical limestone soil is a very shallow (usually < 10 cm) rendzina; another name of Russian origin. These are often stony and well drained, rich in calcium with a high pH between 7 and 7.9 usually punctuated by rock outcrops and screes. The generally low nutrient levels are critical for supporting a high diversity of plants without any one dominating unduly. The soil profile is typically brown-stained from organic matter on the surface, but lacks a B horizon on the steepest slopes, passing straight into fragmented rock and stones. The underlying limestone surface is not uniform, there being many rock crevices and joints, usually widened by solution weathering, into which plant roots can penetrate. These soils can be prone to drying, which was widely visible during the severe 2018 summer drought.

Similar soils called brown rendzinas occur on slopes with other aspects, but here the reduced insolation and increased moisture and leaching result in calcium loss in the upper horizon, leaving a neutral or slightly acidic soil except around rock outcrops. These can therefore support lime-loving plants (calcicoles) and those intolerant of high pH (calcifuges) within the same slope, which makes for interesting botanising. There are bands of chert and/or loess downwash from the plateau in some of these soils. The pH is usually in the bracket 5.8 to 6.3.

Screes are found in many dales, both near the bottom or part way down the slopes (Figs. 11 & 34), sometimes below limestone cliffs, which will have been the source of material during periglacial conditions. These provide skeletal soils termed a protorendzina supporting specialist species tolerant of the dry conditions and stony material.

In the bottom of some dales, a seasonally waterlogged, calcareous surface-water gley develops where the winter water table rises close to or above the surface. These are often deeper soils derived from sediment movements down slopes, which may dry out in summer, but are mottled with anaerobic conditions or completely waterlogged or flooded during wet periods, as in the bottom of Cressbrook Dale in its upper reaches (Fig. 38).

Any of the limestone soils can be interrupted by thin strips of finer, loamy soils overlying igneous outcrops such as the Toadstone layers. These tend to be deeper brown earths of good texture that are slightly acidic (pH 5–6) unless in receipt of calcareous water from higher slopes. Similarly, lead and other metal extraction in the past have left trails of soil disturbance in waste heaps and holes. Metal levels (especially lead and zinc) will be high enough to be toxic to many plants on some of the waste heaps and it is here that our very specialised metal-tolerant flora occurs.

SPECIES DISTRIBUTIONS

We have seen how the accident of geography determines the local weather and climate and how the soils respond to this across different rock types and sediments. These, in turn, dictate where we find many plants and animals. The detail of the assemblages or communities together create a real sense of place. If you can read the signs and signals, you know you are in the Peak District. The plethora of species for which the Peak District lies on the boundary of their national distributions has been highlighted, but why is the interesting question. There are probably two fundamental factors – one is the lack of suitable habitat in lowland England. For example, there is no blanket bog and therefore no blanket bog species to the south and east but they might persist down to south Wales or the southwest of England. The other is some aspect, usually of climate, limiting growth or reproduction. Theory suggests that there would be either or both a reduction in populations owing to depressed reproductive success and fewer favourable habitat patches on boundary edges, thus making these species rarer than in their core areas. Species can also occur within assemblages or communities that differ from those in their core ranges.

FIG 74. Stemless Thistle, a southern species, growing here in a large population on Wardlow Hay Cop, above Cressbrook Dale.

This makes observations particularly exciting as finding the unexpected is always fun.

Few studies of individual species explore these relationships, but some are particularly relevant in the Peak District. Temperature and humidity have been found to be important in limiting several plants. Studies of Stemless and Melancholy Thistles (*Cirsium heterophyllum*) (Figs. 74 & 75) in the Peak District reveal some possible explanations. Stemless Thistle is found on slopes of all aspects on Surrey's chalk hills, but is more restricted to south-facing slopes in the Peak District limestone grasslands; even micro slopes the size of mole hills (Pigott, 1968). Jump *et al.* (2003) found stemless thistle flowers produce less seed in the Peak District: 37 per cent of the maximum seed mass compared with plants in its core range. Experiments suggest this is related to reduced heat and higher levels of infection by the grey mould *Botrytis* in damper conditions, which is partly compensated by growing in as warm a spot as possible on south and southwest facing slopes (Pigott, 1968). Thus, Stemless Thistle is more frequent and more often found in the southern, lower limestone dales but scarce in the

FIG 75. Melancholy Thistle, a northern species, with 'Scotch' thistle type head and white cottony under surface of the serrated but undivided leaves which are just visible between the flowering stems.

more northerly ones in the Peak District. Stemless Thistle also has a lower density in our area than in its central range and exhibits strong isolation by distance, a significant decline in genetic diversity and an increase in genetic divergence (Jump *et al.* 2003, Jump & Woodward 2003). This is considered to be the result of less pollen exchange and seed dispersal and could be interpreted as suggesting poorer fit of a species to its environmental location and therefore a greater vulnerability to change.

Melancholy Thistle's distribution is the opposite to that of Stemless Thistle, being largely northern, reaching its southeastern limits in the Peak District. Having been subjected to the same investigations, it too shows a massive decline in seed production (1.2 per cent of its maximum) here. It also exhibits declining density with populations more isolated from each other at its range edge. However, the same genetic studies did not show the decline in diversity that Stemless Thistle exhibits. Both species can spread vegetatively and thus persist for years within a community, but their spread to new areas will be more restricted. Although Jump and colleagues found evidence for neither of these thistles being found in unusual communities, local differences are evident for the Melancholy Thistle at least. In the Yorkshire Dales for example, it is a typical component of upland hay meadows, it occurs in rock ledge and limestone pavement communities, in light-canopied woodland and on in-washed sediment beside fens (Lee, 2015). In the Peak District, it is more likely to be met on the limestone in damp grasslands, scrub and open dales, particularly on north-facing slopes. The mechanism for its restricted distribution is unknown, but my observations on a flat meadow where it has spread for 20 years from a single plant, showed a substantial decline in patch size (as leaves withered) and flowering stems in droughted conditions in April–June 2017, more dramatically during the longer and more severe April–August 2018 drought and again in spring 2020, suggesting that water balance is a factor.

Similar control mechanisms have been recorded for other species. Dogwood produces less viable seed owing to *Botrytis* infection in the Peak District's cooler and wetter climate (Jarvis, 1960). This changes its ecology compared with the southern chalk downland where it readily establishes from seed. In our area it forms bushes suckering outwards over time, but rarely establishes in new spots in the limestone dales. It is much less abundant therefore than on southern chalkland.

Small-leaved Lime (Fig. 183) shows a different response to climatic controls. It is rare in White Peak dale woodlands and does not produce viable seed except in exceptionally dry and warm summers such as 1975–6, although even then, only some 10 per cent of seeds were fertile, compared with higher percentages in lowland England. Temperature here is the controlling factor preventing growth of the pollen tube to reach the ovule (female germ cell that develops into the seed after fertilisation) and of the development of the seed itself when temperatures are too low. Small-leaved Lime is therefore considered a relict species, establishing at a time when climate conditions allowed seed production and establishment and persisting since as old trees (Piggott, 1981).

Bee Orchid (*Ophrys apifera*), another southern plant (Fig. 200), also seems to be climate affected. My counts in Derbyshire Wildlife Trust's Miller's Dale Nature Reserve since 1977, show flowering plant numbers crashing in 1986 and a cyclical pattern since, but at a lower level, with none in 2018 or 2020. Analysis of the data as a time series suggests that winter cold provides some population control. A cold severity index in the previous November and in February using the accumulative daily minimum below $-2°C$ (taken to indicate severe frost) averaged over the number of days such cold occurred, is strongly correlated negatively with

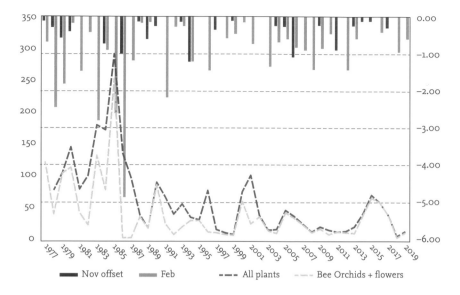

FIG 76. The effect of winter cold on Bee Orchid counts in Miller's Dale Quarry Nature Reserve from 1977 to 2019. The cold severity index (right-hand axis) for November the previous year (Nov offset) and for February (Feb) are correlated with the declines in Bee Orchid counts (all plants) and with numbers of plants only with flowers (Bee orchids + flowers) shown on the left-hand axis.

summer flowering plant numbers (Fig. 76). The overwintering rosette may be sensitive to winter cold. Summer growth provides a food store in the tuber and plants flower only when this is large enough several years after establishment. Winter cold could check this growth (affected plants tend to blacken and die back) and delay flowering. As an essentially southern species generally avoiding the uplands, it is conceivable that it is the severity of winter weather that normally dictates its success.

Northern species like Bird Cherry (*Prunus padus*), Globe Flower (Fig. 202, *Trollius europaeus*) and Mossy Saxifrage (*Saxifraga hypnoides*) may be reacting to other stimuli, although the controlling factors are not always clear. For Bird Cherry (Fig. 77) for example, drier, warmer summers further south cause leaf water stress which young plants cannot tolerate, although living in damp ground can compensate for this (Leather, 1996). Look for this plant in the Peak District on north-facing slopes, near streams (its white blossoms decorate the upper stretch of the Dove Valley in spring), or in the bottom of the Dales as in Cressbrook Dale.

FIG 77. Bird Cherry, a mostly northern species, looking gorgeous in flower on the lane-side between Peter and Monk's Dales.

Globe Flower and Mossy Saxifrage also tend to occupy north-facing slopes within the limestone dales, so they may be equally prone to the effects of water stress. Indeed, Mossy Saxifrage in more open spots suffered badly in the 2018 drought and my Globe Flower counts since 1977 from Miller's Dale Nature Reserve, analysed against various climatic parameters, suggest the number of plants are positively correlated with an index of evapotranspiration or water loss two years previously, suggesting a possible lag effect related to climate.

Jacob's Ladder (*Polemonium caeruleum*) certainly follows the same proclivity for northerly aspects where Pigott (1958) noted air temperatures 2–4°C lower than the regional averages all year, with an even greater disparity of soil temperatures on north-facing ledges or at the bottom of north-facing slopes. This is partly owing to reduced insolation and additional shading from cliffs or slopes. Experimental work shows Jacob's Ladder wilting even in cool shade and needing to regain its water balance overnight. If it cannot, owing to soils being too dry (below 50 per cent soil moisture), it succumbs. This demonstrates how a plant generally

FIG 78. Jacob's Ladder, a nationally rare northern species adopted as Derbyshire's County flower, growing on the north-facing side of Lathkill Dale.

of more northerly climes can survive in specialist niches at the edge of its natural distribution. North-facing or shaded slopes and ledges in Lathkill Dale, the Winnats, Taddington Dale, Wolfscote Dale and Wye Dale are where to search for this nationally rare species (Fig. 78).

The factors that dictate why there are so many northern animals at the southern edge of their ranges in the Peak District have yet to be fully explored. Pearce-Higgins (2010) considers Red Grouse (Fig. 104) and Golden Plover to be northern species ill adapted to warmer climates. Red Grouse, for example are well insulated with extra leg feathers to survive winter cold at high altitude, but are poorly adapted to warmer conditions. In contrast, Golden Plover breeding is closely synchronised with peak emergence of their main prey – adult craneflies in blanket bog (Pearce-Higgins *et al.*, 2009). These are in turn cold-adapted and larval numbers are reduced in warmer summers (although this could also be due to drying peat, which they cannot tolerate). The craneflies *Tipula grisescens*, *T. cheethami* and *T. alpium* are all at their southeast limits in the Peak District. Golden Plover are therefore likely to be restricted more by their food than by direct effects of climate.

A number of other birds are also at the southeastern edge of their natural ranges, including Merlin (*Falco columbarius*), Dunlin (*Calidris alpina*), Common Sandpiper (*Actitis hypoleucos*), Wood Warbler (*Phylloscopus sibilatrix*), Dipper (*Cinclus cinlcus*) and Ring Ouzel. Come north from East Anglia or the Midlands, and the Peak District is the first place you will see any of these during the breeding season.

Of the invertebrates, the Northern Wasp Hoverfly (*Chrysotoxum arcuatum*) has a clear north and west UK distribution (Fig. 79) and is common in the Dark Peak where it is found typically at ground level near woodland and moorland edges. The Hairy Wood Ant (*Formica lugubris*; Fig. 142) is another northern species at the edge of its range in the Peak District (see Chapter 10). It is said to be cold-adapted (Brian, 1977), able to forage at much lower temperatures than the Southern Wood Ant (*Formica rufa*), which it replaces here. The Bilberry Bumble Bee (Fig. 125, see Chapter 9) is also at its southern limits in the Peak District, but being dependent on Bilberry flowers for nectar and pollen at the beginning of the season (Benton, 2006), it is restricted to moorlands and their fringes. It is thought to be cold-adapted.

Southern species will be reacting to different controls. Essex Skipper (*Thymelicus sylvestris*) is a small butterfly that just creeps into the southeast edge of the Peak District. It has a distinct southeasterly distribution in the country correlated with warm summer temperatures. This might relate to enhanced growth rates of larvae in higher temperatures, which could also reduce predation

FIG 79. The Northern Wasp Hoverfly, the only wasp hoverfly with a northern-biased distribution, found feeding on flowers in acid grasslands, woodland and moorland edges in the Dark Peak. (Rob Foster)

risks (as the larval period is shorter) (Pollard, 1988). The Peak District climate may not, therefore, be suitable for some southern invertebrate species where lower temperatures might inhibit larval development rates. For species that overwinter as larvae, reduced survival could also relate to insufficient growth in the previous summer if temperatures are too low. However, for many invertebrates in particular, we just do not know why they are at the edge of their ranges.

Mammals have not been mentioned, but these are in general less restricted by climate than other animals since they are more effective at controlling their body temperature. Some are limited in the Peak District by habitat availability rather than climate, but the Harvest Mouse (*Micromys minutus*) would appear to be largely a southerner, rarely appearing above 152 m, and thus excluded from most of our area (Mallon *et al.*, 2012).

There are fewer species with largely western or eastern ranges. Westerly plants include Rusty-back Fern (*Asplenium ceterach*) (which is in the south as well), Western Gorse (*Ulex gallii*), Ivy-leaved Bellflower (Fig. 111) and Bog Pimpernel – but the reasons are not clear. Wetness is likely to be part of the story.

This account explores why some plants and animals that not only call the Peak District home but are generally at the edge of their ranges here. Some are expanding, some contracting, some are stable. This then provides the background to an exploration of the main habitats and their occupants in the following chapters.

CHAPTER 8

Blanket Bogs

*The cotton grass moors are extensive, dreary and monotonous. One may walk
many miles over the moors of this district without seeing any trace of Sphagnum.*
(Moss, 1913)

[They] *present one of the gloomiest spectacles amongst British Vegetation.*
(Rodwell, 1991)

D ESPITE THESE DREARY PEN PORTRAITS, the blanket bogs of the Peak
District are probably our most important habitat and, happily, their
monotony is now diminishing. Peatlands are national treasures.
Ours provide us with a sense of place, a feeling of wilderness, an expansive open
vista (Fig. 45) or a challenging way-finding exercise, depending on the weather.
They preserve the rich archive of cultural and environmental change and, of
course, they provide a fascinating habitat.

Our peatlands are one of our most extensive habitats – there are over
25,000 ha in the protected areas (SSSIs and SACs – see Appendix 2) but more
outside these. The UK is internationally important for its peatlands, to which
the Peak District makes a significant contribution. In addition, our peatlands,
along with those of the South Pennines, are of European importance for their
breeding birds. Peat also supports some really important environmental
provisions (termed ecosystem services). It represents the single most important
terrestrial carbon store and some 70 per cent of the country's drinking water
supply comes from upland catchments (evidenced by our wealth of reservoirs)
most of which are peat-clad. Additionally, downstream flows are determined
by the catchment quality that can control or at least mediate flood risks. Peat
matters to everyone.

This is our peatland story: celebrating its wildlife, examining how peat degradation developed and applauding the huge restoration efforts being expended to restore and increase resilience to climate change. Our story began in Chapters 5 and 6 where peat development was described. The way the peat drapes itself across the landscape defines it as blanket bog, as distinguished from valley or basin bogs. There are some generally small examples of the latter but most of our peat can be classified as blanket bog.

THE VEGETATION TODAY

It is true that our blanket bog plant diversity is poor. The bulk of the vegetation consists of two cottongrasses on the more intact domes of peat; Hare's-tail Cottongrass and Common Cottongrass. These are neither cotton nor grasses, but a type of sedge with cottony seed heads that provide a profusion of fluttering, white tassels in the wind in a good year. Hare's-tail Cottongrass cushions have wiry leaves and single flower/fruiting heads of short cotton-adorned seeds (Fig. 6),

FIG 80. Common Cottongrass with fluffy cottony hairs on fruits, and red-tinged leaves grows in the wetter parts of blanket bogs, here beside the Snake Pass.

whilst the Common Cottongrass comprises patches (it spreads via rhizomes) of single shoots, often flushed wine-red, with branched, pendulous, fluffier fruiting heads (Fig. 80). Although not a cotton substitute, they have been used in the past for stuffing pillows, paper production and wound dressings. They are adapted to growing in anaerobic conditions and can exude oxygen in a narrow halo around the roots. Indeed, there is a bright metallic-coloured reed beetle *Plateumaris discolor* that, amazingly, is adapted to survive in this zone. It accesses oxygen through two piercing anal spines in the plant root intercellular air spaces. The larvae lacks these spines, but eats a hole into the root cortex (the outer layer) to source its oxygen.

Although bog-mosses, *Sphagnum* are mostly not abundant amongst the cottongrasses, they are increasing in cover and diversity as a result of the downturn in air pollution (described in Chapter 6). There are now scattered clumps and mounds of a number of species visible every few meters (Fig. 46). Ringinglow Bog (above Hathersage), for example, boasts 14–15 species of *Sphagnum* plus Bog Rosemary (*Andromeda polifolia*), which together have not been seen since the 1970s. In general, look out for the wine-red, compact patches of Acute-leaved Bog-moss (*Sphagnum capillifolium*), or Lustrous Bog-moss (*Sphagnum subnitens*), which tends to have bright green capitula surrounded by pinkish outer branches, or the generally paler, ochre-brown inflated leaves of Papillose Bog-moss (*Sphagnum papillosum*). Bog Rosemary is a rare species in our blanket bogs. It might have been affected by peat acidification through acid rain since it has appeared in several locations in recent years and seems to be spreading, such as at the south end of the Goyt Valley, the north side of the Kinder massive and on the top of Bleaklow.

Other species that occur within the Blanket Bog community are Cloudberry and Royal Fern (*Osmunda regalis*). Cloudberry is related to Bramble (*Rubus fruticosus*) but spreads by rhizomes, producing a thin carpet of palmate leaves in the blanket bog vegetation. Its dazzling white flowers (male and female ones are on different plants) develop into amber fruits, which usually eaten by birds before we find them. Grindon (1886) notes they were sold for 2s 6d per quart (13 p) and being particularly abundant between the Isle of Skye (a former public house on Saddleworth Moors) and Holmfirth. They are rich in vitamin C but rather tart, so were usually made into jams or preserves.

Royal Fern was thought to be extinct in most of our area but is atypically appearing fairly randomly. Linton (1903), Lees (1888) and Hawksford & Hopkins (2011) give no records in the Peak District in Derbyshire or South Yorkshire and only on lower ground in the Staffordshire Moorlands, whilst Moss (1900) attributed its demise to collectors for ferneries or gardens. The Pennine Way

FIG 81. Common Haircap, forming large cushions consisting of tough, wiry shoots about 20 cm long, with dark-green spear-shaped leaves spreading away from the stem. Among many such clumps at the upper end of the Goyt Valley on blanket peat.

north of the Isle of Skye road, near the Home Moss BBC aerial, Snake Summit, Dove Stone and Alport Moors are all new locations but there is no clear explanation why this fern should be appearing. Do report any you find.

Much more common are Deergrass and the moss Common Haircap (*Polytrichum commune*) (Fig. 81), which frequently accompany the cottongrasses. Where abundant, the moss often indicates past disturbance, such as burning, and can form a dense, thick mat other plants have difficulty penetrating. Its holly green colour, long stem and thick opaque leaves are distinctive. Deergrass is a wiry, tufted plant which is bright green in summer, with brownish flowers at the end of the stems. Bog Asphodel (*Narthecium ossifragum*) and Cranberry (*Vaccinium oxycoccus*) sometimes flourish in areas of wetter peat. Cranberry often sprawls in a dark green mat over *Sphagnum*, producing tiny pink flowers before red berries develop (Fig. 82). It is the smaller, poorer cousin of American Cranberry (*Viburnum opulus var. americanum*) used for Christmas sauces. Bog Asphodel, which has flattened leaf bases resembling a small iris and pretty yellow flower spikes, is generally more frequent in flushes rather than blanket bogs. Look for it on Axe Edge and Goyt Moss for example in the west. *Ossifragum* literally means 'bone-breaker' and relates to the traditional belief that stock eating it suffer from brittle bones, but this is more closely associated with the calcium-deficient acidic soils than the plant itself. There is, however, more recent concern over grazing asphodel owing to toxicity in some (not all) plants, suggesting production of a secondary substance in response, possibly, to infection, which results in photosensitisation, a serious sheep skin condition. More and larger plants are generally available to lambs and cattle as reduced grazing levels have facilitated its spread and this may be partly causing the problem.

Most of the species described so far are limited to the Dark and South West Peaks, so are missing from the limestone areas, making the flora locally distinctive. Most are also much more frequent northwards and westwards in the UK uplands, with the Peak District being at their southeast limit, although some occur on the lowland mosses as in Lancashire and Thorne Moors in Yorkshire and in boggy conditions in southern Britain. Some of the species, though, may

FIG 82. Cranberry's dark-green evergreen leaves creeping over a *Sphagnum* carpet with developing red fruits, in Abbey Brook, a tributary of Upper Derwent Dale.

FIG 83. Typical drier blanket bog vegetation with more abundant Bilberry and Crowberry growing with cottongrasses, here on Wessenden Moor. Bracken showing as green patches below the peat in the middle distance.

also be found on the limestone heaths and more acidic patches of grassland on the north-facing edges of the Limestone Dales (Chapters 13 and 14).

Drier bog surfaces are, in fact, more common than intact, wet, peat communities. Many blanket bogs are uneven and/or dissected by variously sized gullies. With so many edges, water tables fall in the upper layers, so drying the peat producing an aerobic environment suited to dry-ground species. This encourages plants like Heather, Crowberry (Fig. 42), Bilberry and Cross-leaved Heath, which add to the colour and diversity (Fig. 83). In extreme conditions, Heather has become dominant at the expense of cottongrasses, even on deep peat, but usually where it is dissected by gullies and regularly burnt to support Red Grouse (more of which in Chapter 9). Large areas of such dominant Heather appear on some of the eastern moorlands as on Bradfield and Langsett Moors and on Harrop Moss south of Longdendale.

The Peak District is renowned for the abundance of Crowberry but possibly not for the right reasons. Crowberry occupies the drier spots – gully edges, hummocks and thinner peats. This dwarf shrub has dark-green leaves along a woody stem that often trails across the ground. Its inconspicuous flowers produce rich magenta-coloured anthers yielding clouds of pollen (Fig. 42) and later, black, acrid-tasting berries. The white line on the leaf under-surface marks its folding to enclose glands which if punctured give a distinctive moorland smell on a warm day. Its abundance could reflect the general drying of many of our blanket bogs, to be explored further later. One worrying feature is the extent of apparently regular Crowberry die-back, more particularly in the Dark Peak. I first noticed this in the 1980s but it seems to have increased subsequently. The leaves turn foxy-red in autumn before dying leaving a grey mat of brittle stems (Fig. 84), sometimes on a large scale, sometimes only a single plant. Many plants show some recovery, but not always. Researchers have failed to find any pathogens in dead plants (there are a number of *Phytophthora* related to potato blight threatening dwarf shrubs at the moment). Hypotheses now focus on possible sensitivity of Crowberry to spring drought (which has increased over the last 40 years with the Peak District at the southeast edge of its national range) and/or the synergistic effects of this with nitrogen deposition, but there is no answer currently.

FIG 84. Dead Crowberry, grey and brittle, with limited recovery (the brighter green strands) on drier ground within a Cottongrass community in blanket bog on Wessenden Moor.

One interesting curiosity of blanket bogs is the small, aromatic, bushy shrub, Labrador-tea (*Rhododendron groenlandicum*). Found in a few remote localities (north of Chew Reservoir and near Barrow Stones, for example), this inconspicuous evergreen bush is an enigma. The Derbyshire Flora sees it as a native of Europe and North America, possibly originating from bird-sown seed or a relic of game cover planting, but research reported in Anderson and Shimwell (1981) suggests it might have been brought here by birds from Greenland as our plant is the variety confined to North and West Greenland, Alaska and northern Canada. Greenland Wheatear (*Oenanthe oenanthe leucorhoa*) does migrate from these climes through Britain on its way to Africa and could be implicated in Labrador Tea's occurrence. A third possibility comes from a local keeper who told the story of the original Barrow Stones plant being established there at the turn of the 19th century as part of game management (Capper, 2001). The plant lives up to its name as a tea was made from its leaves in Labrador.

An abundant moss in the drier bogs is Heath Plait-moss (*Hypnum jutlandicum*) (Fig. 101) with its neatly curled branches and washed out pale-green colour. This forms a mat, with branched shoots entwined through the Heather and Bilberry

FIG 85. Fir Clubmoss colonising restored bare peat on Joseph's Patch on Bleaklow.

stems. Other mosses are also beginning to recover now that sulphur dioxide pollution has declined and as peat acidity reduces. Some plots set up on Holme Moss and Alport Moor in about 1980 were revisited in 2005–6 where numbers of mosses and liverworts were found to have increased from 5 to 16 and from 3 to 22 respectively. Although still poor in species diversity compared with more pristine areas, this is a substantial improvement (Caporn *et al.*, 2006).

Much rarer on the blanket bogs is Fir Clubmoss (*Huperzia selago*), another plant reaching its southern limits in the Peak District. Its last South Yorkshire records were in 1800 (Wilmore *et al.*, 2011), and it was reported from Staffordshire 'Mountains' in the same period (Hawksford & Hopkins 2011). But there were no records for the high moors until recently. Clubmosses are a primitive fern ally that reproduces from spores carried at the base of the leaves. The Fir Clubmoss boasts short, tightly packed stems with dark green, evergreen leaves (Fig. 85). It occurs along the Pennine Way, especially south of the A57 Snake Pass, below the west side of Kinder and is also spreading around some of the restoration areas on Bleaklow.

INVERTEBRATES

Research into the invertebrates of blanket bogs in the Peak District is sparse and we are dependent on data collected for studies such as investigating the effects of wildfire in the early 1980s (Anderson, 1986), researching diets of birds or mammals or the effects of drain blocking (Carroll, 2012). Data from these sources is used here integrated with information from Sorby Natural History Society recording schemes.

Blanket bog provides a distinctive habitat for invertebrates. Although many may be inconspicuous, they nevertheless play a vital role in the upland ecosystem in terms of nutrient cycling, decomposition and food web dynamics. Ninety per cent of the invertebrate fauna list are flies (mostly craneflies and their allies), beetles, spiders and harvestmen. Pearce-Higgins & Yalden (2004) add aphids as common on cottongrass bog, more so than in drier Crowberry/Bilberry communities. The keystone group, craneflies and their relatives, comprise a significant 20–30 per cent of the total invertebrate species present and play an important role in litter decomposition and herbivory. The main species are the craneflies *Tipula subnodicornis* and the much smaller Palearctic species, *Molophilus ater*. Imagine a gangly fly with long legs, beautifully marked, brown wing venation and a typical pale brown, segmented body. Cranefly eggs and larvae live in the upper peat layers and are a very high protein food source for breeding birds. The adults emerge in May and June and can form an astounding 75 per cent of the above ground invertebrate biomass. The larvae and eggs are at their most dense in damp peat but they are easily desiccated and die if the peat dries, as in droughted periods, and are less abundant or largely absent from the drier Heather-dominated or more mixed plant communities. Interestingly, adult females tend to lay their eggs in the same area of the moor from which they emerged – so it can take several generations with random movements for changes in density or new colonisation to occur.

Beetles (Coleoptera), especially predatory ground beetles (Carabidae) and to a lesser extent, rove beetles (Staphylinidae), constitute about 25–30 per cent of the insect fauna in terms of the numbers of species, but only a minor 1 per cent of the biomass. The mixed dwarf-shrub/cottongrass bogs tend to support the most species, perhaps due to the mingling of the drier and wetter ground species. This has also been observed in the Northern Pennines so is not unique to our area. *Pterostichus nigrita*, *P. strenuus* and *Agonum fuliginosum*, all of which are known to prefer damp areas, are the commonest on the wetter cottongrass areas, whereas the Violet Ground Beetle (*Carabus violaceus*) and two other *Pterostichus* species are more prevalent on the mixed moor with fewer species and numbers on dominant Heather areas on blanket bog. Some of the species found are fairly catholic,

FIG 86. Heather Beetle adult, well camouflaged, brownish, about 6 mm long. They tend to drop to the ground off Heather plants when disturbed. (Alex Hyde)

found in a wide variety of habitats, whilst others are specialist and some at the southern end of their ranges like *Patrobus assimilis*, which tends to be a high altitude species.

One beetle of economic significance is the Heather Beetle (*Lochmaea suturalis*) (Fig. 86). It is small, brown and unremarkable but its larvae feed on Heather, drop off when mature and pupate before overwintering in wetter *Sphagnum*-rich patches until spring temperatures reach about 9°C. A female can lay up to 700 eggs in *Sphagnum* and larvae, when hatched, climb into the Heather to repeat the cycle. As they are weak fliers, outbreaks tend to spread downwind. With climate change and warm, damp springs, populations can build-up, although they can crash through predation by a small parasitic wasp whose larvae feed on the beetle larvae. The evidence on the ground is dead Heather bushes, sometimes singly, sometimes on a large scale, that turn foxy-red in autumn before becoming grey/brown and dead. Younger plants can show some recovery subsequently. Heather can also die of winter browning after bright sunny days with frozen ground, so it is not always easy to distinguish different causes of death.

There are also some water beetles, particularly *Hydroporus* species (diving beetles), that occur in acid, stagnant water bodies like peaty pools in the Dark and South West Peaks, *H. erythrocephalus* and H. *gyllenhalii* are the commoner ones and not restricted to the moorlands, whilst *H. melanarius, H. morio* and *H. tristis* are more distinctly upland here occurring in bog pools with cottongrasses. The first two of these are Nationally Local (explained in Appendix 2). Another equally restricted species is *Illybius montanus*, which prefers lower altitudes in a similar habitat on shallow peat.

Of the spiders, money spiders, Linyphiidae, are the most important component, but being very small and often black, are easily missed. They are better adapted physiologically to high altitudes and latitudes than other spider groups. They can be active all winter, even under snow, and are therefore important food sources for active moorland species like Pigmy Shrew. Of the spiders caught in my study, 82 per cent were money spiders, with more species and individuals found again on the mixed moor and cottongrass bog than in the Heather-dominated vegetation on peat. There are also some Lycosid or wolf spiders, which are either active ground hunters or retire into silk-lined holes in the ground.

Biting-midges (Ceratopogonidae), infamous in western Scotland, seem to be getting more prevalent in Peak District moorland areas in recent decades. These tiny biting flies, sometimes called 'no-see-ums' are active in dull, still weather and at dawn and dusk. Their tiny larvae live in the peat. They avoid sunshine and wind, and are more common in wet summers, less so following spring droughts.

MAMMALS

Mammals linked specifically to blanket bog are rare, with Mountain Hare the most conspicuous (Fig. 87). It favours mixed Cottongrass moors where it feeds on Heather and other plants, leaving readily identifiable pear-drop-shaped pellets. They are best seen by approaching slowly upwind as they often shelter in little dugouts on gully edges or amongst taller vegetation, or in winter when they turn white, being more conspicuous of course when there is minimal snow. We have already seen that it was native to the Peak District after the Ice Age (Chapter 5), but died out about 6,000 years ago. Naturally only occurring now in Scotland and Northumberland, it was introduced between 1870 and 1882 for sporting purposes to the Peak District and has since spread widely to most of the blanket bog areas in the Dark Peak from Meltham in the north to south of Kinder Scout. It is absent from the South West Peak and Combs Moss and from moors on the eastern side south of Hathersage. Population estimates in 2003 were between 1,500 and 5,000 animals, but sharp declines in severe winters means numbers can fluctuate (Mallon *et al.*, 2012).

FIG 87. Mountain Hare – typical sighting of it disappearing as you approach on Bleaklow in winter. Winter pelage is conspicuous when there is no snow. (Kev Dunnington)

BIRDS

A walk across the blanket bogs in spring is incomplete without the Golden
Plover's evocative, plaintive call (Fig. 88), often accompanied by Dunlin. Golden
Plover are one of our more important northern breeding waders, preferring
open, high, flat or gently sloping bog with short vegetation. According to the
different county bird atlases, the population appears to be fairly stable or
expanding, except in the Cheshire sector. This contrasts to around 20 per cent
range loss across Britain in the last 40 years (Balmer *et al.*, 2013). The Peak District
is the southern Pennine hub for breeding Golden Plover with around 424 pairs
recorded in 2004 (Frost & Shaw, 2013), with only non-breeding flocks in lowland
fields in surrounding regions. Our blanket bogs support some 1.7 per cent of
the British population within the Dark Peak SSSI, which makes them nationally
important without adding those in the South West Peak and Eastern Moors.
This figure increases to 3.3 per cent (some 752 pairs) in the whole South Pennines
which embraces nearly all our moors: a figure of European importance as part of
the SPA (Appendix 2). Golden Plover commute to nearby fields principally to find

FIG 88. Male Golden
Plover establishing its
territory on Dove Stones,
typically standing on a
small plant hummock.
(Geoff Carr)

earthworms and cranefly larvae when they first arrive to breed, but once hatching starts they are dependent on the local supplies of moorland craneflies, although beetles and moth caterpillars are also taken. Chicks may move up to 2 km to feed in wetter areas, where there is more invertebrate food. Adult birds scold loudly when with chicks, and although lovely to watch, careful retreat is recommended to avoid disturbing them and attracting predators.

Less numerous but equally important are Dunlin. Again, the South Pennines SPA holds some 140 pairs which constitutes at least 1.3 per cent of the breeding Baltic/UK/Irish population. The Peak District total is about 67 pairs. Britain holds some 83 per cent of the biogeographic population, which is very significant and emphasises the importance of our contribution. There are signs, however, of severe declines in range and numbers locally, which is in line with national trends. There may be hope for some revival though, as we will see later. Dunlin often seem to shadow Golden Plover, sometimes called 'plover's page', possibly to benefit from detecting danger by their alarm calling. They are much more secretive when breeding, but share the same moorland diet, selecting areas of open vegetation but in wetter patches to breed.

Compared with these high value breeding waders, our next example is not so welcome – Canada Goose (*Branta canadensis*). This is an introduced species that has spread throughout the country since the 19th Century (see Chapter 11). First recorded as breeding on moorland here in 1986, it is increasing rapidly, often well away from water and at high altitude (Frost & Shaw 2013).

DEGRADATION OF OUR BLANKET BOGS

I have already hinted at the condition of the Peak District's blanket bogs: low botanical diversity; dearth of bog-mosses; effects of air pollution; drying of the peat; its extraction for fuel; wildfires and managed burning; high levels of grazing; and the abundance of erosion features. Together, unfortunately, these have resulted in some of the most degraded bogs in the country, ones that leak peat and the carbon in it. They also bear the distinction because of their degraded condition, of being one of the most thoroughly researched. The degradation is an amalgam of multiple overlain events that do not occur everywhere or at the same time. We do not always know when they started, separating natural phenomena from the human-induced ones can be difficult, and the interactions and repeated occurrence of some factors multiply the impacts in a way that is hard to decipher. As peat is a soft material, it is susceptible to headway erosion from natural stream systems, which could increase in periods of higher rainfall. Tallis, for example, found evidence of gullies in the peat, which he dated to *c.* 1450 CE and possibly earlier. At the same time, in drier periods of climate, *Sphagnum* cover would have decreased, before increasing again in wetter times as described in Chapter 6. *Sphagnum* is an ecosystem engineer. The large quantities of water plants hold act as a sponge on the peat surface, some species more effectively than others. This increases the surface wetness and helps maintain higher water tables by drawing water up the acrotelm, whilst facilitating the accumulation of peat as leaves die. Wet surfaces also increase resilience to fire and erosion. *Sphagnum* plants acidify the surrounding water, which maintains their abundance when appropriate conditions prevail. Reduced *Sphagnum* cover therefore is significant in drying the surface and reducing peat formation.

Erosion gullies dry out for at least 1–2 m from their edges and Tallis (1997) has shown that this is where Crowberry predominated from about 1100 CE onwards as the first, possibly naturally developed, gullies developed. Recent observations show that the peat often cracks, vertically but in parallel, some 0.3 to 1 m or more away from a dry gully edge. Overland water flow then slips down the cracks, frost-heave expands them, thus precipitating mass collapse that is evident in many gullies. Calculations based on gully floor peat losses suggest that many of the gullies up to one metre deep could have been formed within the last 200–250 years (Tallis, 1997a). This coincides with the period of high aerial pollution and the loss of *Sphagnum* from our blanket bog surfaces. However, as this has been all pervading and erosion features vary locally, other drivers must also be involved. An obvious one is fire.

As anyone can see walking across fire-affected peat – and the severe wildfire on Stalybridge Moor that reached the national news in 2018's droughted summer is a case in point – there are considerable variations. Blown by a strong wind, a fire might be large, but scorch the vegetation and leave the peat and root mat intact. In contrast, a fire that moves slowly, back-burns in intensive patches for days or even weeks, can destroy the root mat, and eat into the peat exposing it to erosive forces (Fig. 89). The heat a fire generates will depend on the amount of accumulated dry vegetation, especially if it is highly flammable (such as those containing abundant woody material or natural oils), the wind, and small-scale topography. Fire can tear up a gully for example with great heat compared with a slower rate of spread on a flatter surface.

The time of year is also important. After a period of drought in hot conditions when the vegetation and peat are extra dry, damage can be much greater than in winter when vegetation and peat are wetter. In this respect, it is important to distinguish between wildfires and managed burning. This is more pertinent now owing to the restrictions placed on managed burning, with the open season in

FIG 89. Typical damage caused by wildfire on blanket peat. The fire has burnt through the vegetation (a mixed vegetation with Heather, Bilberry and cottongrasses) and the root mat and resulted in loss of peat as well. Ashed peat and vegetation is visible in surface grey patches. Stalybridge 2018 wildfire site visited in 2019.

upland England from 1 October to 15 April. There were no such restrictions in the past, but managed burning would still mostly be undertaken when the moor was damp. Wildfires, on the other hand are those that are unintended, although occasionally managed burns get out of hand.

The incidence of wildfires in particular has been much studied in the Peak District, perhaps due to it being one of country's most fire-prone upland areas owing to large visitor numbers and the tendency for the moors to be drier due to our geographical location. The Peak District has a long history of fire. It could be a natural phenomenon through lightning strikes, although Lindsay (2010) suggests that the return period for such an event could be in the order of 500 years, which does not make it a key factor. Burning in Mesolithic and Neolithic times has already been described (Chapter 5). Little is known about the incidence of burning in the intervening period, although Radley (1965) considered that fires would have been relatively rare prior to the enclosure period due to the closely protected manorial system. Lindsay (2010) identifies pressure to increase productivity in the 18th century and of the rising sport interest in the 19th century as factors increasing the use of burning. The former links well with Farey's (1815) description:

> … the firing of the heath … in dry weather had at different periods set fire to the peat, and into which it had continued to penetrate, and make large and irregular holes, apparently, until heavy rains fell to extinguish it: the source of unevenness of the groughs and gullies and of local dead black places on the surface of these mosses, is perhaps more common than has been supposed.

Farey describes shepherds going out all day on horseback with their tinder boxes to set fire to the moors. There would have been little control of fires then and they would have probably been large as this used to be the tradition for sheep grazing management. The description suggests a well-established practice by this period and erosion features of some age. It is difficult to believe that land owners could wantonly destroy their livelihoods through injudicious burning, but Farey's description suggests this was not an isolated phenomenon and that burning was being carried out in dry conditions and not necessarily in winter.

Yeloff et al. (2006) found increased frequency of burning from 1920, with a peak during the 1950s on the Marsden Estate on our northern moors. This would accord with increased sporting interests. The extent of managed burning on blanket bogs increased significantly with the first Environmental Sensitive Area agri-environment scheme, the North Peak ESA in the 1980s, which included a requirement for a 20–25-year-cycle burning programme. This stimulated much more extensive burning than had been conducted previously. Current scientific opinion views this as damaging. Lindsay (2010) for example, suggests that at least 200 years are needed for complete recovery after burning wet peat habitats. In addition, some of the burning carried out post 1980 used hot burns which can kill Hare's-tail Cottongrass and even retard Heather's re-establishment– the main desirable plant for grouse moor management. This more regular burning, especially where too hot, has resulted in Heather-dominated peatlands as described earlier, especially on dry peat. There is a general move to cool burns now to reduce damage.

It is the major wildfires that can inflict huge damage. Records of conflagrations appear in the literature and in the media. Tallis (1997a) lists one across Holme Moss in the early 1700s thought to have produced most of the bare peat that was evident until recently. There is a record of another wall of fire on Holme Moss in 1895, and 1921, 1948, 1959, 1976, 1980, 1988–90 and 1995–6 have all been identified as particularly bad years for moorland fires in the Peak District linked to droughted periods (Radley, 1965, Anderson, 1986, Anderson *et al.*, 1997). Radley was of the opinion that the 1959 fires were unequalled up to that date and resulted in large-scale retreat of the blanket peat edges and significant erosion. The National Park Rangers' fire log lists over 400 wildfires just from 1976 to 2009 (Moors for the Future website), although not all will be on bogs, and there have been more since, particularly in 2018.

If early fires were largely lit deliberately by shepherds, damaging fires now are mostly a product of carelessness (usually involving cigarettes, barbeques or camping equipment) or, alarmingly, arson. Other culprits in the past have included sparks from steam trains in Longdendale and Bilberry pickers. Radley recounts how the latter were blamed for a fire in 1826 that made a 'scorched and arid plain' of Broomhead Moors and again in 1868 when Bilberry pickers were alleged to seek revenge for being ejected from traditional picking areas by setting fire to 1,214 ha around Moscar. More recently, aeroplane crashes, usually with tragic loss of life, have contributed to burning peat. Dove Stones and Shelf Moor were affected by fires after such crashes in or just after the last World War (Fig. 90).

FIG 90. Shelf Moor, above Glossop, site of the crash of B-29 Superfortress of the Photographic Reconnaissance Squadron of the USAF, November 1948 with the loss of 13 lives. Most of the bare peat all around is now being restored, but this area has been left as a memorial.

FIG 91. A peat pipe running underground at the bottom of the peat becomes a deep gully with mass peat collapses, extending the gully over time, on Wessenden Moor.

Damaging fires are those that are large, hot, destroy the vegetation and root mat and consume peat, sometimes burning for weeks or months. For example, in 1976, 79 fires burnt across 6,268 ha, which was the highest number and area in any year. Such fires are drawn up the cracks and gullies, enlarging them as they go. This can result in what are called peat pipes – cracks and holes of all shapes and sizes, some large enough to crawl into, which criss-cross the peat mass. They seem to be particularly plentiful where the peat is most degraded – possibly related to the frequency of severe fires. Walk up any peat gully and you will find weeping spots where water appears, often at the foot of the peat, but sometimes between the catotelm and acrotelm. Other larger peat pipes have collapsed leaving a gully or parts of gullies with intervening peat bridges or apparent swallow holes along the pipe alignment. There are places where it is obvious that water spouts forcefully out of peat pipes, spraying fine mineral material on top of the peat during storm events. Peat pipes are everywhere across our more degraded peatlands (Fig. 91), adding to the drying process as water escapes from the system.

Where the top layer of peat is removed, the exposed surface may be caramelised by the heat, changing its chemical signature and become hydrophobic, making it difficult for plants to re-establish. Removing shallow layers of peat also re-exposes the polluted layers with heavy metals that also pose re-establishment problems. Once exposed to the elements, bare peat is prone to frost-heave and wind erosion when dry as well as water erosion on slopes. Peat is puffed up by frost-heave hindering new plant growth. Air pollution and the acidification of the peat has made growth very slow from a restricted variety of species too, all combining to retard re-colonisation and peat stabilisation.

Other factors interplay with the bared peat. If, as frequently happens, there is a wet period after a drought and fires, then storm events can wash away the peat or generate new gullies in bared peat very quickly. September 1976, for example, was wet and storm events washed away more than a metre of peat on Burbage Moor after fire had exposed the surface. Careful observation on different moors will reveal lowered peat levels, sometimes down to mineral soil, often with pedestals of peat and toupees of vegetation, marooned within them. The edges will be very wavy and uneven (distinguishing them from old peat cuttings) and gullies radiate out from them. Look on top of Holme Moss and Tooleyshaw Moss to the west, as well as Bradfield or Langsett Moors. There are many more.

Over-lubricated peat can result in peat flows, although these are very rare in our area, probably owing to our generally dry climate. Nevertheless, Anderson *et al.* (1997) describe one dramatic example reported in the local press that affected Cabin Clough above Glossop in 1834. Two gamekeepers were sheltering in a shooting lodge near the A57 and described the event:

> ... *immense quantities of peat-earth in several different places round them, thrown into the air to the height of thirty or forty feet, accompanied by fires and by large quantities of water, which appeared to boil out of the earth ... the ground ... appears to have been ploughed and torn up in a very singular manner.*

The torrent of water was said to be impassable by horses and to wash away hares, rabbits, grouse and many other wild animals that were drowned. Many fish were reported to be caught up in the turnpike road. Several hundred trees were uprooted beside the stream, torn up by their roots and carried away. 'The quantity of peat, earth, ling and bilberry bushes carried from the moors to the lower part of the valley is almost incredible.'

Grazing is another important factor. Sheep (the predominant animal) are highly selective grazers, seeking out the sweeter grasses (like bents and fescues) in preference to other species. Cottongrass flower buds provide a much sought after supply of phosphorus and calcium at lambing time; important for adequate lactation. Mat-grass, Crowberry and Cross-leaved Heath are avoided, the first owing to the high silica content that would grind down teeth and provide little nutriment, the others probably owing to unpalatable glandular products. Heather is taken in the autumn when its high copper content is useful, but cannot alone support sheep growth rates. Sheep also like to graze in open or short vegetation, so do not push through tall shrubs like old Heather, or tussocks of Purple Moor-grass. This means that any young plants trying to re-establish on bare peat will be grazed hard, particularly after a fire since this releases a minor fertiliser effect in the ash. High levels of sheep grazing alter the vegetation. Woody shrubs are not adapted to this and Heather, especially, declines or disappears. There are still grassy cloughs for example where the only Heather plants are hanging over the streams out of reach of hungry mouths. Crowberry can be damaged through trampling, but Cross-leaved Heath tends to survive in a degraded form as it is not eaten. Bilberry, too, can survive as short browsed stems for many years owing to its underground rhizome system that provides some resilience to hard grazing.

Intensive grazing prevents flowering and seeding thus also reducing the seed bank, food for many invertebrates, and new plants to colonise bare peat. As the more favoured plants are grazed more intensively, they decline in the sward, whilst those that are avoided tend to increase. On the blanket bogs, this includes Purple Moor-grass, Heath Rush (*Juncus squarrosus*) and Common Haircap (Fig. 81). Mini bare spots develop in the heavily trampled ground and the surface becomes uneven – thus more prone to frost and rain damage. Large scale bare peat is probably not directly attributable just to sheep grazing, although this has occurred on some mineral slopes in the recent past (Upper Heyden Brook near Holme Moss and the west facing side of Kinder Scout in the 1980s for example). Storms, fire or localised heavy trampling by visitors might expand this small-scale bare peat too.

This all sounds traumatic for the moor, but there is ample evidence for this sequence of events. Yeloff (2006) for example, has shown increases in sheep numbers in the Colne Valley Parish during the period 1937–2000 apart from dips in 1947 and 1969 when many animals were lost in snowy winters. The rise was particularly sharp after farm subsidies from the EU started in the mid-1970s as this was based on headage rather than area payments, and there was little incentive to reduce numbers. Although sheep farming has been a feature of the area since about 1550 CE, sheep stocking levels increased four-fold from the 1930s in some areas (Phillips *et al.*, 1981). These are parish statistics based on agricultural returns, and some of the sheep will have been housed on enclosed fields and other areas, not just on the moorlands. This also indicates the high level of agricultural intensification in these fields. Derek Yalden and I compared the vegetation on the ground with vegetation maps produced by Moss in 1913 and showed a loss of some 50 km² of heather-dominated and co-dominated ground, attributed to increased sheep grazing (Anderson & Yalden, 1981). Early experimental work to restore bare and damaged ground on the high moors also showed conclusively that without sheep exclusion fencing, re-vegetation was impossible, and even then, the acidity and mobility of the peat thwarted some attempts (Anderson et al., 1997).

One last factor, which is less important in the Peak District than further north in the Pennines and Bowland, are grips. These are drains, often arranged in a herringbone pattern, across rather than with the contours, that were supported by Government funding in the middle of the 20th Century. They often add to or extend the dendritic pattern of gullies, increasing their drying effect. Their objective was to drain the bogs and increase the suitability for grazing or sport by encouraging Heather expansion. At that time, carbon, natural flood defences and the importance of the blanket bog habitat were not on the agenda. The grips both capture overland water, thus drying out the land below them, and drain the excavated route as well. They are visible in the Goyt Valley, on some of the moors north of Longdendale, and in small patches and sparsely elsewhere as well. They have been responsible mostly for further peat drying rather than initiating erosion.

Blending these factors together into a coherent story is not easy. They are best used as a palette of possibilities for any piece of ground. This is what Yeloff et al.,(2006) did for an area of blanket bog on March Haigh, part of the Marsden NT Estate. Since 1840, he found that first *Sphagnum* disappeared in the mid-19th century, possibly as a result of a drying climate, but then remained absent owing to air pollution; dwarf shrubs then became dominant tied in with increased managed burning regime from 1880 to 1900 on something like a 20 year cycle. This was followed by a severe moorland fire in the period 1910 to 1930 from which the dwarf shrubs failed to recover. The severity could be related to accumulating biomass owing to a dearth of keepers during the First World War and a wildfire that coincided with droughted summers such as in 1925. Sheep numbers on the area increased substantially after this period, possibly preventing the recovery of the site. Dwarf shrubs failed to regain dominance and were replaced by grasses (mostly Purple Moor-grass) and an increase in mosses that are pioneers of bare peat and ashed surfaces after fire. There was then another catastrophic wildfire, probably in 1959's droughted summer, and the charcoal and documentary evidence points to more severe fires in the following years after the onset of peat erosion. In the meantime, rising sheep numbers (more than three times the number in 1958–61 and seven times by 1984 compared with 1937) maintained the bare peat, which in turn precipitated gully erosion. A peak in accumulated sediment in the reservoir below the moor indicates increased

FIG 92. The very extensive bare and eroding peat appearing as a mass of anastomosing gullies (groughs) of bare peat separated by upstanding haggs. This type of eroding peat formerly covered large parts of the Kinder Scout plateau (and elsewhere) prior to large scale restoration undertaken by Moors for the Future.

erosion between 1976 and 1984 reflecting more severe summer wildfires. Dried out peat shrinks and cracks, facilitating its removal in rainstorms.

On the ground the pattern of bared peat varies with slope. On flatter areas, peat pans only a few metres across are considered to form after fire through differential burning of more flammable material like Crowberry on drier hummocks, leaving a hollow where the unburnt mound had been (Anderson, 1997). Although this interpretation lacks empirical evidence, it makes sense on the ground. The hollows hold water in winter but dry out with a cracking cover of dried algae in dry periods. Where there is a slope, water can drain between pans, redistributing peat and initiating rills and gullies. Peat pans are common on several watershed moors. A feature of the particularly severely eroded peat summits are the anastomosing channels separating peat mounds a few metres across (Fig. 92). This could possibly represent an older stage in erosion of peat pans, or could be derived from a more severe fire that has consumed peat and then been dissected by gullying. On more sloping ground, the peat gullies tend to be regular and parallel, sometimes quite dense and therefore with a greater drying effect. Even without erosion, the peat surface will have shrunk as the peat dries as well.

This sorry story is not unique, but exemplifies the layers of events that have, to varying extents, affected most of our blanket bogs. The condition of the Peak District blanket bogs was probably worse than anywhere else in the country (Fig. 92). The importance of the different factors vary between sites. Happily, the problem has been recognised and restoration projects form the next, more constructive part of the story.

RESTORATION PROJECTS

The first research into how to repair moorland wildfire damage was undertaken in the North York Moors National Park after the 1976 fires, but the Peak District National Park Authority initiated a major restoration project in 1979 that still continues. The Moorland Management Project first assessed the extent of moorland damaged ground, analysing the main contributory factors before undertaking experiments to revegetate the bare ground and improve habitat condition. The key findings of the first stage were that 8 per cent of the National Park moorlands were bare or severely eroding (out of 52,000ha), largely in the Dark Peak, and that 74 per cent of the peatland was eroding along with 81 per cent of its margins, mostly at altitudes above 550 m. The causative factors were those teased out above. These analyses were the beginning of experimental and landscape-scale trials to identify the most appropriate means of restoring the moorlands, but focused on re-vegetating and stabilising bare peat and mineral ground (Anderson et al., 1997).

Key findings were that bare, mobile peat needed to be anchored to prevent frost heave and a material called Geojute was found to help. This is a biodegradable, open, jute mesh that absorbs water once on the ground thus increasing its weight so that it stabilises bare peat (Fig. 93). It is now used widely on the steepest slopes to stabilise peat whilst plants establish. It was also discovered that dressings of lime and fertiliser were essential to enable plants to counteract the polluted acidic peat, along with a nurturing cover of grasses, which enabled establishment of more locally native species within their matrix. At first, Heather seed was added and new techniques were developed to collect and stimulate their germination (which benefits from smoke and heat as they are adapted to fire). Heather seed can be collected on plants (called brash), cut between October and January before seed dispersal, or brushed from plants with a harvesting machine. Geoff Eyre from William Eyre and Sons in Brough was a major player in the development of techniques and machines. Early in the project, the North Peak ESA started to have a significant impact through

FIG 93. Geojute was used to stabilise the steep gully side before adding a nurse grass seed mix, fertiliser and lime as well as Heather seed on Dove Stone. The remains of the biodegradable Geojute is visible in the centre. The bare peat beyond this shows the lack of re-vegetation where no stabilising material was added.

supporting reduced grazing levels and with, at first, small-scale restoration works. New agri-environment schemes have continued and expanded in scope to support the large scale restoration projects that have been implemented since, particularly on SSSI land.

The Moors for the Future programme (which evolved out of the Moorland Management Project), United Utilities' Sustainable Catchment Management Programme (SCaMP) more recently in collaboration with RSPB, the National Trust's High Peak Vision management work, Yorkshire Water's restoration programme, the National Park Authority on its Warslow Estate, along with work by private estates have all been instrumental in taking the earlier smaller scale restoration projects forward on what is now a grand scale, but with a change in emphasis. The earlier restoration work was geared to re-establishing vegetation on bare ground as a matter of urgency, but this ignored the hydrology which we

FIG 94. The River Etherow at the east end of Longdendale showing strong brown colouration from excessive dissolved organic carbon derived from peat decay in the surrounding catchment.

now know is the key for healthy condition and abundant *Sphagnum*. The more recent, larger scale projects have been restoring hydrology too to slow down or reverse the drying that has taken place over so many centuries. This has not occurred in isolation in the Peak District as similar approaches are being applied in many upland areas elsewhere, sharing data and techniques.

From the Water Company's perspective, increasing levels of colour derived from dissolved organic carbon from the peat pose major problems (Fig. 94). The colour is generated from drying peat in aerobic conditions and has to be removed at the treatment works at considerable financial and environmental cost. Levels are increasing internationally, possibly owing to reduced sulphur dioxide pollution. Therefore, any action to rewet the peat and slow down colour production is beneficial. At the same time, such action enhances blanket bog condition, stimulating recolonisation by mosses like *Sphagnum*, as well as holding more water and attenuating downstream flooding. Raising water tables also increases peat's resilience in the face of droughts and wildfires.

Techniques to hold more water on blanket bogs have therefore been developed. These are essentially various dams: wood (rather vulnerable to wildfires), piling (made of recycled plastic), peat dug from beside the gully, coir rolls, heather bales, and more recently, stone (Fig. 95) have all been deployed. Different types suit different gully depths, slopes and water pressures. You can see them on many moors now – Cutthorn, Wood Moss and Danebower Moors in the southwest, the Goyt Valley north of these, large areas of peatland to the north and south of Longdendale, the top of Kinder Scout, both sides of the Snake

FIG 95. Different kinds of dam materials for blocking gullies and raising water table levels on blanket bog. (a) Stone being dropped by helicopter into gully, Chew. (Andy Keen). (b) Recycled plastic dams in gully, Higher Nether Moor, Kinder Scout. (c) Coir rolls being placed to trap mobile peat in peat pans, Chew. (d) Stepped wooden dams in gully, Buckstones.

Pass A57, Upper Derwent Valley moors and many more. To raise the water table to support wet bog vegetation requires water to be within about 15–20 cm of the surface. Thus once dams are filled with sediment, new dams will be built on top: this whole water table re-building process might take many decades but at least progress is in the right direction.

As well as re-wetting, re-establishing blanket bog species rather than the dwarf shrub cover is essential. Thousands of cottongrass plants have been planted (usually with volunteer help), and most importantly, methods for introducing *Sphagnum* have been trialled. Initially instigated by Moors for the Future working with Micropropagation Services, various forms of *Sphagnum* introduction products have been tried, including bead-sized capsules, a gel and plug forms for different situations and scale. RSPB has favoured using handfuls of *Sphagnum* harvested sustainably (with permission) to plant 70,000 with volunteer help on Dove Stone. The better developed the plug, the faster the plants establish and spread, but many are needed for large areas (Fig. 96). Growth can be good, for example, in 4 months in 2017, plug clusters expanded from 3 cm to 15 cm across and one plug established in 2009 had enlarged to 1.05 m in

FIG 96. *Sphagnum* plug inserted (with many others) into revegetated peat on Kinder Scout.

six years on Holme Moss. Monitoring by Manchester Metropolitan University on Featherbed Moss 18 weeks after establishing *Sphagnum* plugs showed that 70 per cent of those in gullies and 88 per cent of those in peat pans were still alive and many were spreading well, being more successful in already revegetated rather than bare peat.

Between the projects, nearly all the main moorland areas within the Dark Peak and in parts of the South West Peak have undergone treatment. This is a massive achievement in a short period exemplified by some statistics. The Moors for the Future after 15 years had restored 2,428 ha of moorland, established 250,000 plants, spread 40 million beads of *Sphagnum* moss, laid 140 miles of Geojute and, for just one of its projects on Kinder, installed 1,060 stone or timber dams and spread 8,500 bags of Heather brash to stabilise bare peat. There is much still being done. You might have seen the brash bags ready on the moors for helicopter transport to the bare peat areas, or the piles of dam material awaiting installation.

Although huge progress has been made, restoration to fully functioning blanket bog will take many decades or more, but improvements are already evident. Grip blocking has been shown by SCaMP in the Goyt to elevate water table levels and reduce their perturbations, thus facilitating natural increases in *Sphagnum*. There are also early signs of slow increases in other bog species and reduced Purple Moor-grass (Penny Anderson Associates, 2015). Elevated water tables are helping cranefly survival, more particularly in dry seasons (Carroll *et al.* 2011). Monitoring by the RSPB at Dove Stones has shown breeding birds benefiting significantly. Dunlin in particular, after a rapid decline, has shown a substantial increase from a low 7 pairs in 2004 to 43 pairs by 2014 on the 2,500 ha monitoring area. This increase coincided largely with where rewetting took place. Concurrently, Golden Plover numbers trebled on the same areas, with a critical increase in chick survival from 0.57 fledglings per pair in the 1990s, which is consistent with maintaining a stable population, to 1.22 fledglings per pair over just three years. Curlew (Fig. 118, *Numenius arquata*) also showed signs of increasing with rewetting on Dove Stones and other species such as Red Grouse and Skylark (Fig. 156) have also increased on the restored areas, critically despite reduced predator control (Carr & O'Hara, 2015; O'Hara & Carr, 2017).

Some of these positive signs could be related to the new pools created behind dams. Brown *et al.* (2016) in a study including some new Peak District pools, show that these quickly provide habitat for large populations of non-biting midge larvae (Chironomidae) with smaller numbers of specimens and species of water beetles (Coleoptera) and bugs (Hemiptera): an assemblage similar to that found in long-established peat pools. The predominance of midge larvae was attributed to their ability to adapt their diet as they consume detritus, which will be abundant in peaty pools, thus enabling them to occupy even the smallest of water bodies. The emergence of the midges especially significantly increases the available wader food. This is particularly important since the future of some moorland species is under threat through climate change (Pearce-Higgins *et al.* 2010, see Chapter 15). The rewetting therefore increases resilience to this.

Several other benefits have followed from the restoration. Monitoring of SCaMP and Moors for the Future projects shows that landscape-scale revegetation of bare peat reduces sediment loading to streams and reservoirs. Bare ground reduces by around 88 per cent within four years of revegetation treatment and, although not a blanket bog vegetation at that stage, the plant cover significantly reduces loss of peat in water or wind. On the other hand, colour reduction in the runoff water takes more time. Although reductions in dissolved organic carbon were detected after only two years in outflowing streams after grip blocking raised peatland water tables in the Goyt Valley as part of SCaMP, decreases are more difficult to achieve from severely eroded moorland (Penny Anderson Associates 2015, Pilkington *et al.*, 2015).

Of special significance to those vulnerable to downstream flooding, such as in Sheffield and Rotherham in 2007 when the River Don burst its banks, is the evidence of flood attenuation from blanket bog restoration. A Defra-funded project on Kinder in the River Ashop catchment examined the effect of both revegetating bare peat and blocking gullies and showed that storm-flow is slowed down through reduced lag times and decreased peak and total storm discharge. This was from restoring only 12 per cent of the Upper Ashop river catchment, but even when modelled for the wider catchment, suggested a possible reduction in peak discharge (the highest flow at a given point) of 5 per cent on average at a 9 km^2 scale after both gully blocking and revegetation (Pilkington *et al.*, 2015). Translated into flood risks in Derby on the main River Derwent into which the River Ashop flows, blanket bog restoration is set to be a critical future friend.

We have seen how the enormity of the scale of blanket bog degradation is now matched by restoration efforts by numerous partners. The results are positive at a landscape scale and auger well for the future of these fragile ecosystems. It is time to turn now to the rest of the moorland environment which forms an integral part of the overall upland system, although one that has not been subjected to the same degree of investigation or degradation.

CHAPTER 9

'Desolate, Wild and Abandoned Country'

'T'HE MOST DESOLATE, WILD and abandoned country in all England.' So wrote Daniel Defoe about the Peak District in 1726, further damming the moorland as a 'waste and howling wilderness'. Fortunately, attitudes today have changed and we can delight in the splendid expanses of Heather (Figs. 50 and 97) the deep cloughs with hidden flushes, Bracken and woodland, as well as the intermixing of grasslands up to the blanket bog edges. Moorland is essentially the land above the agriculturally improved fields, generally over about 300 m altitude, dominated by low vegetation – grasses, dwarf shrubs, Bracken, sedges and rushes, including the blanket bog already described. It consists of low summits and upland plateau together with the intervening valleys, thus distinguishing it from mountainous terrain. In the Peak District, most moorland lies within open access land (as shown on the OS 1:25,000 maps) within the Dark and South West Peaks.

DWARF-SHRUB HEATH

There is little better than being immersed in flowering Heather with pollen puffing and bumblebees humming in warm summer sunshine. Heather moorland is more correctly classified as dwarf-shrub heath, or heathland/ heath for short. Heather is just one of the characteristic dwarf shrubs, along with Bilberry, Crowberry, Cowberry (*Vaccinium vitis-idaea*), Bell Heather (*Erica cinerea*) and Cross-leaved Heath. Each has its preferred niche and are not equally common. Intermingled with these are moorland grasses like the fine-leaved Wavy Hair-grass or the more wiry Mat-grass tussocks.

FIG 97. Heather moor on Nether Moor, east end of Kinder, with patches of Mat-grass and Wavy Hair-grass, overlooking more Heather moorland on the Derwent Moors in the distance.

Bilberry and Cross-leaved Heath come into their own on the heathlands. Look out for swathes of shorter Bilberry on drier soils within, often, a Mat-grass matrix, or taller, bushier and more dominating plants mixed with other dwarf shrubs where grazing pressure is lower. Bilberry's deciduous leaves unfurl in swathes of lovely peachy/pale green colours in spring (Fig. 98) before producing profuse, bell-shaped pink flowers which ripen into appetising fruits rich in vitamin C and anthocyanins (naturally occurring antioxidants). Eating them has been a long tradition here, but spare a thought for foraging ring ouzels and flocks of young starlings when collecting them.

Cowberry, a Bilberry relative sometimes called Lingonberry, is distinctive with evergreen leaves and rather tart red berries attracting feeding birds. It often accompanies Bilberry and represents another mostly north and western species. It hybridises with Bilberry, with the only British locations occurring in Staffordshire, Yorkshire and Derbyshire, possibly related to overlapping flowering periods here of parent plants. Most of the records lie on the eastern

FIG 98. Rich peachy and green colours of Bilberry as leaves emerge in spring, above the Woodlands Valley.

side of the Peak District, along the lower moors, especially beside disturbed track edges where seedlings can establish.

Much rarer is another similar shrub, Bearberry (*Arctostaphylos uva-ursi*). Again we are at the southern edge of its circumpolar distribution. Plants occur in Upper Derwent Valley (Fig. 99) and on Ladybower Tor, but are difficult to find possibly as herbalists were gathering armfuls from North Derbyshire valleys in 1911 to treat kidney complaints and as a diuretic and disinfectant. Recent pharmacological research suggests that long term use may cause chronic liver dysfunction, vomiting and constipation, so hopefully there is little threat from further collecting.

Cross-leaved Heath is more characteristic of wet heath, growing with *Sphagnum*, some cottongrasses but usually Heather too. Wet heath is rather restricted, but occurs in perched wet areas above the cloughs. Cross-leaved Heath's pale pink flower clusters are much prized by bumblebees. The plant is avoided by sheep, but can appear grey and lifeless after an attack by moth caterpillars.

FIG 99. Bearberry in the Upper Derwent, showing its dark-green evergreen leaves which have a net pattern rather than the glandular dots of Cowberry.

The closely related Bell Heather blooms earlier than Heather, helping to extend the heathland flowering season for bees and other pollen or nectar-dependent invertebrates. Moss (1913) described this species as abundant or co-dominant with Heather, yet few sites now match this description. Look instead on dry, unburnt banks, as along the A57 east of Glossop and around Curbar Gap (Eastern Moors). Scattered plants occur in many cloughs, but only where grazing is light.

FIG 100. Western Gorse on Revidge in the South West Peak, with less deeply furrowed spines than European Gorse, flowering late summer.

FIG 101. Typical mosses of dwarf-shrub heath from which bilberry is emerging, including Heath Plait-moss with metallic-toned curled ends of the shoot branches and the upright narrow leaves of Broom Fork-moss, growing on a steep bank overlooking Woodlands Valley.

The two Gorse species: Western and European Gorse (*Ulex europaeus*) also extend the heathland flowering season for invertebrates. Easily separated when in bloom as Western Gorse flowers in late summer (Fig. 100), whilst European Gorse is at its best in spring accompanying any walk with a pungent burnt-biscuit aroma. However, both are quite restricted within the moorlands. Being taller and thicker than the other dwarf shrubs, they are also a useful bird breeding habitat. Hucklow Edge is particularly spectacular when the European Gorse is flowering and some of the southern cloughs below Kinder Scout support patches of Western Gorse.

Just as integral to the heathland vegetation, though not as noticeable perhaps, are the mosses and lichens. Heath Plait-moss (*Hypnum jutlandicum*) is the most plentiful, but others include the upright Broom Fork-moss (*Dicranum scoparium*) (Fig. 101), the pale-green flattened shoots of Waved Silk-moss (*Plagiothecium undulatum*) and the ubiquitous Springy Turf-moss (*Rhytidiadelphus squarrosus*). On north-facing slopes in the cloughs, *Sphagnum* may be abundant drooling down banks in the sheltered, damp conditions. Small upright mosses are dominated by the *Campylopus* species, Heath Star-moss (*Campylopus introflexus*) is conspicuous with white hair-tips on reflexed leaves, but others are more difficult to differentiate.

Lichens are also common heathland species, often with bright red or dusty grey spore tips of *Cladonia* species of which there are several, including the tangled mass of a Reindeer Lichen (*Cladonia portentosa*) (Fig. 102). A more lichen-rich habitat, often within the heathland environment or in cloughs, are the many rock surfaces. You will see lichens established as flat patches (crustose) forms or those that are leafier in appearance – the foliose species. They vary from blackish, to grey, cream, brown or khaki, often with different coloured fruiting

FIG 102. The lichen *Cladonia portentosa* growing around bilberry bushes on the edge of Featherbed Moss, Snake Summit.

bodies. They represent a symbiotic relationship between a fungus and an alga or cyanobacteria. The fungus benefits from carbohydrates produced by algae and cyanobacteria can also fix nitrogen from the air or water. Lichens generally grow very slowly as they shut-down when dry and can be very long-lived. The most conspicuous species on the rocks are *Acarospora fuscata*, which can form extensive brown patches with a surface resembling cracked mud; the yellow crustose *Rhizocarpon geographicum* with black fruits (the map lichen); and *Pertusaria corallina*, which forms bright white patches with a surface resembling a miniature garden of coral reef polyps. The commonest foliose lichen is *Parmelia saxatilis* with its grey leafy appearance that can cover quite large rock areas.

The general lichen coverage on exposed rocks is still suppressed by historical industrial pollution but is increasing. Some rarities are also still recovering, such as the beard lichen *Usnea*, or several upland north and west species like *Ophioparma ventosa*, which has vivid purple-red fruiting bodies dotted on a creamy-white thallus, or *Umbilicaria* species that are brown and leafy looking rosettes with a central 'holdfast'.

The extent of dwarf-shrub heath in the Peak District at 11,350 ha within SSSIs and SACs (Appendix 2) is second only to that of blanket bog, with most also in the Dark Peak. Seemingly common in Britain, it is perhaps surprising that 75 per cent of upland Heather-dominated ground in the world lies in the UK (BASC, undated) and it is actually rarer than rainforest (Moorland Association: www.moorlandassociation.org). This emphasises the importance of this habitat nationally and regionally.

Anyone visiting the sweeping vistas of heather moorland will see small blocks of burnt or cut vegetation, often in a regular pattern (Fig. 103). This is part of their management for grouse and such grouse moors along with the Red Grouse subspecies (Fig. 104) are unique to the UK. Historically, shooting began in the mid-17th century but gained popularity in the 18th century. Indeed, the shooting season of 12 August to 10 December that still applies was set by a 1772 Act. It was only with the successive improvement of guns in Victorian times that driven grouse over butts, rather than walk-up shooting, became usual practice, Hey (2014), prior to the tightly managed heather moors that we see today.

FIG 103. The typical appearance of a heather moor that is managed for Red Grouse, in this case by cutting Heather in rectangular-shaped blocks to provide young growth for Grouse feeding and taller cover adjacent.

FIG 104. Male Red Grouse setting up its territory in November on heather moorland.

Moorland management for sheep and grouse are closely tied to Heather's life cycle of about 40 years which is divided into four main growth phases. There is an early pioneer period of small plants establishing after managed burns or cutting, followed by the building phase when plants increase their density and canopy. Reduced vigour, shorter growth increments, greater woodiness and less flowering mark the next mature phase culminating in the degenerate state after about 30 years when the plant's central frame opens out. The young shoot tips in the pioneer stage are the prime food source for Red Grouse, although sheep and Mountain Hares also favour them, sometimes resulting in growth suppression. Heather management for Red Grouse focuses on 12–15 year cyclical cutting or burning of Heather before it reaches the mature, woody phase as older plants recover more slowly. Prevailing weather conditions usually thwarts this target though.

Most of the Peak District heathlands are maintained as shooting estates for Red Grouse, although varying numbers of sheep (and sometimes deer) are also accommodated. Red Grouse require a territory with a good supply of young nutritious Heather and taller, leggy plants for cover and nesting. An optimum balance of Heather growth phases is therefore needed in each territory, which could be as low as 4 ha per pair or up to 12 ha for a high or low density population respectively. Research shows that predation is reduced and access to cover is optimum for Red Grouse if the managed patches are no more than about 30 m wide, although they can be much longer. Management on scheduled sites (SSSIs) also seeks to retain significant amounts of old Heather for species that depend on it. Cutting is encouraged rather than burning where the vegetation is on any depth of peat to avoid damage and loss of its carbon store. Ideally, diverse blanket bog should not be burnt, and it has already been shown how some of our upland heaths are Heather-dominated on deep peat. Exposing the dark peat to sun increases surface temperatures and aerobic conditions, thus speeding up peat decay and subsequent loss. Cutting is less weather dependent and requires less expertise compared with burning, so is a useful alternative on accessible land.

This account explains the vegetation patterns on a moor (Fig. 103), but also probably accounts for Heather's dominance. Bilberry is usually mixed with the Heather, but other plants are often scarce once the canopy becomes dense. This type of grouse moor management was instigated after the Committee of Inquiry on Grouse Disease in 1911, so Bell Heather's past abundance in the Peak District probably diminished under this more intensive management. Rabbits (*Oryctolagus cuniculus*) too may have played a role since they readily graze young Bell Heather plants, which could have given Heather (although this is also grazed) a competitive advantage. Rabbit numbers would have grown and peaked prior to decimation by myxomatosis in the 1950s, but there is very limited local information on this.

Additional grouse moor management practices merit mention since they affect other heathland animals. Even if you find shooting an anathema, it is worth understanding the management system, and how it fits into the regional economy. Grouse are normally driven over guns in butts, which are horse-shoe shaped bunkers or barriers such as walls or boards behind which the shooter is screened. A few moors adopt walk-up rather than driven shooting (although this is less profitable). Grouse shooting generates economic inputs, labour and investment which are positive in a rural area where making a living from the land is difficult. Certainly in the past, moorlands without grouse shooting interests were more often heavily sheep-stocked resulting in degraded or lost heathland. This is now largely reversed through more recent agri-environment schemes and nature conservation management objectives being applied to protected areas.

Grouse moor management also involves legal predator control of Foxes, Carrion Crows, Stoats and Weasels, but the methods used are regulated.

In addition, Red Grouse suffer from the debilitating diseases, louping ill and Strongylosis. Louping ill is transmitted by sheep ticks, which can be reduced by inoculating the sheep. Strongylosis, in contrast, has a greater effect when Red Grouse densities are high and has been implicated in the regular cyclical population fluctuations. It is caused by a nematode worm, *Trichostrongylus tenuis* and controlled by offering the birds medicated grit (Game & Wildlife Conservation Trust). Grit is needed to help grind food into a more digestible state in the bird's gizzard. The grit trays usually have a flap for covering the medicated side prior to the shooting season to avoid drug by-products entering human food-chains.

You might also see grouse drinking pools on Stanage Moors. Hey (2014) describes how its owner commissioned a young stonemason, George Broomhead, in 1907, to excavate artificial drinking troughs fed by carved runnels collecting water on a range of gritstone boulders in an attempt to prevent 'his' birds moving to neighbouring moors seeking water and natural grit. Altogether 108 numbered troughs can still be found, many along Stanage Edge, though often obscured by Heather and Bracken, with the largest 1.50 m wide and 60 cm wide.

Burning for sheep management used to be in large blocks to produce a new bite of younger, more nutritious growth. However, where repeated too frequently, this led to dominance by more fire-resistant grasses and the loss of many dwarf shrubs. This was the origin of some of our larger areas of Purple Moor-grass. Heavy grazing alone also had major impacts. Comparing Moss's 1913 vegetation map with moorland vegetation in 1980 showed a major 36 per cent decline of the Heather or Bilberry-dominated communities and their conversion to acid grassland or Cottongrass dominated blanket bog (Anderson & Yalden, 1981). Moss described the Heather-Bilberry and Bilberry-dominated vegetation as characteristic below gritstone edges like those fringing Kinder Scout or Derwent and Bamford Edges. Bilberry also formed hill crest cappings. This description still holds for Taxal Moor in the Goyt Valley, where there is a luxuriant moss-filled Bilberry/Heather vegetation, but no summit matches Moss' description now– even though the vegetation would be more stunted in such exposed situations. Crowberry tends to more frequent instead, mixed with Mat-grass.

Recent restoration is reversing some of these earlier trends. Heathland has been re-established on Purple Moor-grass in large parts of the Howden Moors (Fig. 105), Pigford, High Moor and Shutlingsloe, parts of the Eastern Moors and patches within the Longdendale catchment since the 1980/90s by land managers. Large scale trials into the most effective restoration methods found that herbiciding or cutting dense Purple Moor-grass low to the ground, sometimes burning off the accumulated litter to provide bare ground, and then adding heathland seeds worked well. The west face of Kinder is another showcase. The

FIG 105. This moorland on Howden Moors was dominated by Purple Moor-grass until the 1990s when it was gradually restored to heather moorland (the dark areas) by Geoff Eyre. Some grassy areas (pale colour) remain and other species were added to produce a more diverse vegetation.

National Trust purchased the estate in 1983 and set about restoring its very severe condition. After removal of sheep, recovery was substantial by 1992. Bare ground declined as Wavy Hair-grass increased significantly. The gradual expansion of Heather cover was probably limited by lack of local seed and Bilberry only slowly spread from underground rhizomes (Anderson *et al.* 1997). This recovery has been spectacular and continues today. Dwarf-shrub heath has also recovered from overgrazing on Lantern Pike, Gun Moor (now a lovely Staffordshire Wildlife Trust reserve, Fig. 50), Gradbach Hill, and the Eastern Moors along with many other sites, although not everywhere is in ideal condition yet. There are still stunted Bilberry patches, Mat and Purple Moor-grass grasslands and cloughs with little Heather or its relatives.

Some large heathlands, like the Eastern Moors and Roaches Estates, are managed principally for nature conservation objectives. Here burning or cutting creates firebreaks for firefighting access and to subdivide the heathland

FIG 106. Invasion of Silver Birch into heather moorland after a wildfire near Surprise View car park, Hathersage.

to reduce potential damage from wildfire. Heather's accumulating woody framework with age is at high risk of destruction in dry summers through wildfires, as witnessed repeatedly in recent decades. Short, young Heather burns less intensely or rapidly than old plants, thus also damaging soils less. On these sites, stock grazing is geared more to the required habitat condition rather than to agricultural or game outputs.

Without management, dwarf-shrub heath would not persist. This is clear on Blacka Moor and Greave's Piece on the eastern moorlands, as well as several wider road-side verges and ex-wildfire sites where trees colonised (Fig. 106) and grow amongst the remains of dwarf-shrubs. Blacka Moor was part of the Duke of Rutland's grouse shooting estate, but was sold in 1927, eventually to the Graves Charitable Trust to be preserved in its natural state to avoid development. It was then gifted to Sheffield City Council in 1933 who were to maintain it whilst providing access. Grazing was largely withdrawn from the 1950s until the current managers, the Sheffield and Rotherham Wildlife Trust, reintroduced it in 2006. Although there were woodlands around the fringes and on the lower sides of the moor, the absence of stock (Red Deer are present), resulted in a substantial spread of trees forming scrub patches and extending the woodlands. Given more time, the heathland would have largely disappeared: a common story on unmanaged heathlands in Southern Britain. Since dwarf-shrub heath has its own specialist flora and fauna and is of cultural interest as well as being beautiful, protecting and restoring it is an important element of our heritage. This does not necessarily mean removing all trees and shrubs, but containing them so that wide, open spaces for the heathland are maintained.

SPRINGS, FLUSHES AND DAMP PLACES

Anyone visiting the moorlands interested in flowers will be drawn to the cloughs, where the greatest diversity and many of the rarities live related to the general lack of burning, the variable soils and hydrology, and to diverse physical structures like rock ledges, waterfalls, steep banks and streams (Figs. 28 & 107). In many cloughs heathland adorns the steep slopes, often with more varied dwarf shrubs nestled in mosses. Many clough banks of this kind are luxuriant, bushy and colourful as in Black Cloughs at the upper end of Longdendale. Others lack this heathland mixture and are more Mat-grass/Bilberry dominated, such as Linch Clough in Upper Derwent Valley and Oyster Clough off Lady Clough.

FIG 107. One of the larger, very rich flushes in Fair Brook Clough, where groups of trees including willows and patches of Bracken (brighter green patches) are also visible between Heather. The acid grassland covered slopes of Kinder Scout in the distance.

In all cloughs of any size, wetlands are varied, matched by a diverse flora. Springs bubbling, or oozing, from under rocks or at seepage lines are frequent where the shales and grits meet. Water may be flushed through slopes – this is where it passes through soils below the surface except in very wet weather. This, therefore, provides oxygen and minerals and reduces acidity, which many plants favour. The detailed chemistry of the shales gives the springs their chemical signature, but the lie of the land will dictate how fast the water flows and spreads out, whether it sinks into the ground, sometimes re-emerging or not, sometimes becoming a surface stream or running below (often audible). All these factors control the habitat conditions for the plants and animals.

In recent sampling of clough flushes, I have found 183 different plant species, excluding mosses and liverworts. This is remarkably high compared with the natural general dearth of heathland and blanket bog plants and is linked to the more base-rich conditions. I can mention only a few, so exploring is the best way to appreciate them. Floristically, these wet patches often support plants that also grow in other habitats, often more abundantly – such as White Clover (*Trifolium repens*), Creeping and Meadow Buttercups (*Ranunculus repens* and *R. acris*), and those that are rarely seen elsewhere, like some of the sedges and marsh plants, resulting in unique communities. There also tends to be differences between habitat altitude, with some species more often found on lower ground, such as Angelica (*Angelica sylvestris*) and Sneezewort (*Achillea ptarmica*).

In general, the flushes at the upper end of cloughs are less diverse than those further down, having lower dissolved minerals, resulting in classic mixtures of Soft Rush (*Juncus effusus*) with the Flat-topped Bog-moss (*Sphagnum fallax*), Common Haircap (Fig. 81) sometimes with few other plant species. There are two classic spring communities, one dominated by mosses and the other by Opposite-leaved Golden Saxifrage (*Chrysosplenium oppositifolium*), often accompanied by the inconspicuous Blinks (*Montia fontana*). The moss community usually contains Fountain Apple-moss (*Philonotis fontana*), a lime-green, upright moss with red stems that forms cushions over a wet stony matrix. The little, star-like Marsh Bryum (*Bryum pseudotriquetrum*) is a common associate, sometimes accompanied by the reddish liverwort, Water Earwort (*Scapania undulata*), which also grows on rocks in small streams. There may be the odd stem of Sweet Vernal-grass (*Anthoxanthum odoratum*) or Creeping Soft-grass (*Holcus mollis*); typical lowland grass species we will meet again later, but often little else in a community that might be less than a metre across.

Where water is trapped and sits in little pools within the flushed slopes, Round-leaved Crowfoot (*Ranunculus omiophyllus*) can grow, often with Blinks, although this is also found in other situations but only in the Dark or South

FIG 108. A typical mixture of Bog Pondweed (broad, smooth-margined leaves), Bulbous Rush (with clusters of flowers along the stems), and Round-leaved Crowfoot in a small runnel. Carnation Sedge and Soft Rush on either side of the flush in Oyster Clough, off Woodlands Valley.

West Peaks. Bog Pondweed (*Potamogeton polygonifolius*) might accompany these (Fig. 108). Look out too for water dribbling through golden orange-brown mats of Lesser Cowhorn Bog-moss (*Sphagnum inundatum*), with horn-like branches or where an equally orange-brown moss, but one with side-twisting, curled ends to leaves, Curled Hook-moss (*Palustriella commutata*) occurs. Neither are common, but the latter indicates a greater concentration of calcium in the spring water and occurs in a few of Abbey Brook's and Green Clough's springs for example, on either side of the Upper Derwent.

Soft Rush is a ubiquitous flush plant, whose abundance may be related to its prolific seeding and seed persistence. There are also short, sedge-rich patches occasionally, usually with the contrasting grey-green (glaucous) Carnation Sedge (*Carex panicea*) and the bright green of Common Yellow-sedge (*Carex demissa*). Slightly more peaty conditions sees Common Cottongrass joining these, whilst drier edges are marked by Mat-grass and wetter runnels by Bulbous Rush (*Juncus bulbosus*), a much smaller relative of Soft Rush (Fig. 108). Common associates include Tormentil (more usual in drier acid grassland) and other sedges.

More plants typical of less acidic conditions accompany Soft Rush communities in the richer flushes. Marsh Violet (*Viola palustris*) is common, Lady's Smock (variously also called Cuckooflower, as it flowers at about the time cuckoos return, *Cardamine pratensis*), Common Sorrel (*Rumex acetosa*), Marsh Willowherb (*Epilobium palustre*), Bog Stitchwort (*Stellaria alsine*), Marsh Thistle (*Cirsium palustre*) and Star Sedge (*Carex echinata*) are regularly encountered. Marsh Thistle is particularly important as a source of nectar and pollen for bumblebees and butterflies.

Some plants show associations with specific cloughs. Less common sedges, Smooth Stalked-sedge (*Carex laevigata*) and Flea Sedge (*Carex pulicaris*), the more widespread Greater Bird's-foot-trefoil (*Lotus pedunculatus*), Marsh Pennywort (*Hydrocotyle vulgaris*), Selfheal (*Prunella vulgaris*), Autumn Hawkbit (*Scorzoneroides autumnalis*) and Yellow Pimpernel (*Lysimachia nemorum*) are conspicuous in some cloughs but absent or nearly so in others. Such patterns must relate to water chemistry, but could be a product of past grazing history or other factors too. Other associations can give unexpected surprises. There is at least one flush in the South West Peak, for example, that has both Primrose (*Primula vulgaris*) and Meadow Vetchling (*Lathyrus pratensis*) growing together next to more normal flush plants, which is very unusual.

Often beside the springs, sedge communities and rushes, small *Sphagnum* carpets show the increased acidity and sometimes slowing of flow rates. Bog

FIG 109. Round-leaved Sundew with remains of flies still in on its leaves. Above William Clough, near Hayfield.

FIG 110. Heath Spotted-orchid growing in flushed soils on Houndkirk Moor. Note the small central lip lobe compared with the larger wings.

Asphodel, Common Cottongrass, Cranberry and Cross-leaved Heath sometimes mark these changes. It is always special to find Round-leaved Sundew – we have few sites and most are in these boggier flush patches (Fig. 109) largely in the Dark Peak. The plant so fascinated Charles Darwin that he conducted many experiments on it. Its sticky, red, long-stalked glands fringing the leaves trap small insects (just visible in Figure 109), which are digested by enzymes excreted to extract the nutrients, especially nitrogen, which are then absorbed through the leaf surface enabling sundews to grow in very low nutrient environments. Numbers vary from a handful to several hundred plants and can be seen in the Upper Derwent Valley, Heyden Brook, Jagger's Clough (Edale Valley) and several other cloughs. They are very susceptible to trampling damage so do take care when exploring.

Heath Spotted-orchid (*Dactylorhiza maculata*), which is not common, is striking in flower (Fig. 110), when it can be easily separated by its smaller central lip lobe from Common Spotted-orchid (*Dactylorhiza fuchsii*), which is rare in the higher moorlands. The specific name *maculata* means spotted which relates to the black blotches on the leaves. It is more abundant in north and west Britain and increasingly restricted moving east and south. There are other specialities which are a joy to find. Ivy-leaved Bellflower is one with delicate pale-blue flowers resembling harebells but angular-lobed leaves and a creeping habit (Fig. 111). Recorded in only a few cloughs in the Dark Peak, usually at the edges of wetter areas, it occurs in Heyden Brook, Oyster Clough (in abundance) and in some cloughs on the eastern side of the Upper Derwent Valley. One record in Audenshaw Clough (Longdendale) was noted in 1650 according to the Cheshire

FIG 111. Ivy-leaved Bellflower in base-rich flush with Marsh Violet, Marsh Thistle and Lemon-scented Fern (*Oreopteris limbosperma*).

Flora – and nearly 400 years later, very satisfyingly, is still there. There could be new sites as I discovered several in 2018 in Heyden Brook and Jagger's Clough. It might be increasing after reduced sheep grazing or possibly from reduced acidity as soils recover from historic pollution.

Equally inconspicuous are several other rare flush plants. Lesser Skullcap (*Scutellaria minor*) has pale salmon-pink flowers on slender stems. It is more widespread in Wales and Southern England in marshes or wet bogs. We have only a few sites, all in the Dark Peak, including Heyden Brook and the Eastern Moors. Marsh Arrowgrass (*Triglochin palustris*) is a second rarity only readily visible when fruiting as these elongate and are held vertically up the stem. It is more frequent at lower altitudes in the Staffordshire Moorlands and even grows occasionally alongside roads, as do Marsh Valerian (*Valeriana dioica*) and Ragged-robin (*Silene flos-cuculi*). Both are rare in Dark Peak flushes but more frequent in the South West Peak. Bogbean is a third rare species, more likely to be found at lower altitudes such as on the Eastern Moors or rarely in the South West Peak moors. All these rarities were also labelled as such by Moss in 1913, suggesting their distribution has not altered significantly.

Other marsh plants prefer wet ledges beside clough streams such as Marsh Hawksbeard (*Crepis paludosa*) and Common Butterwort (*Pinguicula vulgaris*) (Fig. 204), which is on the England Red List. The Hawksbeard, with yellow dandelion-like flowers, has tall stems with distinctive clasping leaves. It grows on wet rocky ledges, often in partial shade, particularly in the Upper Derwent Valley and in the South West Peak. In contrast, look for Butterwort on steep, dripping, shaley banks above streams. Its pale lime-green leaves are sticky and curl over slowly to trap insects thus supplementing their nutrient requirements in the same way as sundews. Its name may have originated from the belief that if leaves were rubbed onto cow udders it would protect them against evil and bad butter! The upper half of Alport Dale supports one of the larger populations on several wet banks tumbling into the stream, but it also occurs sparsely in other cloughs. It is very rare in the South West Peak. More spectacular perhaps, but not native here, is the Large-flowered Butterwort (*Pinguicula grandiflora*), possibly introduced before the First World War from its native Southern Ireland, which has not only persisted but multiplied since. In some years there are over 200 flowers on banks in Grindsbrook.

Where dripping rock outcrops are more shaded they can be curtained with mosses and liverworts. The flat dividing thallus of Great Scented Liverwort

FIG 112. Beech Fern growing on a damp, north-facing rock ledge in Blackden Clough. The cover of Great Scented Liverwort can be seen on the vertical face behind the fern fronds.

(*Conocephalum conicum*) (visible in the background of Fig. 112) or the more wavy *Pellia* liverworts can form sheets over the rocks, sometimes interspersed with patches of mosses like *Fissidens*, which has two ranks of flattened leaves. *Sphagnum* Bog-mosses sometimes hang in great curtains dripping from rocks above, sprinkled with a few grass leaves. All these provide invertebrate habitat and food for small birds like Wrens (*Troglodytes troglodytes*), which can be heard in every clough. The plants on the drier rock outcrops beyond reach of grazing animals might surprise you. They include Rosebay Willowherb (*Chamaenerion angustifolium*), Ivy (*Hedera helix*) and Honeysuckle (*Lonicera periclymenum*), all commoner in other habitats. The lovely yellow flower heads of our native Golden Rod (*Solidago virgaurea*) reach into the highest sections of cloughs on rock ledges and Woodsage (*Teucrium scorodonia*) is not uncommon lower down.

The specialities to look out for though are Oak and Beech Ferns (*Gymnocarpium dryopteris* and *Phegopteris connectilis*), both of which are rare. They are small, usually growing separately in small patches with perhaps 10 to 50 or more fronds on sheltered rock ledges beside streams (Fig. 112). The Peak District is at the southeast edge of their national distributions. They occur in a few Dark Peak Cloughs around Kinder Scout and the Upper Derwent Valley and Oak Fern only more rarely in the South West Peak where it has been recorded in Lud's Church (Fig. 31) since before 1872. Other ferns are much more abundant. The commonest smaller one is Hard-fern (*Blechnum spicant*), conspicuous because it has spore-bearing fronds with narrower leaflets (pinnules) than the non-reproductive ones. Larger ferns are more difficult to differentiate perhaps, but the lovely lime-green Lemon-scented or Mountain Fern (*Oreopteris limbosperma*) sometimes smells lemony – try rubbing young fronds between your fingers. It has distinctive short stubby pinnules at the bottom of the frond and is common in damper sections of cloughs. It hardly ventures here outside the Dark and South West Peaks though. Lady-fern (*Athyrium filix-femina*) prefers a damper environment, although this could be provided in woodlands, flushes, or damper rocky hillsides as well.

WOODLAND

There are remnants of woodlands or just small groups or isolated trees within the upland cloughs (Figs. 28 & 107). Many are open to grazing and show little signs of regeneration, although some are now enclosed as in Upper Derwent Valley and Abbey Clough. The contrasting case is exemplified by several of the eastern moors where woodland, individual trees and scrub are plentiful, either as a product of removal of grazing or establishment after wildfire when the bared ground was colonised – often with a dense mass of Silver Birch (Fig. 106). The commonest species are Silver Birch and Rowan, both able to establish from seed out of sheep's reach. Many cloughs also boast specimens of Sessile Oak (*Quercus petraea*), although few younger ones are evident. Sessile refers to the stalkless acorn cups compared with those of Pedunculate Oak (*Quercus pedunculatus*), which have a clear stalk (the peduncle). The leaves also differ but examining any oaks in the cloughs may be confusing as they hybridise and these are also common. Ash is, perhaps, surprisingly regular in the cloughs, often quite old specimens, usually occupying more base-rich soils near streams. Some are already showing signs of Ash Dieback disease (caused by the fungus *Hymenoscyphus fraxineus*) and may not survive long term. Occasionally, as at Three Shires Head, in lower Alport Dale and the Upper Derwent Valley, Aspen occurs, often in groups as they sucker. The lovely near circular leaves trembling in the slightest breeze are distinctive and sometimes support Poplar Hawk moth caterpillars (*Laothoe populi*). Also scattered in damper places are different willows (*Salix* spp.) and some lovely old, gnarled Alders. Being palatable, the willows are generally out of reach of grazing animals. Holly (*Ilex aquifolium*) and Hawthorn (*Crataegus monogyna*) are the other occasional contributors to clough woodland, often clinging onto rock ledges. Indeed, Hawthorn bushes have a long history and some may be very old where they remain, gnarled and crooked, leaning with the wind along clough edges.

There could once have been a more extensive scrub or woody edge to the blanket peat. Many peat edges are receding and evidence from elsewhere suggest that a woodland or scrubby vegetation against these margins would reduce this (R. Lindsay, pers. com). So are these scattered Hawthorns the remnants of a now missing peat-fringing habitat? We do not know, but such ghost or shadow woods are explored further in the area, although mostly at lower altitudes, by Rotherham (2017).

The sparse wooded patches in grazed cloughs have a depleted ground cover, many of the species being also associated with the open moorland such as Wood Sorrel (*Oxalis acetosa*), which also appears under thin Bracken canopies. Species variety is greater in the more wooded environments as on Blacka Moor or in the enclosed woods, and those tolerant of the increased shade often flower and seed more profusely without grazing. One particularly palatable species is Greater Woodrush (*Luzula sylvatica*), which thrives alongside fenced streams or on inaccessible shaded rock ledges.

There are several clough woodland creation/restoration schemes operating with significant, landscape-changing ambitions, especially by the National Trust and Moors for the Future working with private and public landowners. Over

840 ha of new or restored clough woodlands is the impressive target, which focus on Sessile Oak, plus the other native species described above, designed to produce patchy new woodland, with scattered trees, denser areas and a better complement of species thus creating more varied habitats. This will lead to changes in the invertebrates, small mammals and bird life as the trees mature. The most diverse existing flushes and heathland patches are carefully avoided to maintain their value. Some of these new woods will be open to grazing once established and will provide more shelter over time. They will also help slow the runoff into streams thus helping reduce downstream flood risks and keeping stream water cool in hotter conditions (essential for some upland aquatic invertebrates). The trees capture and store carbon too thus assisting moderation of climate change and they enhance the landscape when established in the right place.

BRACKEN

Much of the new planting is being established in Bracken beds. This fern is one of the most successful and widespread species in the world, occurring in every continent except Antarctica and in all habitats except deserts, although moorlands and heathlands are a typical location (Fig. 107 and the distant green patches in Figure 83). It is one of the more extensive vegetation types in our moorlands, but rarely extends onto deep peat. Bracken is historically important as it was cut for winter bedding, which, once composted, was a valuable fertiliser. It provided thatch, packing materials for earthenware, or could even be burnt as a fuel. It was cut and burnt in potash kilns, with the ash used for removing the grease from woollen cloth, or for making lye, used as the alkali in soap making. None of these past uses persist, although some cutting is undertaken to control its vigour. Unlike other larger ferns, Bracken spreads through rhizomes. On richer and deeper soils, Bracken beds are tall, dense and impenetrable, with little growing underneath. In less ideal conditions, on shallower soils for example, Bracken is thinner, shorter and with other vegetation persisting underneath. It avoids wet areas, so fringes flushes rather than venturing far into them. Droughted conditions, as in 2018 summer, lead to a shorter, less dense canopy, especially on south-facing slopes.

Bracken is a plant that is loved or hated. It contributes significantly to landscape colour and texture, changing from lime-green as it emerges to rusty-red after frosts. In the cloughs, it can provide important vertical structure and therefore cover for breeding birds like Whinchat (*Saxicola rubetra*) or Stonechat (*Saxicola rubicola*) (Fig. 113). The latter has been increasing recently, possibly linked to warmer

FIG 113. Male Stonechat perched on an old Bracken stem in Burbage Valley. A species that is increasing in the Dark Peak. (Kev Dunnington)

winters associated with climate change, whilst Whinchat, formerly in every clough, has declined significantly and is now included on the Red List. Ring Ouzels and Twite (*Linaria flavirostis*) both Red Listed, often select breeding sites containing Bracken, whilst it also plays host to a range of invertebrates that feed exclusively on it. Common Lizards (*Zootoca vivipara*) and Adders (Fig. 122, *Vipera berus*) may hibernate in Bracken patches. There is also a common microfungus, Bracken Map (*Rhopographus filicinus*), growing on dead Bracken, forming a series of irregular black lines down the stems which are more visible when the fronds are dying.

On the negative side, the plant is poisonous, producing a carcinogen ptaquiloside and the enzyme thiaminase, which destroys thiamine (vitamin B1) causing a possibly fatal disease in animals like horses. These are naturally occurring chemicals evolved to deter grazing animals and which are at their highest concentration in the young fronds as they unfurl from their 'shepherd's crook'. Stock and wild grazing animals usually avoid Bracken unless there is nothing else to eat. With increased sheep numbers in the moorlands and concomitant reductions in cattle, Bracken has spread because sheep avoid it and do not trample it significantly, to which Bracken is susceptible. Historic high sheep numbers also heavily graze plants that are competing with Bracken, thus accelerating its expansion. Cattle would have trampled Bracken more and being cut for winter bedding controlled its growth and vigour significantly. Additional factors are probably climate change with fewer frosts and possibly air pollution – Bracken is quite competitive so responds positively to nitrogen deposition. Thus, the lack of cutting, reduced trampling, addition of nitrogen and ameliorated climate have probably all contributed to its expansion.

Recent comparisons of Bracken cover in sample areas of the Peak District using aerial imagery from the 1960s to 2005 found a 1.8 per cent per year expansion since 1990, more on north and northeast-facing slopes and at higher altitudes, suggesting a climate change effect (McAlpine, 2014). Moss (1913) for example, considered Bracken above 472 m was exceptional, but McAlpine found it regularly, although not commonly, and at up to 533 m around Kinder Scout. This expansion is despite efforts by many land owners to control it, mostly using herbicides.

ACID GRASSLANDS

These are the grasslands typical of acidic soils on the grits and shales in the Dark and South West Peak. They have not been subjected to liming or fertilising to improve their agricultural productivity, although basic slag (an early waste material from Sheffield's blast furnaces and elsewhere used as a lime-rich fertiliser) may have been applied historically to more accessible patches. The main grassland types are those dominated by either Mat-grass or Purple Moor-grass, their abundance related to the soil wetness and past management. There are smaller areas of more diverse patches, which are usually short, brighter green in colour and populated by Common Bent (*Agrostis capillaris*), Sheep's-fescue (*Festuca ovina*), and Wavy Hair-grass; all of which also occur sporadically amongst Mat-grass.
Centuries of sheep grazing and moorland management, including burning, determine the character of these grasslands. Swathes of bleached Mat-grass-dominated swards cling to steep slopes below the blanket bogs in many places in both the Dark and South West Peaks. This very widespread grass naturally occurs in snow-bed communities, but owing to its very high silica content – like sandpaper if chewed – sheep generally avoid it in favour of more palatable and nutritious grasses like Sheep's-fescue, Common Bent and Wavy Hair-grass. This gives it a competitive advantage and results in its historic spread in the uplands to dominate large areas nationwide. Moss (1913) describes the abundance of this grass in the Peak District (Fig. 114) and highlights its cover in the Edale

FIG 114. Alport Castles landslide covered in Mat-grass-dominated acid grassland, plus a pond trapped between hillslopes with a Soft Rush fringe and green carpet of the Floating Hook-moss *Warstorfia fluitans* in shallow water. Black Darters are numerous in this habitat.

Valley on the southern flanks of Kinder Scout and in Longdendale, where it is still widespread, showing this is not a recent phenomenon. Indeed, it will have declined as grazing pressures reduce.

Mat-grass' companion, Wavy Hair-grass, is particularly frequent in the Peak District compared with other upland areas where Sheep's-fescue replaces it. Heath Bedstraw is a regular companion, its tiny white flowers and whorls of leaves being distinctive. The yellow four-petalled flowers of Tormentil and Springy Turf-moss *Rhytidiadelphus squarrosus* are also common, but other species may be scarce. The moss is not grazed, but sheep often pull it out to find tastier morsels below, and its abundance reflects a history of high sheep grazing pressure. One rare interesting fungus is Horn Stalkball (*Onygena equina*) growing on putrefying hooves and horns where it digests the hard keratin. It has small, white fruit bodies on a thick stem (stipe). It is mostly a northern species, but has been recorded in the Kinder Scout area, Bleaklow and Upper Derwent Valley.

It is only in more diverse patches that you will find plants like Common Dog-violet (*Viola riviniana*), Hard-fern, Harebell (*Campanula rotundifolia*), Green-ribbed Sedge (*Carex binervis* – the rusty orange leaf ends are diagnostic in autumn) and Barren Strawberry (*Potentilla sterilis*), which looks like a rather hairy, diminutive strawberry plant. It is also worth looking out for waxcaps and other fungi, especially in the autumn, although the range of species may be inferior to those in some of the enclosed grasslands, see Chapter 11.

Comparing this sparse plant list with Moss' 1913 inventory, which included Bird's-foot-trefoil (*Lotus corniculatus*), Ribwort Plantain (*Plantago lanceolata*), the tiny fern Moonwort (*Botrychium lunaria*) and Quaking-grass (*Briza media*) amongst others, begs the question why most of these are rare in our acid grasslands now. If already acidic soils with little natural buffering became more acidic through acid rain, then plants preferring slightly more base-rich conditions, like many highlighted from Moss' list, could disappear. Simultaneously, increased sheep numbers could also preferentially supress these same more palatable species. The outcome is that many of our acid grasslands exhibit limited diversity.

Purple Moor-grass is the other major grassland dominant. Extensive sheets of this tussocky grass can clothe whole hillsides on the gentler, wetter shale-based slopes. It is conspicuous in the upper parts of Lyme Park near Disley for example, in the upper Longdendale valley and its tributary Heyden Brook, in the Upper Derwent valley and in places on the Eastern Moors. Tracts of dead leaves (they are deciduous) turn the landscape creamy-brown in winter, whilst the purple anthers in summer bestow a purplish haze over the dull green leaves. Purple Moor-grass attains its greatest vigour on slopes with lateral water movement that provides sufficient oxygen to the root zone and shallow peaty soils, where it develops a monotonous monoculture of robust tussocks that are particularly arduous and hazardous to traverse. In more waterlogged, anaerobic conditions, as in some wet blanket bog and boggy spots in flushes, the grass is much less tussocky and competitive as it is unable to pass oxygen to its roots effectively as cottongrasses can for example. There is some evidence that the high nitrogen deposition in the Peak District (see Chapter 6) has also contributed to its increasing dominance and vigour.

Purple Moor-grass has always been a component of our wetter moorland vegetation, but it has probably spread significantly over recent decades mostly through too frequent burning to which it is adapted, its tussocks protecting buried buds. Such burning was used to produce a flush of more palatable growth in spring, but sheep usually prefer other forage at that time, so it actually leads to increased dominance. Moss (1913) describes Purple Moor-grass as much more local compared with Mat-grass and includes a long list of associated, largely wetland species, most of which are rarely found with it now. In its dominant mode, Purple Moor-grass vegetation tends to be remarkably homogenous with Wavy Hair-grass a minor component and Soft Rush a sporadic companion. Cattle forage the grass much more readily and would have kept it under better control in the past when they were more widespread on our moorlands. Cattle are being used again, for example by the Eastern Moors Partnership and heathland is being restored as described earlier.

WATER

Every clough has a stream, often with tributaries, dripping rocks and waterfalls (Figs. 28 and 115). Most flow all year, even in droughted periods. During extreme rain events, these streams can overtop their normal banks, flooding adjacent river terraces, ripping out fences, eroding banks, picking up plant material as well as boulders, and redepositing them across the flood plain. These forces are accentuated where clough-sides are covered in short vegetation with little woody component as water can then flow off more rapidly. Flows will be better moderated as the new clough woodlands develop and more gullies are dammed on the peat above.

FIG 115. A typical clough stream, flowing over rocks in the deep-sided valley, here in Hollingworth Clough, near Hayfield, with Soft Rush patches, Bracken, Heather and the odd tree adjacent.

For the most part, the aquatic and splash zone plants have to be tolerant of both the acidic conditions and to periods of severe scour or desiccation, and only a limited suite of lichens, mosses and liverworts fulfil this role. Purplish patches of Water Earwort (*Scapania undulata*) often cling to rocks, whilst the bright green River Feather-moss (*Brachythecium rivulare*) covers natural stone steps in the stream bed. Alpine Water-moss (*Fontinalis squamosa*), another north and western species, has long stems that stream from underwater stones. The main lichens of this habitat are *Ionaspis lacustris*, *Porpidia hydrophila* and *Rhyzocarpon lavatum*. The first is a yellowish-ochre to reddish-brown crustose (crust-like on the surface) thallus. It grows on stable rocks and in the splash zone, but can withstand prolonged periods of inundation. *Porpidia hydrophila* grows in similar situations, but has a whitish crustose thallus with black sessile fruiting bodies. *Rhyzocarpon lavatum*, another splash zone species, has a grey-brown thallus with black sessile fruiting bodies. It is worth searching rocks standing proud in the streams or along the edges for patches of these and other distinctive lichens.

Although these are common species in this habitat, Gilbert & Giavarini (1997), whose study included several Peak District streams, found a sequence of species both across and down a stream. They separated those that occur in the pebbly headwater or stony flushes from the torrent zone where scouring prevents many species from establishing, compared with the colluvial zone where flow rates are gentler and large stable boulders provide ideal lichen habitat. The richest lichen flora occurs in the middle reaches of upland streams owing to the extent of rock exposure and generally slightly higher pH. Here different lichen assemblages could be identified on submerged rocks, in the splash zone, the zone of occasional submergence and the drier edge zones.

It is worth exploring stream invertebrates here as they are mostly absent from the other habitats. Allan (2004), who sampled many Peak District moorland streams, showed different assemblages of invertebrates related to the acidity, aluminium levels and calcium content in particular. The higher altitude, more acidic upland streams, largely on the gritstones, often with high levels of aluminium, supported fewer species than the lower altitude streams with a higher pH and calcium levels, mostly overlying the shales. These are the same factors that influence the flush plants.

Groups like midge larvae (Chironomidae) – aquatic filter feeders   are universal and Nemouridae – a stonefly group – are similarly widespread, with *Nemoura cambrica* more likely in the more acidic high-altitude locations and *Amphinemoura sulcicollis* in the lower altitude, more base-rich waters. There is a singular lack of most of the stonefly and mayfly larvae in the highest, most acidic streams, whereas they flourish more in lower, more base-rich waters, more often found in the South West Peak and Eastern Moors for example. Other more acid-sensitive groups included Gammaridae (Freshwater shrimp family) and Hydropsychidae (a caddisfly family), whereas the caddis family Polycentropodidae were found in more of the acidic streams. Unlike the White Peak stream fauna, Crustaceans (such as fresh water shrimps) and snails were extremely rare or absent from all the sites owing to low calcium levels. Moreover, water beetles are also relatively scarce, with only some diving beetles (Dytiscidae) present in the more acidic waters. One Nationally Local species rarely found in a few streams in the Dark Peak is *Oreodytes davisii*, which hides under stones, in pools or amongst waterlogged Common Haircap in the Rivers Alport, Ashop, Derwent, Little Don and in Hollingworth Clough (Merritt, 2006).

General diversity and numbers of species increase as you pass downstream. This is where to watch out for the beautiful Golden-ringed Dragonfly (*Cordulegaster boltonii*), for example at the southern end of the Eastern Moors on Barbrook or beside Agden Bog. This large dragonfly (considered the longest species in Britain) is an upland acid stream specialist largely confined to uplands in general plus the south and southwest of Britain. The larvae live buried in the stream bed for up to five years where they ambush passing prey. The Peak District is an outpost for it.

In general terms the range of fauna found are typical of soft water, nutrient-poor systems and the organisms found match those found in other studies in Northern Britain, but it is suggested that the acidification of the upper streams in the Dark Peak in particular from historic sulphur dioxide pollution (Chapter 6) is still impacting the aquatic fauna. It is also possible that the release of lead and other heavy metals (see Chapter 8) could be having an effect. Certainly, Ramchunder *et al.* (2012) showed that high dissolved organic carbon levels in moorland streams in the Pennines suppresses the diversity of freshwater invertebrate diversity downstream of peat erosion gullies.

Dipper is the main bird associated with the upland streams (Fig. 116), but it occurs at low densities, probably owing to the low food abundance compared with the lower and White Peak rivers, where it is much more numerous. Its delightful habit of bobbing on rocks before searching for prey underwater makes it distinctive.

Ponds are rare in the unenclosed moorlands. There are several of significant size trapped in landslides, such as below Alport Castles (Fig. 114), or Mermaid's Pool below Kinder and others on high moorlands in the South West Peak, such as Blake Mere and Doxey Pool. Most of these are said to be home to mermaids, although their stories vary. A young man seeing the mermaid in Kinder's Mermaid's Pool at Easter is said by some to offer the gift of immortality, but

FIG 116. Dipper in Fair Brook Clough. Densities of pairs are much higher on the limestone rivers where there is more available food.

others translate this as being fortunate in money and goods but never in love. Blake Mere's mermaid is either one brought back from the sea by a local sailor who fell in love with her and then haunted the lake after his death, or it is the ghost of a girl said to have been labelled a witch and drowned in the pool after spurning a young man's love. This pool is also said to be bottomless and cattle will not drink from it or birds fly over it (untrue). Doxey Pool's mermaid is known as the blue nymph called Jenny Greenteeth (who appears in myths elsewhere too!), who entices unsuspecting victims into the pool to a watery grave. Whatever the myths and legends, these pools are generally rather peaty, acidic and lacking significant plant life. The pool at Alport Castles is shallower than others and its floor is covered with a carpet of Floating Hook-moss (*Warnstorfia fluitans*) (Fig. 114). *Sphagnum* patches, Soft Rush and some sedges provide marginal vegetation in shallower water.

Other pools are more recent and without legends. Several reservoirs surplus to requirement were re-designed to avoid the liabilities of the Reservoir Act by breaching the main dams and creating new pools and other habitats. Barbrook and Ramsley Reservoirs in the Eastern Moors, Lightwood (Buxton) and Stanley Moor Reservoir below Axe Edge were subjected to this treatment. Some of the new pools on the Eastern Moors, which are favoured by large Toad breeding populations (*Bufo bufo*), were designed to safeguard the existing population of the rare Shoreweed (*Littorella uniflora*), which is characteristic of the reservoir draw-down zones. These new pools, as well as some of the larger ones and those in ex-bomb craters on Leash Fen are home to the moorland specialist dragonfly, the Black Darter (*Sympetrum danae*), easily identified by its small size, black-waisted body (males) and black legs (Fig. 117).

There are many other small pools or puddles within the flushes, trapped in the river terraces, or in backwaters in the streams where flush species thrive like Bog Pondweed, Round-leaved Crowfoot and Water Starwort. These will be very important breeding sites for Common Frogs (*Rana temporaria*), which are regularly encountered throughout the Moorlands at higher altitudes than Toads seem to venture. Even vehicle tracks puddles are sometimes filled with frog spawn, suggesting a probable dearth of suitable breeding habitat.

FIG 117. Black Darter dragonfly, which has a short and erratic flight, appears in mid to late summer. The waisted, black abdomen of this male is clear. (Steve Orridge)

OTHER ANIMALS

Many of the other moorland animals occupy a wider suite of habitats. The bird life is special, the full assemblage being one of the qualifying features for the National and European importance of a large part of the whole area. The mammals are limited in variety and number, whilst the reptiles and other invertebrates are distinctive.

The moorland birds are special – not just for nature conservation but because they are a joy to see and hear, their presence intensifying the upland experience. The curlew, with its large downward curved bill, has to be highlighted as the most quixotic, the harbinger of spring, its spine-shivering liquid calls one of the most emblematic icons of wild places (Fig. 118). It is not just a moorland bird, but one found throughout the Peak District, more especially in marshy pastures as well as in the lower wetter moorland places. Highlighted by local communities as a symbol of the South West Peak HLF Landscape Partnership project, it means

FIG 118. Curlew, with its long, downward pointing beak and rippling call, is a summer breeding bird of the blanket bogs, lower moorlands, wet rushy pastures and some parts of the limestone plateau. (Kev Dunnington)

much to local people irrespective of any interest in birds. This is significant in the context of national declines in range and its elevation from Amber to the Red List (Appendix 2).

Snipe (*Gallinago gallinago*), another wader, also seeks out larger flushes and rush patches where soft ground makes feeding easier. We will return to these species in the discussion on rushy pastures in Chapter 11, but there is also a few Redshank (*Tringa totanus*) on rushy ground in particular. Their populations peaked in the 1940s and declined in the 1960s so that pairs are only found now in Longdendale, Beeley Moor and Middleton Moor in Derbyshire, with no confirmed breeding in the West Yorkshire section of the region. The cause of the decline is unknown, but could be related to drainage and high levels of grazing at that time on many marshy grasslands at the edge of the moors. Oystercatcher (*Haematopus ostralegus*), a magnificently distinctive vocal species, has started appearing in the Peak District to breed with the first breeding record only in 2004, but it is spreading, possibly due to a change in behaviour, and occupies lowland areas outside the Peak District too.

Red Grouse (Fig. 104) has already been highlighted as our most iconic bird of heather moorlands, but others of interest in the core assemblage are Reed Bunting (*Emberiza schoeniclus*), Skylark, Meadow Pipit (*Anthus pratensis*), Cuckoo (*Cuculus canorus*), Nightjar (*Caprimulgus europaeus*), Ring Ouzel, Wheatear (*Oenanthe oenanthe*), Grasshopper Warbler (*Locustella naevia*), Twite, Whinchat and Stonechat. These all have separate niches within the upland environment. Reed Bunting favours wet patches and scrubby moorland edges, so are more abundant in the South West Peak and eastern moorland fringes. They have declined significantly in lowland Britain, so our populations are becoming increasingly important. Twite, in contrast, breed often in Bracken beds (and now in some limestone quarries near Buxton) but feed mostly on seeds in enclosed fields. Their numbers have also plummeted nationally as well as in their Pennine stronghold, including the Peak District, probably caused by agricultural improvement of flower-rich hay meadows, switches to silage cutting and the loss of plentiful seed sources in the landscape. The 2004 Moors for the Future survey found only 10 pairs compared with 131 pairs at many more sites in 1990 (Wood & Hill, 2013). Their future here is dependent on projects to increase wild flowers in the landscape and thus a varied seed source.

Skylarks, another Red List species, have declined by over 58 per cent nationally, mostly in agricultural land, but our moorlands are still a stronghold, where you can enjoy their seemingly never-ending aerial display and distinctive song. There are signs of declines in the Peak District too though. They select territories on the grassier or mixed moorland vegetation rather than dense

Heather and feed mostly on insects and spiders. More widespread and abundant than Skylarks, Meadow Pipits occupy similar habitat but are easily distinguished by their smaller size, outer white tail feathers and parachuting display onto rocks, walls or the ground. The peak of cranefly emergence is an important food source, along with beetles, moths and spiders.

The Cuckoo (also Red-listed) is another species that is declining nationally and locally. Gosney (2018 b & c) found reductions of over 50 per cent along the eastern Peak District flanks in just 30 years. Meadow Pipits are the main host, and these are still common, though declining in places (Gosney, 2018 b & c). Poorer survival on Cuckoo's westerly migration route through Africa compared with their easterly route is a possible reason for their decline.

Nightjar and Dartford Warbler (*Curruca undata*) are possible beneficiaries of climate change. Nightjar numbers have increased 43 per cent on the Sheffield side of the Peak District and also breed in the Goyt and below Curbar Edge. Gosney (2018a), for example, counted 26 different calling birds in 2018 just along the drivable roads between Strines and Langsett. The upper edge of Wyming Brook Nature Reserve (Sheffield and Rotherham Wildlife Trust) is specially managed for them, with small coupes removed in the tall Heather for nesting and retention of scattered birch for perching. They also utilise the cut coupes in conifer plantations as in Matlock Forest. Nightjar could be increasing due to milder winters, as is Dartford Warbler, which has also appeared at a few sites. This warbler was once limited to Dorset heathland after the severe 1963 winter but after many milder winters has spread north. It prefers gorse patches within heathland to provide cover and adequate food.

We are very much at the southeast edge of Ring Ouzel's breeding habitat (Fig. 119); another of our iconic species. Its white bib stands out along with its song, which is slightly reminiscent of the Mistle Thrush's (*Turdus viscivorus*). Look for it in cloughs and below edges. Unfortunately, this is another Red Listed declining species, with losses recorded in all Peak District county atlases. The generally reduced grazing pressures and concomitant enhancement of invertebrate populations and berry crops, along with increased cover of fruiting trees like Rowan and Hawthorn should all help to increase food availability, but the declines have been linked with migration issues and chick survival.

Unfortunately you will no longer see Black Grouse in the Peak District. They declined around the Eastern Moors in the 1960s and 1970s and were eventually lost, as they were from the South West Peak by 1998, despite considerable conservation effort. This matches widespread trends across England and Wales, with 29 per cent reductions in range over the last 40 years attributed to habitat loss, fragmentation and degradation of habitat mosaics that should form

FIG 119. Male Ring Ouzel, another northern and western species. It breeds amongst Bracken and Heather in some of the Cloughs and below the Edges, but is sensitive to disturbance. (Kev Dunnington)

transitional zones between open moorland and inbye farmland (Balmer *et al.* 2013). Climate change may also be a factor (Wood & Hill, 2013). Black grouse were re-introduced into the Upper Derwent Valley in 2003 with birds released over three years. Some birds are known to have dispersed widely and breeding was recorded. However, they are not persisting as a sustainable population. Another attempt in the future may be possible when habitat conditions improve so that we can see this fascinating bird again in our moorlands. Other species' declines such as for Whinchat and Wheatear are unexplained. The latter could be related to reduced grazing levels on the moors since they prefer short vegetation and populations have remained fairly stable on the limestone area, where vegetation is often more closely grazed.

Where there is more tree cover the bird assemblage changes. Tree Pipit (*Anthus trivialis*) is a bird of open country with scattered trees or woodland edges. It is another Red-listed species having declined by 73 per cent from 1970 to 2008, nationally and locally. The Peak District is now a hotspot for it, particularly the upland valleys, the Goyt Valley, and eastern moors where trees are more frequent. Recent tree planting will provide good new habitat for it. Similar woodland/woodland edge and scrub species include Linnet (*Linaria cannabina*), Lesser Redpoll (*Acanthis cabaret*), Common Redstart and Wood Warbler, although none are restricted to this habitat. Except for Redstart, these are all Red-listed. Linnet is a moorland edge species, whilst Lesser Redpoll has disappeared from many adjacent lowland areas so the Peak District uplands now provide critical habitat. They breed mostly in birch woodland and conifer plantations and feed extensively on the small seeds of Birch and Alder.

Turning to birds of prey, there have been some spectacular recent successes but also some troubling failures. Ravens were abundant here up to the late 1860s and returned naturally probably in 1992 to become a regular sight. The Common Buzzard story is similar, with first breeding recorded in the 1975–80 period and subsequent spread throughout the area, possibly related to reduced persecution. Similarly, Peregrines (*Falco peregrinus*) were absent from the Peak District for most of the second half of the 1900s owing to persecution plus the effects of organochloride persistent pesticides and their allies. They re-appeared in 1981 as part of their national recovery and bred successfully for the first time in 1984, after which there was a slow increase (Melling *et al.*, 2018). There are around 33–44 pairs

FIG 120. Short-eared Owl with Field Vole prey gliding silently over the moors. (Kev Dunnington)

across the Peak District and youngsters begging food or calling adults are regular delights. Most breed in quarries or rocky edges in all areas, including the White Peak, but numbers have declined substantially across the Dark Peak in recent years. Only nine pairs were recorded in 2018 in the Dark Peak and only four of these laid eggs. There were none in 2017.

The Goshawk story is similar. They were exterminated as a breeding species in the late 19th century, but re-established after the escape or deliberate release of falconer's birds in the 1960s. By 1980 about 60 breeding pairs were estimated nationally, increasing to between 437 and 616 pairs in 2015 (Melling *et al.*, 2018). The population in the Dark Peak moorland edge woodlands was deemed of national significance (Wood & Hill, 2013). However, by the late 1990s, the range and numbers of territories had declined locally, particularly on the eastern side of the Dark Peak, at least partly attributed to illegal persecution (Wood & Hill, 2013).

Hen Harriers (*Circus cyaneus*) have always been rare in the Peak District (and in most of England except Bowland, supporting their Red List status) and have bred successfully only six times in the last 23 years (to 2019) after a long absence despite suitable habitat being available. Radio-tagging of a recent fledged brood ended in one being lost, assumed shot, nearby and a second disappearing over the North York Moors. This was the tenth satellite-tagged Hen Harrier to disappear in the UK during the 2018 autumn: a shocking statistic (PDNPA, 2018). Illegal persecution is a major issue for this species, not just here, but more widely.

Short-eared Owl breeding pairs tend to fluctuate with Field Vole numbers; their chief prey – (Fig. 120), but benefit from reduced grazing and ranker vegetation which the voles favour, like tussocky Purple Moor-grassland. It is a real treat to watch these large, silent owls drifting over the moors, particularly at dawn or dusk when they are most active. They are, however, not immune to persecution. One was found shot dead in the Saddleworth moors area in 2018. At least one Red Kite (*Milvus milvus*) had a similar fate. This is not a regular sight in the Peak District yet, but they are increasing.

Merlin (also on the Red List) is the smallest raptor on the moors, difficult to see, usually flying close to the ground with rapid wingbeats interspersed with glides. It nests mostly amongst old Heather, but also uses old crow nests. Like so many species, the Peak District is at the southeast edge of its national breeding range. Meadow Pipits are their main prey but they can also take Skylarks and Wheatears. Their territory numbers have remained fairly stable at about 24 since 2015, with 3.46 young fledging per successful nest. The total is lower, however, than in the 1990s when around 37 pairs were present.

As a result of these sharp reductions and after some years attempting to reverse their fortunes, a Birds of Prey Initiative, a partnership between the NPA, Natural

England, Moorland Association, RSPB and National Trust backed by local raptor groups, was set up in 2011. The project works with land managers and owners to try to reverse the alarming downward trends described and is endeavouring to promote a safe environment for Hen Harriers too. Work is focussing on Peregrine, Merlin, Goshawk and Short-eared Owl within the Dark Peak and South West Peak, where populations are below expected levels. So far there has been some recovery in 2018 after a very poor year in 2017 in those areas where relationships between land managers and the project developed well. However, the results are still far from the agreed targets (PDNPA, 2018). Indeed, RSPB has withdrawn from the partnership frustrated by lack of progress. Analysis of data for Peregrine Falcon and Goshawk by RSPB demonstrates a clear correlation of raptor persecution incidents, reduced occupancy rates and low probability of successful breeding in the Dark Peak with heather burning for grouse shooting (Melling *et al.* 2018). Until illegal persecution is addressed, possibly at a national scale, future progress is problematic. Part of the equation has to address high intensity management of grouse moors to maximise numbers for shooting. This is considered again in Chapter 15 in the face of climate change prognostications.

In March 2019, the National Heritage Lottery Fund grant-aided the development of 'Upland Skies': a partnership project aimed at raising public awareness, educating and engaging with children and young people about this precious wildlife, inspiring people to take action to help increase the numbers, and championing positive land management techniques to help provide better habitats for Peak District birds of prey. Only time will show the degree of success this can achieve.

The significance of these issues lies in the overall importance of the main moorland breeding bird assemblage. Their dearth or absence in lowland areas around the Peak District makes our region special. They feature in the designation of the SSSIs across the Dark and South West Peaks and are reflected in the European importance of much of this area in the SPA. Table 7 lists their significance at the time of notification, the percentage being of the national or international population (above 1 per cent is nationally significant). The overall assemblage is also nationally important and '+' shows the contributing species that do not reach the one per cent level. The South Pennines SPA stretches from Ilkley in the north to Leek and Matlock in the south (Fig. 3).

Some very contrasting mammals merit attention here, although none are unique to the unenclosed moorlands. Pigmy Shrew, our smallest mammal, is distinguished from the Common Shrew by its longer tail (65–70 per cent of its body length) and by the clear demarcation between darker top and paler underside giving it a two-toned as opposed to Common Shrew's three-toned

TABLE 7 The importance of the moorland bird assemblage (per cent of British population)

Species	Dark Peak SSSI 1993	Eastern Moors SSSI 1998	Leek Moors SSSI 1988	South Pennines SPA
Golden Plover	1.7	+	+	3.3
Dunlin	0.9		+	1.3
Curlew	+	+	+	
Redshank		+		
Snipe		+	+ especially	
Lapwing		+		
Merlin	3.3	1.5–2.4	+	5.9
Short-eared Owl	1.1	+		2.5
Red Grouse			+	
Twite	Nationally important		+	
Peregrine	0.8			1.4
Ring Ouzel	0.7		+	
Reed Bunting		+		
Stonechat		+	+	
Whinchat		+	+	
Wheatear		+		
Nightjar		+		
Whole assemblage	Important	Important	Important	

appearance. They might be rarely seen, unless you search abandoned bottles or cans in which they sadly get trapped, but they are often heard when two interact, squeaking loudly in territorial disputes. They have been recorded at 580 m on top of Kinder Scout and the high moors are an important habitat. Indeed, Pigmy Shrews outnumber Common Shrews in the moorlands by a factor of about 8 to 1 attributed to the scarcity of earthworms (which Common Shrews favour) and the abundance throughout the year of smaller prey (in the 2–6 mm range), particularly money spiders (mostly *Centromerita* spp.), Carabid beetles, harvestmen and, in summer, moth larvae, which Pigmy Shrews predate (Yalden, 1981).

Water Voles (*Arvicola amphibius*) have declined here as elsewhere in the country. Formerly widespread and common on waterways, they were lost throughout the 20th Century, with a further sharp decline from about 1980 linked to predation by the non-native American Mink (*Neovison vison*), which are not only aquatic but can also penetrate Water Vole burrows. However, although numbers are still low, small populations of moorland Water Voles have been discovered associated with rush patches and water runnels down to only 20 cm wide but deep enough to dive into, away from Mink. They have been found as high as 571 m on the side of Bleaklow and the moorlands are becoming an increasingly important refuge as the decline in the lowlands continues. Look out for their latrines, feeding platforms and runs alongside tributary streams where rushes and Bracken lie in the Upper Derwent, Upper Alport Dale and Etherow catchments and the Eastern Moors especially. In these areas Soft Rush, Heath Bedstraw and even the tough Common Haircap (Fig. 81) form their main diet. Water Voles and their burrows are protected under the Wildlife and Countryside Act 1981 (as amended).

Wild Red Deer died out in the Peak District at the end of the 18th Century but were maintained in deer parks or collections, from which they subsequently escaped or were released. Now, there are three centres of population: the Eastern Moors, the Goyt Valley and the Staffordshire Moorlands. The Red Deer on the Eastern Moors (Fig. 121) are presumed to have originated from Chatsworth, and are now frequently seen with the largest count in 2010 of 126 on Big Moor (Mallon *et al.* 2012). Red Deer in the Goyt Valley inhabit the conifer plantations, but frequent the higher moors too. Around 60 to 100 animals may occupy this area, although numbers are controlled by the Forestry Commission to reduce tree damage. These animals may have originated from Lyme Park to the north, where there are Red and Fallow Deer herds, or from the private collection which included a small number of Red Deer once held at Roaches House, near Leek. The southern population centred around Swythamley was thought to develop from this menagerie when maintenance became impossible during the last World War (Yalden, 2001). Numbers vary owing more to illegal poaching, but are calculated to have reached 100–120 deer. Occasional animals are seen elsewhere in the Peak District. Counts in 2019 co-ordinated by the Cheshire Wildlife Trust suggest a total population in the order of 350 now in the South West Peak.

We only have one wild population of Fallow Deer on Stanton Moors, which are the dark form rather than the normal brown and cream colouring visible in captive herds in Lyme Park and Chatsworth (Fig. 52). The original animals were brought to Stanton Park from Chartley Park in Staffordshire in 1800 and escaped possibly in the 1920s. They have expanded, utilising woodland, conifer plantations, open moorland, small fields, scrub and Bracken, and local gardeners also complain of their marauding. The natural barriers of the limestone plateau on two sides and the open river valleys on the others seem to be containing their spread (Yalden, 2001, Mallon *et al.*, 2012).

FIG 121. Red Deer stag and hind on the Eastern Moors in winter with Purple Moor-grass background amongst managed Heather. (Kev Dunnington)

Anyone familiar with the South West Peak will know about the Wallabies that lived there in the 1970s and 80s. Five were released in 1939–40 from the private menagerie at Roaches House. At that time, the Roaches Estate was closed to the public and the Wallabies (Bennett's Wallaby [*Macropus ruficollis rufogriseus*]), developed a population of 40 to 50 by 1962. Heavy mortality occurred in the severe winters of 1962–3 and 1978–9, although numbers subsequently partially recovered. However, the population declined in the 1990s and early 2000s. Increased disturbance on the Roaches Estate after access was encouraged post-1981 resulted in their dispersal into inferior habitat with additional losses from traffic accidents. The cessation of Heather burning on the Roaches at this time also failed to generate more patches of younger plants on which the Wallabies depended and they were considered lost to the Staffordshire Moorlands by 2013 (Yalden, 2013). However, reports of their demise may be premature since occasional sightings of this enigmatic animal are still reported, although they are spread further than the Staffordshire Moorlands (http://roaches.org.uk).

It might be surprising to include Moles in the moorland mammal assemblage but their hills are regularly found on even very small patches of better soils where earthworms might be found. Many are adjacent to roads and tracks across moorland. Others follow river valleys where alluvium enriches the soil. Young Moles, dispersing from their birth sites in summer, cross inhospitable habitat on the surface until they find suitable soils, unoccupied by other Moles and these surface-running animals are frequently seen in July. There are other mammals on the moorlands – Rabbits in some places, Stoats and Weasels (although these are usually controlled on grouse moors so rarely seen) and Woodmice, which are common amongst the walls and Bracken in particular.

The Peak District is not generally a reptile-friendly place. The high altitude, exposure and cold winters (at least until recently) deter most. However, there is a good, though well dispersed and low density population of Common Lizards throughout the moorlands, including blanket bog, although glimpsing a disappearing tail may be all you see. Females often bask to help egg development, which burst as they are deposited giving the impression of delivering live young. Lizards feed mostly on insects and spiders. Kestrels take lizards if they can, revealed by the scales found in their pellets. Slow Worms (*Anguis fragilis*) are legless lizards with a highly polished appearance. They are very rarely seen, but worth looking out for in the Upper Derwent Valley where two recent records are known.

Compared with other upland areas, the distribution of Adders in the Peak District is extremely restricted to in or near the Eastern Moors, centred on Big Moor (Fig. 122). Why is not known, but higher altitude and colder climate with longer snow-lie could be factors. Using old records and museum specimens indicates that the current distribution was not so different in Victorian times, although there has been some range shrinkage suggested by the comment in 1901 that 'Adders were more common than Grass Snakes in the Bakewell District' (Whitclcy, 1997). Sunny spring mornings from February (or even late January) to May are ideal for Adder-watching. They bask on southeast-facing slopes, often in Bracken, whilst tussocks provide shelter and wet rushy ground is essential for finding prey. They are best approached slowly as they quickly slide away, but do avoid disturbing them. The Eastern Moors Management Plan incorporates Adder's requirements, thus avoiding grazing that is too heavy or vegetation control where sensitive hibernation or basking sites are located. All these reptiles will be especially vulnerable to wildfires in the moors and could take many years to recolonise large damaged sites. Managed burning may also play a role in their distribution, particularly for Slow Worm and Adder.

FIG 122. Female Adder in the Eastern Moors area. Note the larger size and much browner colouring compared with the more silvery hue of the males which has a blacker zig-zag pattern. (Derek Whiteley)

There are no overarching data on invertebrate groups favouring the different moorland habitats, so information has been extracted from different reports to furnish the general picture. There are also some specialities that merit our attention. My study of moorland fires shows that Heather-dominated ground (although on peat rather than thinner podsols) was much poorer in invertebrate species than the blanket bog and mixed peatland vegetation (Anderson, 1986), and this is also noted elsewhere in the Pennines. There are, though, different assemblages utilising the different stages of Heather growth, with a group of pioneer species and scavengers moving into newly burnt Heather patches, where they are probably an important source of food for moorland birds. The dominant species of moths also change with Heather age. More ants, hoverflies, springtails, parasitic wasps and various caterpillars were trapped on the drier heather moor compared with the wetter mixed and blanket bog vegetation in my study. The predatory ground beetles and, to a lesser extent, rove beetles (Staphylinidae) comprised some 25–30 per cent of the insect fauna in terms of species number. I found only 11 ground beetles based on spring and summer pitfall trapping to be associated with the Heather habitat, of which *Carabus problematicus*, *Trechus obtusus*, *Calathus micropterus* and *Trichocellus cognatus* accounted for over 70 per cent of the animals captured. These are all typical of open country, including dry heather moorland, although *C. micropterus* is also a litter dwelling forest species. *Carabus problematicus* is the predominant Violet Ground Beetle, with the nominate Violet Ground Beetle *Carabus violaceus* being more localised. These are amongst the largest of our beetles, growing up to 30 mm long. *Carabus problematicus* is a heathland and woodland species, a nocturnal hunter that hides under stones, litter or logs in daytime.

Oil beetles are a group of nationally important beetles. The Violet Oil Beetle (*Meloe violaceus*) is another northern and western animal but one of only five oil beetles that occurs in our moorlands as in the Upper Derwent Valley, around Kinder, on the Sheffield moors and at Three Shires Head in the west. Look out for it where there are sandy or friable soils such as along footpaths but where there is some botanical diversity (Whiteley, 2011). This beetle has an interesting association with solitary bees in that once hatched, the ant-sized larvae crawl up the nearest flower stem and hitch a lift from a visiting bee back to its nest where it feeds on bee eggs, before pupating and emerging the following year to repeat the cycle.

Water beetles are not just found in water but also associated with flushes and springs. *Anacaena globulus* and *Hydrobius fuscipes*, for example, are common and widespread in flushes, drainage ditches and margins of streams, sometimes in *Sphagnum* mosses, but not restricted to the uplands, whilst *Hydroporus nigrita* frequents flushes and springs in the Dark and South West Peaks (Merritt, 2006).

My post-wildfire investigation found the most abundant spider on drier heather moorland was *Pelecopsis mengei*, a small money spider with a limited upland distribution, followed by much more widespread and catholic *Walckenaeria acuminata* and *Centromerita concinna*, which is a more typical moorland species. Web-spinning spiders are also quite common in the moorland environment where there is adequate vegetation structure. The specific plants involved are less important as all spiders are predators.

A wide variety of other invertebrates inhabit our moorlands, although many are inconspicuous and more the target of the specialist rather than the casual observer, but there are some which attract attention. Amongst the many moths, the large day-flying species are conspicuous. The Northern Eggar (*Lasiocampa quercus* f. *callunae*), the northern and western race of the Oak Eggar (Fig. 123), is a large common orangey-coloured moth appearing from late May to early July. It is slightly larger, the female browner and it has a two-year life cycle compared

FIG 123. The Northern Eggar moth with a wingspan of 68–96 mm. Males are frequently seen flying erratically over heather moors on a sunny day and the large hairy caterpillars are often seen resting on vegetation in autumn prior to hibernation and in spring and summer in their second year.

to one year for the Oak Eggar. The distinctive brown hairy caterpillar feeds on Heather and Bilberry, overwinters as a small larva growing quite large in its second year with rings of white and black, before pupating for the second winter. The Fox Moth (*Macrothylacia rubi*), which flies in May and June, also has a large hairy brown caterpillar without pale markings when mature (but with yellow rings when young), which is quite common from autumn to spring.

The Emperor Moth (*Saturnia pavonia*) is another large, colourful moth of open moorland and heathland, the caterpillars of which, like Fox Moth, feed on Heather. These larvae are green with black stripes when fully grown, dotted with yellow or purple warts growing out of hair tufts. Spent cocoons of these and Northern Oak Eggar are often seen. Nationwide, larger moths, like so many groups of invertebrates, have suffered a significant 28 per cent decline in total abundance between 1968 and 2007 (Butterfly Conservation, 2013), but this is most severe in southern Britain (40 per cent loss), whilst the north (North Lancashire and North Yorkshire northwards) has been less affected. The large, interconnected scale and variety of habitat within out moorland environment is critical in ensuring the success of our moths as well as many other invertebrates that specialise in these habitats.

Several common and widespread butterflies are also at home in the moorlands, with Green-veined White (*Pieris napi*) and Small Heath (*Coenonympha pamphilus*) most regularly seen. The food plant of the former is probably Lady's Smock in the flushes, whilst Small Heath caterpillars utilise a variety of finer leaved grasses like Common Bent and Sheep's-fescue. This species seems to be increasing in the uplands as lowland habitats are lost. Green Hairstreak (*Callophrys rubi*) is our real moorland gem (Fig. 124), although it also lives in the

FIG 124. Green Hairstreak butterflies mating on Heather on Combs Moss. (Steve Orridge)

limestone dales. Moorland larvae feed on Bilberry and adults are often seen bobbing around prolific bilberry patches from mid-April to late June. The Peak District is a stronghold, particularly in the moorlands where colonies are larger, on account of its 41 per cent national decline since 1976. Populations crash following wet summers when damp conditions seem to affect underground chrysalis survival. There is also an association with ants – the chrysalis is said to be able to emit clucking and chirring sounds (even audible to humans), which attract ants to take them into their nests (Duncan *et al.*, 2016). Look out for them anywhere with Bilberry in the cloughs and the eastern moors, but also at Teggs Nose Country Park (near Macclesfield), in the moorland sections of the Goyt Valley, and in Swineholes Wood in the Staffordshire Moorlands (Machin, 2018).

Widespread bumblebees, like some butterflies, also collect nectar and pollen from a variety of moorland plants. Red-tailed Bumblebee (*Bombus lapidarius*), White-tailed Bumblebee (*Bombus lucorum*), Buff-tailed Bumblebee (*Bombus terrestris*) and Common Carder (*Bombus pascuorum*) are the commonest, but like the butterflies, we also have a moorland speciality, Bilberry Bumblebee

(Fig. 125). This is another northern and western species usually living above about 300 m altitude mostly in moorlands. The queens emerge in April and the colony expands to 50–70 workers, observable from May with a lull in July before increasing again in August. The nests are built on or just below the ground in tall, tussocky grassland. Peak District-based research showed that this is essentially a moorland edge species (Yalden, 1982, 1983). Bilberry has a fairly short flowering period, so to sustain colonies, bees must forage more widely using adjacent hay meadows, European Gorse, as well as flushes wherever there are suitable flowers. Main species visited other than Bilberry are Cowberry, Willows, Bird's-foot-trefoil, White Clover, Red Clover (*Trifolium pratense*), Bell Heather, Cross-leaved Heath and Heather. They can be seen throughout the Dark Peak and South West Peak, but hotspots with particularly high densities include the Roaches and Swineholes Wood in the west and Blacka Moor and Agden in the east (Yalden, 1983). Since Bilberry plant height and flower abundance depend on grazing levels, Bilberry Bumblebee numbers strongly correlate with low grazing intensities and recent reductions will support their numbers into the future provided moorland edge habitats like flower-rich hay meadows are also retained.

FIG 125. Bilberry Bumblebee on willow catkin. Note the extended red colouring of the abdomen and yellow stripe at the top of the thorax compared with Red-tailed Bumblebee. (Steve Orridge)

There is a wealth of other moorland invertebrates, with space to highlight only a few. Many are common and widespread in a variety of habitats, others more local and confined to the uplands, and some are at their range edges, like so many already described. One good example is a shooting moth, one of the caddisfly order, *Rhadicoleptus alpestris*, which is a Palearctic species. It is rare in the Peak District, found originally in the late 1800s on Dunford Moor, which is where, after over 100 years, it was re-found by Crofts (2011). It has a very restricted national distribution, largely in western and northern England. Hoverflies have been well studied with up to 50 species discovered on Peak District moorlands. Characteristic ones include the elegant large wasp mimic *Sericomyia silentis*, which breeds in wet peat. Adults appear from May to October sunning on bare ground, rocks and vegetation, or visiting moorland flowers. *Chrysotoxum arcuatum* is another typical moorland wasp mimic, easily recognised by its unusually long porrect (long, pointing forwards) antennae and by its smart waspish appearance (Fig. 79). It is associated with ants' nests, probably the Small Black Ant (*Lasius niger*). Adults start flying on sunny April days and appear until early October.

Very few grasshoppers and crickets (orthopterans) contribute to the musical soundscape of the moorlands. The typical widespread species is the Common Green Grasshopper (*Omocestus viridulus*), whilst the Meadow Grasshopper (*Chorthippus parallelus*), an abundant species of southern England, is quite localised to sites with lush damp vegetation on lower moorlands. Agden Bog has a good population. A more recent colonist, spreading from southern England in the past three decades is the Slender Ground-hopper (*Tetrix subulata*) now found increasingly in moorland wet flushes with patches of bare mud in sunny spots.

Owing to a shortage of calcium, slugs rather than snails are the molluscs of the moors, with rarely a flush on a damp day without its Large Black Slugs (*Arion ater*). Equally common, but less noticeable, is the Hedgehog Slug (*Arion intermedius*) and only slightly less common is the Dusky Slug (*Arion subfuscus*), easily identified by its orange body mucus that readily stains fingers. It is tolerant of acidic conditions but prefers grassy moorland.

We turn now to the enclosed moorland environment within the Dark and South West Peaks. Many of the species we have seen in the unenclosed habitats occur lower down too, but there are also distinctive differentiating features, as we shall see.

CHAPTER 10

Woody Habitats in the Dark and South West Peaks

W E NOW LEAVE THE SWEEPING landscapes of blanket bog, dwarf shrubs and steep cloughs to explore the more enclosed, relatively sheltered land. Lower in altitude, with a more equable climate, often based on clay soils derived from shales, this embraces the larger woodlands, various grasslands like rush pasture, reservoirs and rivers – more mature and meandering than in the steep cloughs, along with their associated habitats. The land is enclosed, mostly by gritstone walls, with access largely via public rights of way restricting their exploration, but there is still plenty to see. The shelter, scale and enclosure lead to a different character and some new players. This chapter focuses on the woody habitats, whilst the rest follows in Chapter 11.

WOODLAND

Although woodland and trees are very important in the Peak District, it is not renowned as a wooded landscape. Our best estimates of extent come from Forestry Commission statistics combined with the NPA's priority habitat mapping in support of its BAP (Appendix 2). Neither are fully complete and Forestry Commission data omits the smallest woods, nevertheless, woodland cover is estimated to be around 12,175 ha or 9.2 per cent of the whole of the Dark and South West Peak areas, excluding new clough planting. This general dearth is attributed to past losses, to the amount of high, peat-covered land where woodland is not a natural feature, and to the many years of high sulphur dioxide pollution levels making growing trees commercially uneconomic.

The woodlands are about 55:45 split between broadleaved and conifer, with mixed woodland only a minor component. The habitat inventory for the Dark and South West Peaks Natural Character Areas suggest a total of about 3,953 ha of broadleaved woodland, but of this only some 1,262 ha is regarded as ancient woodland with a further 657 ha replanted ancient woodland. These are the most valuable woodlands for nature conservation, but they only total some 1.45 per cent of the Dark and South West Peak landscapes. This is tiny and less than the national average (which is still very low at 2 per cent according to the Woodland Trust).

Ancient woodlands are those that have persisted since 1600, a rather arbitrary date linked to the first reasonably accurate maps. The trees may well have been intensively managed for different products, but the soils and ground cover generally reflect this long period of wooded stability and the tree and shrub species are mostly locally native. This does not mean that all the trees are old, but more that the same species have been present, renewing themselves over the last 400 plus years. Ancient woodland is special, with complex communities of plants, fungi, insects and other organisms, many of which are relatively poor dispersers and therefore less able to colonise newer woodlands. Some of these generally immobile species are useful ancient woodland indicators, but these vary regionally owing to different responses to soils and climate. An example is Dog's mercury (*Mercurialis perennis*), which is a good indicator in the gritstone and shale environment but probably not on the limestone where it occurs in grasslands. Some mosses and liverworts and a few lichens are also useful indicators of ecological continuity in old Oak forest and stand-alone parkland trees.

Patterns of distribution

Different woodland types display a distinctive pattern. Often large conifer plantations dominate hillsides above the larger reservoirs, as in the Upper Derwent Valley (Fig. 5), the Woodlands Valley astride the A57, Longdendale, Holme, the Sheffield Reservoirs, Goyt Valley and Macclesfield. In contrast, broadleaved woodland is more scattered in smaller blocks, often narrow and snaking down the valleys. Particularly wooded landscapes lie on the western side of the South West Peak (Shell Brook valley), from Hathersage to Curbar (Fig. 126) in the mid-Derwent Valley (Derwent comes from an ancient Celtic word meaning 'valley of thick oaks', which it still is), on the Peak District's eastern fringes and around Stanton in the southeast. The Chatsworth Estate is well endowed with broadleaved woodland, and there are further sites in the National Trust's Lyme Park near Disley. Many of these woodlands are derived from planting and contain trees and shrubs that are not locally native like Beech and Sycamore. The

FIG 126. One of the Peak District's more wooded areas in the mid-Derwent Valley.

semi-natural ancient woodlands can sometimes be engulfed by larger conifer plantations. Fragments persist for example within the conifers in the Goyt Valley and in Parkin Clough above Ladybower Reservoir. Others are isolated and small, such as in Long Clough south of Glossop. Some are incorporated within broadleaved planting as in Clough Wood east of Birchover.

Recent woodland history

Historic management has moulded woodland character; itself dependent on economic demands. Prior to about 1600, there was little woodland planting, so woods shown on the first maps are potentially derived from the original forest cover with a long period of continuity through the centuries. People were dependant on them for fuel prior to coal being mined, along with a wide range of building and household needs. The prime management would have been by coppicing whereby trees or shrubs are cut to near the ground and then grow back. This regrowth is said to 'spring' back and there are several Spring Woods locally. Uncut or pollarded trees would have been included in many of the woods for larger scale building or other requirements, particularly in lower-lying areas with better growth. Grazing animals were excluded from coppiced woods after cutting to avoid stools being browsed. Thus walls, hedges or dead hedges were essential. Not all ancient woodlands survived from the 1600s. Hay (2014) noted the loss of major oak woodlands in the Rivelin valley during the 17th century for example, as the fashion for hunting waned and Red Deer were removed from the forests and chases before many were destroyed.

Plantations began to be established for timber and for landscape objectives on the larger estates in the late 18th and 19th centuries. Farey (1813) and Pilkington (1789) provide Derbyshire accounts, whilst Moss (1913) gives us a more ecological view. From these, we know that the landed gentry and cotton mill owners established the first major hardwood plantations in the late 18th century. Numerous such plantations were listed by Farey throughout the Peak District such as at Chatsworth, Dinting Glossop, Little Hayfield, Stanton-in-the-Peak and Great Hucklow, planted with mixtures of Scots Pine (*Pinus sylvestris*) Larch (*Larix*), Norway Spruce (*Picea abies*), Fir (*Abies* spp.), Oak, Ash, Elm, Sycamore and Birch. In Glossop, for example, Lord Barnard Howard was planting 50–60,000 Larch, Scots Pine and Beech annually. Larch was a valuable tree at the time, and Major Shuttleworth of Hathersage had a roomful of furniture made from Larch only 59 years old cut from Upper Padley Wood. The plantations seemed to be mostly grown to maturity and harvested rather than being managed as coppice. Few are still extant. Some near Glossop were certainly stripped during the Great Depression (1929 to 1939) for cooking or fire wood when people were desperate.

Many of the tree species planted had specific uses. Alder, for example, was grown in Edale for the Manchester dyers who paid the considerable sum of £6 to £6 10 s for a ton of poles. Bark, cones and leaves produce a range of browns used for dying. Alder grown near Chapel-en-le-Frith was made into clogs for mill workers as it is resilient and extremely durable when wet – and mill floors were usually wet, making clogs essential. However, being relatively soft, it was not the best for dancing clogs as the sound was poor. Mature birches were tapped for their sap to make wine and their wood turned to make bobbins and spindles. Farey commends the beautiful grain of Hawthorn wood used for veneering furniture in Ashford-in-the-Water, but interestingly, Hazel was considered to be unprofitable underwood owing to its slow growth and to the 'temptation it affords to idle and mischievous persons to trespass on the fields and woods, and break down their stems' to collect the hazel nuts! This could account for its general lack in many Dark Peak woods.

Rowan was described as a soft but tough and solid wood used for tables, spokes, shafts and chairs. Even the roots produced knives and spoons, whilst the ground dried berries produced a wholesome bread, perhaps accompanied by a spirit distilled from the berries. Rowan was once widely planted close to houses as a protection against witches. Blackthorn (*Prunus spinosa*), was used for a dye and a cure for agues making a tea from tender leaves, whilst the wood was hard and tough enough for rakes and walking sticks, although Farey was very dismissive of its value as timber.

The greatest value of our Oak was for tanning for which it was regularly coppiced with the bark stripped and sent to the tanneries as far away as Stockport and Sheffield. There were also many smaller local tanneries in the 1830s dependent on this supply at, amongst others, Wirksworth, Bakewell, Baslow, Grindleford, Bradwell (which had three), Edale Mill prior to its conversion to a cotton mill, and three in Hope and Ecton. Leather for shoes and saddlery were essential in a period of horse-based transport and agricultural use. Farey expresses indignation at finding Oak stools with the bark stripped but the wood uncut, as in Nether Padley, predicting that such practice prevents regrowth of the stools possibly causing their demise. We have numerous Oak woods that survive from this management.

Farey also describes the management of Spring Woods as containing much underwood, sometimes with taller trees as well. Examples were given in Chisworth, Nether and Upper Padley, Haddon Hall, Shire Hill in Glossop and woodlands near Kinder. The management of the underwood was on a coppice cycle said to be due to the inadequate prices given for large timber at the time. A 21 – 28 year cycle is described, with annual cuts providing a continuous supply of wood, though Farey thought that as the return from wheat and pasture was greater, spring woods on the best soils should be cleared and cultivated. It is fortunate that this recommendation was not implemented. Apart from Oak for the tanning industry, the main use of the coppiced wood was puncheons (the local name for pit props for the coal mines), stemples and fails (equivalent props for lead mines) and for ladders, sough and gate timbers. Smaller poles provided hurdles, broomsticks and hedge stakes with the rest being used to produce charcoal, usually in or near the woodland site. A number of old charcoal platforms persist, for example, in the Upper Derwent Valley within the conifers above the reservoirs.

Woodland management has changed significantly. Coppicing went into a steady decline during the 19th century, mostly as coal replaced charcoal. Woods that were not destroyed were converted to high forest whereby singling of multi-stemmed stools, clearance of undergrowth, planting with timber trees or natural regeneration changed the previous coppice structure completely. Woods were managed instead for public recreation, game coverts or for timber. Many are now neglected which tends to result in the loss of the understory layer as the canopy develops and there is little commercial use of some older plantations.

FIG 127. Conifer removal in the foreground in Alport Dale by the National Trust, with natural colonisation by Birch. Large blocks of mature conifer plantation remain in the background.

Many of the woods were opened to grazing, as tree harvesting became uneconomic, trees have aged and fallen and with continuous grazing, there is often little regeneration, leading to thinning woods. Reversal of this in places like Padley Gorge has been dramatic.

More recent conifer establishment was designed to protect the reservoir catchments. They consist mostly of Norway and Sitka Spruce (*Picea abies* and *P. sitchensis*), European and Japanese Larch (*Larix decidua* and *L. kaempferi*) and Scots, Corsican and Lodgepole Pines (*Pinus nigra*, and *P. contorta*). Many are still managed commercially, although there are exceptions. In Alport Dale for example, the National Trust is felling some plantations, where acccss for extraction is too environmentally damaging, and replacing them with broadleaved woodland, partly through natural regeneration but with supplementary planting where needed (Fig. 127).

Similarly, some of the water companies are converting parts of conifer plantations into broadleaved woods, especially of Oak, adding more native species and developing better woodland structure through continuous cover forestry, as along the Upper Derwent Valley, in Longdendale, Macclesfield Forest and Goyt Valley. Other conifer plantation conversions are taking place in the South West Peak where the Staffordshire Wildlife Trust is restoring upland oak woodland and moorland habitats from Black Brook Plantation, and on Bradfield Moors where a large part of Holling Dale Plantation is being reverted to heathland.

There is little information on recent changes in semi-natural woodland cover in the Dark and South West Peaks, but a comparison I made of Moss' 1913 northern vegetation map (covering the Dark Peak core area) and the situation in 1982 showed a 55 per cent loss from 108.5 ha in 69 separate small sites down to 60 ha. Only 20 sites remained semi-natural woodland covering 31.6 ha. Fifteen woods were engulfed in conifer plantations, eight were completely lost, some to reservoirs, some had shrunk in size and a further 18 had degenerated into scattered trees in a grazed environment. These losses are being partly stemmed through different projects. The clough planting (Chapter 9) will help. A recent Heritage Lottery funded project successfully addressed management of significant areas of ancient and semi-natural woodland in the Dane Valley, including thinning, stock exclusion, 10 ha of Rhododendron (*Rhododendron ponticum*) control and creation of 45 ha of new woodland. Much more woodland creation and tree establishment is still needed, but a good start has been made.

Woodland character

Woodland character is closely related to soils and therefore the underlying geology, as well as history. Those on the gritstones are distinctive in being dominated by Oak, usually Sessile Oak, but Pedunculate Oak and their hybrids are also frequent. Pedunculate Oak are generally more abundant at lower altitudes on shale-based soils. Abundant Birches and Rowan accompany the Oaks. Silver Birch prefers the drier ground whilst Downy Birch (hairy, more rounded leaves) frequents wetter soils. Birches also hybridise, complicating identification. Good examples of Oak–Birch woods lie in the mid-Derwent Valley (Fig.128), Shire Hill in Glossop, Kinder Bank in Hayfield, in the Derbyshire Wildlife Trust reserves on opposite sides of the A57 at Ladybower and in many examples along the eastern side of the Peak District in South Yorkshire.

In the Oak woods, regeneration can be slow and research in Padley Gorge demonstrates the severe effect of grazing and impressive regeneration that follows when sheep are excluded. A 1955 enclosure established near the top of the wood, the effects of which are still visible, witnessed first a flush of birch, but few could penetrate the subsequently deepening layer of Wavy Hair-grass and Bilberry turf as these bounced back from heavy suppression. On the other hand,

FIG 128. An Oak-birch woodland with Rowan in the mid-Derwent Valley, with typical Bracken and Bramble understory.

Sessile Oak recruitment was intermittent as a bumper acorn crop depends on mast years. The acorns did best when buried under falling leaves, hidden from the main predator, Wood Mice, and in more light-filled gaps than under denser canopies (Pigott, 1983). Also interesting were the success of a few Hazel and Small-leaved Lime planted into what would normally be seen as unsuitable soils. This work shows the dynamics of our oak woodlands, how changes in grazing, and hence ground flora density, height and bare ground can influence the ratio of different trees which could persist for years, and how our impoverished woodland tree range may be more closely related to past history and management than to ecological controls.

The most distinctive feature of some of these upland Oak woods is the blocky nature of the ground with a jumbled mass of boulders at all angles as it tumbled from above (Chapter 4), especially below the edges from Hathersage to Curbar (Fig. 129). Only skeletal soils have developed on the rocks supporting lichens such as the common *Parmelia saxatilis*, whilst increased litter accumulation heralds a luxuriant cover of Bank Haircap (*Polytrichastrum formosum*), Cypress-leaved Plait-moss (*Hypnum cupressiforme*) and Swan's-neck Thyme-moss (*Mnium hornum*). The pale–green flattened shoots of Waved Silk-moss contrast with the darker mosses. More rarely in some oak woodlands, the glaucous green, swollen cushions of Large White-moss (*Leucobryum glaucum*) turn dirty white when dry, forming cushions that bowl around in the wind.

With deeper soils on the gritstones, Bilberry is more abundant, sometimes accompanied by Common Cow-wheat (*Melampyrum pratense*), a semi-parasitic annual. Find this in gritstone oak woodlands below the edges or on Kinder Bank, Hayfield. This is a declining lowland species, increasing the importance of upland reservoirs as here. On deeper and loamier soils between the rocks Bluebells (*Hyacinthoides non-scripta*), along with Creeping Soft-grass, Honeysuckle and Bramble all appear (Fig. 129). One special species is Climbing Corydalis (*Ceratocapnos claviculata*), a pale green annual with creamy flower sprays straggling over other plants (Fig. 130). It is best seen in the Derwent Valley woodlands and the lower sites in the Staffordshire Moorlands. Other common associates include

FIG 129. The very diverse ground conditions in some Oak-birch woodlands derived from the large rock boulders with little soil and a mossy cover, contrasting with soils on shales or head materials in-between where Bluebells can thrive, as here in Haywood, part of the National Trust's Longshaw Estate.

FIG 130. Climbing Corydalis sprawling over dead Bracken fronds in woodland north of Hathersage.

Wood Sorrel in more sheltered spots and Bracken is often densely dominating on more acidic soils.

Some of the woodlands exhibit greater tree, shrub and ground cover diversity. Holly, often in quantity, provides welcome winter cover and shelter for various birds. It is a northwestern European species and true Holly woods are a British speciality, increasing the value of ours. In the more loamy or heavy soils, Hazel and Hawthorn are more conspicuous, Aspen, a rare associate, and Ash occasionally present too. Crab Apple (*Malus sylvestris*), Wych Elm and Dog Rose (*Rosa canina*) are infrequent but Small-leaved Lime is very rare, known in Abney and Bretton Clough woodlands plus a single tree in Parkin Clough near Ladybower. Sycamore, Beech and more rarely, Sweet Chestnut (*Castanea sativa*) are often planted or have invaded, some of which are magnificent large, old specimens. There may be a few planted conifers like Larch and Scots Pine persisting too.

Significant changes in the ground flora mark differences in soils. More base-rich spots support Dog's Mercury (*Mercurialis perennis*) and Enchanter's Nightshade (*Circaea lutetiana*), both spreading by underground rhizomes. The early delicate white flowers of Wood Anemone (*Anemone nemorosa*) and the later ones of Greater Stitchwort (*Stellaria holostea*) may join them. The Stitchwort also often grows under fringing scrub or hedges. In lower altitude woods, large clumps of Pendulous Sedge (*Carex pendula*) with its graceful pendulous flower stems, are conspicuous but not common, although it can escape from gardens and become naturalised. Ungrazed or lightly grazed (perhaps by deer) woodland ground flora is generally more diverse and much more exuberant with greater tree and shrub regeneration.

Wet woods are extremely important and a priority Biodiversity Action Plan habitat. They are, though, in short supply in the Peak District. Most are buried within larger woods, although some patches occur adjacent to streams or reservoirs where water tables are high. Distinctive trees are Alder and various Willows, and more rarely, Bird Cherry. Herbaceous plants of this wet ground include Opposite-leaved Golden Saxifrage, Common Celandine (*Ficaria verna*), Yellow Pimpernel, which is equally at home in base-rich flushes in the cloughs, and, more rarely, Large Bittercress (*Cardamine amara*) especially in seepages (look out for its distinctive violet-coloured anthers) (Fig. 131). These seepages

FIG 131. Large Bittercress in wet woodland above Shell Brook in South West Peak. Note the four-petalled flowers and violet-coloured anthers.

are very important habitats for invertebrates, particularly flies like some scarce soldierflies and craneflies. Taller wetland plants where there is more light include Meadowsweet and Water Mint (*Mentha aquatica*), identifiable by its strong minty smell. Marsh-marigold or Kingcup (*Caltha palustris*) brightens wet woods in spring with showy buttercup-yellow flowers. Most of these wetland species are not obligate woodland plants but also grow in open areas within wet pastures or near to streams or ponds.

Ferns can be abundant in many woods. The commonest are Male Buckler Fern and Broad Buckler Fern (distinguished by the broadly triangular frond shape of the latter and maximum width in the middle of the frond for Male-fern). Two other locally common species – Lemon-scented Fern (see Chapter 9) and Golden-scaled Male-fern (*Dryopteris affinis* agg) can grace woodland stream banks. The frond stem (rachis) of the latter is splendidly clothed with gingery brown scales. Male-fern was valued in the past as a remedy for various conditions, including liver fluke that affects cattle and sheep grazing on wet, acidic soils, while both Male and Golden-scaled Male-ferns were considered powerful medications for tapeworm, but beware that ingestion of Male-fern can be toxic. Although often abundant and widespread here, some of these ferns are quite rare or absent elsewhere, more particularly to the south and east, making our woods and cloughs special (Fig. 43).

The ground flora in plantations tends to be less diverse. Low light levels in the densest prevent colonisation, although a mossy cover may prevail. With a little more light, Ivy can festoon the ground and trees too. Its flowers (in late autumn) are important sources of food for late butterflies and other nectar feeders, whilst the berries (in spring) attract birds like Blackcap (*Sylvia atricapilla*). Many animals find protection and shelter amongst its dense tree cover. The commoner ferns also feature in many of the plantations, along with Bramble, patches of Common Nettle (*Urtica dioica*) and common shade-tolerant plants like Herb-Robert (*Geum urbanum*) and Red Campion (*Silene dioica*), both of which frequent hedgerows and White Peak woodlands too.

In more open plantations, beautiful carpets of Bluebells have developed as in the Upper Derwent Valley. A May trip is truly uplifting with the blue haze and delicate faint honey-smell along the roadside and in the plantations (Fig. 132). There are other, older Bluebell woods, particularly on the fringes of the Peak District, towards Sheffield and Chesterfield for example, or near Pott Shrigley, Disley and in the Dane Valley to the west. Bluebells also persist along roadside banks and under Bracken some distance from woods as along Shell Brook. Some of these sites might have once been woodland, but Bluebells can colonise new sites relatively easily here. Britain is special for Bluebells,

FIG 132. Bluebells have colonised *en masse* under plantations in the Upper Derwent Valley beside Derwent and Howden Reservoirs.

holding around half the world's population. It is a western European species and Bluebell woods are mostly a British phenomenon. Its root sap was used in the Middle Ages to glue feathers onto arrows and to stiffen ruffs in Tudor times. However, the bulbs are toxic, containing glycosides rather like those in Foxgloves (*Digitalis purpurea*); a species that explodes into a colourful mass after timber removal in many upland plantations.

There are some invasive species in the woodlands. Rhododendron (mostly *Rhododendron ponticum*) was historically widely planted on estates and later in landscaping schemes around reservoirs (Rotherham, 1986). Sites include Chunal Plantation south of Glossop, Beeley Moor, around Park Hall in Little Hayfield, and in the Goyt Valley around the now derelict Errwood Hall where 40,000 specimens were planted in the 1850s. Rhododendron was introduced into Great Britain from Gibraltar in 1763 but was first recorded in the wild over 100 years later. Although attractive in flower, servicing a range of early nectar-feeding invertebrates, it spreads readily by seed from mature bushes into woodland, heathland and dry blanket bog. At some million seeds per average sized bush, its invasive powers are significant. It can then spread vegetatively, swamping the ground flora as on Strawberry Lee Plantation on Blacka Moor and nearby Longshaw Estate. It is now known to be one of the hosts for relatively new fungal diseases *Phytophthora ramorum* and P. *kernoviae* (related to Potato Blight), which can affect dwarf shrubs, Oak and Beech. Rhododendron is much more invasive in the west (Lake District and Snowdonia for example) than in the Peak District, but it is nevertheless being controlled as in the Wessenden Valley near Marsden and Upper Derwent where it was first established in the late 19th century.

In County Wildlife Trust Reserves, some control of other invasive species like Sycamore is undertaken to ensure the perpetuation of native species. Sycamore has a much denser and longer-lived canopy than native oaks and birches, so can exclude ground cover and take over from native trees. Although valuable as a source of aphid prey for some birds, it does not in general support as many invertebrate species or numbers as our native trees, which are therefore preferred.

Woodland Fungi

Fungi are very important woodland occupants here, although there are no comprehensive data. Fungi are in their own kingdom, more closely related to animals as their main cell walls are made of chitin, as in insect exoskeletons. In our woodlands, they are a critical and integral part of woodland biodiversity. They have long, branching filamentous structures called hyphae, which permeate their preferred environment. One group feeds saprophytically on dead material on the ground or on dead wood, effectively recycling it by releasing nutrients back to the soil and supplementing the organic layer. There is often a sequence of different fungi as wood decays, sometimes with one fungus living on another.

FIG 133. Common Stinkhorn in broad-leaved woodland. Its sticky, honeycombed head is clear along with the stout, white, hollow spongy stem. (Steve Clements)

One particularly striking saprophyte is Common Stinkhorn (*Phallus impudicus*) associated with dead wood especially in many of our deciduous woodlands (Fig. 133). Common and widespread, it is easily detectable by its distinctive, pungent odour simulating the smell of carrion. Follow the smell to find troops of phallus-shaped fruiting bodies up to 30 cm tall with maybe up to 100 together. The smell attracts flies, some beetles and other insects expecting to find a carrion meal, but inadvertently collecting spore masses on their legs and transporting them elsewhere. Common Stinkhorn was used in medieval times as a cure for gout and as a love potion – considered an aphrodisiac owing presumably to its suggestive shape.

Another fungal group develops mycorrhizal associations with trees, either internally in the roots (endomycorrhizal) or as an external wrapping (ectomycorrhizal). These are fundamental to woodland functioning as they are more effective at extracting nutrients and water from soils and pass these to the trees, whilst sugars produced by photosynthesis pass from trees to fungus. Many fungi are associated with specific tree species while others are more catholic. It is their fruiting bodies which are many of the familiar Toadstools in the woods. An individual tree, though, can have 15 or more fungal partners at one time. Recent research has demonstrated that the interlinking of fungal hyphae between trees of the same or different species enables chemical signals and nutrients to be passed from one to another across a woodland, with suggestions that these facilitate 'communication' between trees signalling, for example, herbivore or parasite attacks. Only a small proportion of fungi are parasitic; attacking trees or other plants, and resulting in their eventual death. Some species seem to hasten demise, but only attack once a specimen is already damaged or diseased; others do not kill their host.

Autumn is the best time to find many fungi fruiting bodies when damp conditions promote their production, although some, like the bracket fungi, persist for many months or even years. Birch, Oak, Pine, members of the rose family (Hawthorn, Blackthorn, Rowan as well as Roses) and Orchids have ectomycorrhizal fungal associations. Familiar fruiting bodies in our woods are the highly toxic Fly Agaric (*Amanita muscaria*); the iconic red and white, magical fairy Toadstool. This is linked to a range of tree species, more often Birch in our woods (Fig. 134). It has a long history of use in religious ceremonies, but causes hallucinations and psychotic reactions if eaten. Other common mycorrhizal fungi include *Russula* (Brittlegills, often with brightly coloured caps), and *Lactarius* (Milkcaps, usually whitish or autumn colours but with white flesh and a milky exudate). Different Milkcaps are associated with different trees such as Oaks, Beech, Birches or Pines. Look out for Penny Buns (*Boletus edulis*), one of a fungal family displaying pores instead of gills, making the cap appear more spongy.

FIG 134. Fly Agaric under birch in Alport Dale. Cap can be 5 to 20 cm across, spherical at first, covered with a white veil which breaks up to give the white patches when the cap becomes convex and then flattened.

The Longshaw Estate and other mid-eastern sites are regularly surveyed by the Sorby Fungus Group who have amassed over 1,150 species records on the Longshaw Estate alone for example; a number that continues to increase, demonstrating fungi's often fickle appearance and the survey effort required. Not all are in the woodlands and, although this sounds high, it is not unusual in the national context. Only invertebrates will exceed this total species number per unit area. Longshaw's richer fungal habitats are the mature woodlands with a clear forest floor, meaning that boulder-filled woodlands covered in Bracken, for example, are poorer than some plantations and other areas. Remarkably, more fungi are associated with Beech, a planted species not native here, than any other tree, with far fewer associated with Birch and Oak for example. Indeed, in one survey 70 different species were found on a single Beech log in Padley Gorge! The fungal assemblage is truly significant. Look out for small brown mushrooms like mottlegills (*Panaeolus*) usually growing on dung, Bonnets – small delicate mushrooms on soil, wood or other plant debris with bell-shaped caps

(*Mycena* spp.), and different brittlegills. In general, the more dead wood there is, particularly large branches, stumps and fallen trunks, the more fungi there will be, showing how important it is to retain dead wood in woodlands. Although the typical cap and stem mushroom-form dominates the overall fungal list, there are also large numbers of Crust, Bracket, Bolete, Coral or Club and Jelly Fungi. This is just the larger ones, a third again will be micro-fungi including mildews, rusts, moulds and slime moulds, many of which when magnified are stunning (Fig. 135).

There are several rare fungi in our woodlands. The little, easily overlooked Earpick Fungus (*Auriscalpium vulgarum*) with teeth instead of gills and a hairy dark-brown cap, grows in pine cones, mostly on the ground. It has so far been found in Upper Derwent Valley woods and at Longshaw. It is the only member of its genus in Britain and is localised throughout the country. Another oddity, *Ophiocordyceps ditmarii*, parasitizes insects, changing their behaviour as it does so. This is one of only about 10 records in Britain and Ireland at the time it

FIG 135. Some stunning micro-fungi to enjoy. (a) *Lachnella alboviolascens* on rotting deciduous wood; (b) *Melastiza contorta* on soil under Scots Pine; (c) *Craterium minutum* on rotting deciduous wood; (d) Dewdrop Bonnet (*Hemimycena tortuosa*) on rotting dead beech. (John Leach)

FIG 136. The Hoof Fungus on Silver Birch. (Steve Clements)

was found. Finally, the Spidery Moss Ear (*Rimbachia arachnoidea*) grows on the common Swan's-neck Thyme-moss. It has a tiny white ear-shaped cup connected to the moss by spidery threads. When found in Birch woodland at Longshaw in 2015, this was also a new species to Derbyshire.

Bracket fungi form shelves sticking out from dead wood or trees. Some are large and woody, others more delicate and pretty. Some change colour if bruised, while others have a long history of use. Their names can be graphic – Chicken of the Woods, Hen of the Woods, Turkeytail, Blushing Bracket and Dryad's Saddle (*Laetiporus sulphureus*, *Grifola frondosa*, *Trametes versicolor*, *Daedaleopsis confragosa* and *Polyporus squamosus*), all of which occur on trees in our acidic woods, some commonly like Turkeytail with its velvety muted rings on a small bracket surface. Some are parasitic and often long-lived so can be seen throughout the year on stumps, logs or fallen branches. Around 40 different species have been located so far on the Longshaw Estate. The rare Oak Polypore (*Piptoporus quercinus*), largely a southern species, lives on oaks in Chatsworth parkland.

Two more are especially noteworthy. The Hoof Fungus (*Fomes fomentarius*) looks rather like an extended horse's hoof clutching the side of a Birch tree (Fig. 136). It is mostly restricted to Birch, although it grows on Beech too in Europe. It has recently colonised the Dark Peak, although it is more widespread east of Chesterfield. Birch Polypore (*Piptoporus betulinus*) is also restricted to Birch, but the Fungi Group found Hoof Fungus usually associated with Birch Polypore, although not the other way round, suggesting a functional relationship. Remarkably, both these fungi were found on Ötzi's body found in the Alps in 1991, the Copper Age iceman living 3400–3100 years BCE. Hoof Fungus, also called Tinder Fungus, was used for starting fires, but has medicinal value to stop bleeding and has been converted into a fibre for clothing such as hats. The Birch Polypore also once had anti-inflammatory and anti-bacterial medicinal uses.

Other woodland sites merit fungal searches. There are Beech and Pine plantations in the Upper Derwent Valley which are good sites, Alder carr and mixed wood in Abney Clough and Clough Wood near Darley Dale, and other woods on the edge of our area like Linacre south of Cutthorpe and Holmesfield near Sheffield. Note that access to some is only by public rights of way. There are many more woods where regular fungal forays could find more species and assemblages of significance, but this selection provides a snapshot of just a few of the many fungi, common, unusual, weird and rare in our upland woods. Best leave them for others to see so that the species can persist into the future as well as providing an important food source for woodland animals like slugs, springtails and fly maggots.

HEDGES, SCRUB AND TREES OUTSIDE WOODS

Although stone walls are the norm, you may be surprised at the generally woody feel of some of the Dark and South West Peaks owing to the large numbers of trees and shrubs along field boundaries, beside streams, along road-sides and around farmyards and villages (Fig. 137). Ash is abundant and many are old or mature trees. Alder and Willows, sometimes with Bird Cherry, are more frequent closer to water, while Ash, Hawthorn and Sycamore are common along roadsides, mostly as self-seeded pioneers. Groups of Sycamore were frequently planted many years ago to provide shelter around farmhouses and farmyards, a pattern typical of the Pennines as a whole. Some of these trees are now large and old. Closer to villages, there tends to be more non-native species, like Beech and Horse-chestnut (*Aesculus hippocastanum*), and a variety of conifers. Some trees have

FIG 137. Wooded feel to the countryside in the South West Peak with hedges, scattered and field-side trees, looking across Warslow Brook northeast to Revidge moorland.

been planted for landscape purposes, like the lime avenue along the Bakewell to Calver road, or the landscaped grounds of larger Estates (Chapter 6).

There is evidence for past planting or at least nurturing of Holly and gorse. In an era where it was difficult to keep stock alive over winter and when pastures on acidic soils at moderate altitudes provided little winter keep, farmers would have used browse cut from trees to sustain livestock. This is mostly in or before the 1600s and 1700s, prior to more extensive liming of acidic grasslands to increase productivity. There are numerous references to feeding Holly to sheep (especially) but also to cattle, horses and deer; to the number of Holly trees a farm must have to provide adequate browse; to the cultivation or rental of 'haggs' or 'hollins' (Holly clumps held within an enclosure, the hag(g) or portion of woodland), to the fines or agreements related to Holly; and to ensuring the Hollies survived and regrew after cutting on land rented out as pasture (Hey 2014, Spray 1981, Spray & Smith 1977). By overlaying the historical evidence with place names, people's names (using telephone directories!) and Holly's ecology and national distribution, Spray (1981) concluded that Holly used as fodder was more prevalent in the South Pennines on the shales and grits (not the limestone) in the Dark Peak, as well as in additional areas in the North West and Yorkshire Dales. The practice seemed to have been much scarcer elsewhere in the country in general, although with some exceptions.

The reasons for this unique association in the South Pennines are possibly related to the poorer pastures at middle altitudes on the Pennine Fringes, resulting in increased demand for alternative winter forage; the greater abundance of Holly as a native species in the north and west (although altitudinally limited by severe winter frosting affecting its growth layer, the cambium); and the lag in agricultural improvements such as turnip-growing for winter fodder being adopted. That this dependency lasted for several centuries is reflected in the widespread place names incorporating hollin or hag(g) such as Hollinsclough, Hollinwood Road (Disley), Hollingworth (Longdendale) and the rash of 'hag' names in the Derbyshire Hope Woodlands. This traditional use here was dying out in the 18th century, with the last rental traced for the Sheffield region to 1737, but a much earlier 1632 in Derbyshire.

Holly's particularly prickly nature (and Hawthorn's thorniness) reflect natural defences against browsing animals which can also be indicative of high palatability and nutrients levels. Gorse is similar and European Gorse was also widely planted to provide winter fodder – a use that has lasted in the collective memory much longer than that for Holly. Gorse could also have been planted to use for firing bread ovens as it burns hotly, with little ash. Ivy was also widely used for fodder. Winter grazing in forests and woods was once common and they were carefully managed in the Middle Ages, for example, for cattle as well as wood products, so animals would have been able to browse freely then.

It is well known that browse cut from Ash and Elm, historically, but possibly prehistorically, were the mainstay of many stock animals, foliage being collected through pollarding throughout the growing season and winter feed being collected in autumn. Leaves of both are highly palatable and nutritious. The adequacy of fresh browse must have added greatly to the value of land and to the ability to overwinter stock. The extent to which stock search out trees and shrubs in pastures now is clearly visible from the browse-lines. The question is whether the current abundance of old Ash along field margins and former abundance of elm (mostly Wych Elm here) before it was lost to Dutch Elm disease were a product of planting or nurture for stock browse. Wych Elm was frequent on non-calcareous soils in Derbyshire and Cheshire, particularly in the Peak District, and is now rare (Clapham, 1969, Newton, 1971). Thus, my thesis is that many trees and shrubs in the agricultural landscape were once potentially primarily valuable as browse for stock and deer and that their presence is a relic of this former value;

one which has essentially been lost from collective memories. Many of the older Ash trees still gracing our countryside show signs of being pollarded in the past. Are these representative of how our countryside was used several centuries ago? There is better evidence that the many Holly hedges in the western and eastern edges of our area, and perhaps the solitary specimens clinging to the clough sides, are vestiges of former use. Farey (1813) mentions scattered Holly pollards which were used for lopping in severe winters to feed sheep to good effect in the Hope Woodlands around Rowlee, until the agent cut them for charcoal and making birdlime (for trapping birds) and their demise subsequently as sheep ate the young regrowing shoots. The historic evidence suggests there were many more Hollies than now. There are still Hollies in fields near Charlesworth and on the eastern side nearer to Sheffield. Some Gorse patches may also be remnants of once larger ones used for winter fodder.

We have many hedges containing Holly and some where it dominates. Hawthorns too are a constant, with many mature or younger Ash, and even some hedges incorporating European Gorse. However, hedges as a whole are not surveyed or mapped comprehensively across the Peak District, so information is lacking on total amount or historic changes. Some of the hedges, particularly Hawthorn ones, are rather thin and tightly cropped, usually supplemented with wire. Others are regularly managed but well structured, whilst many no longer have a stock-proof function and have grown tall, contributing significantly to the area's wooded feel (Fig. 137). Hedge surveys (Anderson & Shimwell, 1981) show distinct variation in species content across the Dark Peak related to the hedge's age, soils and history. Work in lowland Britain, confirmed in Southern Derbyshire by Willmot (1980), suggests that you can age a hedge roughly by counting the number of woody species in 30 m samples, with one additional species per 100 years. Although not precise, this method allows separation of Saxon hedges (which are often old parish boundaries) from those of Tudor age or later enclosure-period hedges. No work on this has been undertaken in the Peak District, providing future opportunities for interesting investigations. Some new hedges may confuse the surveyor though having been planted with many different species.

Most of our hedges probably date from the enclosures but there are a few seemingly older examples. Hazel and Field Maple (*Acer campestre* – we are on the edge of its distribution so it is not common here) are regarded as poor colonisers and more typical of older hedges. They are certainly quite rare in the Dark and South West Peaks' hedges. Look out for the abundance of Hazel with Holly along the A54 heading out of the Peak District for example, or for Hazel in hedges in the Coombs/Chinley area near Chapel-en-le-Frith and around Charlesworth. Field Maple grows in a few hedges in Harthill and Birchover to the southeast and in the South West Peak. Hawthorn is the ubiquitous species in the younger enclosure hedges and has many uses, not just for stock. Its spring leaves have been eaten universally as 'bread and cheese', i.e. food of basic value, but with a pleasant nutty taste. The buds were made into a spring pudding and the flowers make a fine liqueur. That the white flower scent is overwhelmingly heavy is an added bonus. Blackthorn is more abundant in sample hedges from around Harthill and Birchover, along with Elder (*Sambucus nigra*), the latter also featuring near Charlesworth. Hedge oaks are more likely to occur in the west than the southeast, but other species are scarce. There are lines of willows where ditches run alongside, which may contain Bay Willow (*Salix pentandra*), which is uncommon in the South West Peak, with its lovely shiny leaves.

In general, lichen growth on trees in the Peak District was seriously inhibited by two and a half centuries of industrial pollution, but reductions in sulphur dioxide emissions has allowed colonisation of younger trees. Older trees in exposed situations are generally still suffering pollution effects. The increased nitrogen deposition from vehicle emissions and intensive agriculture has increased significantly the coverage of some lichens, particularly on trees. Most notable are the common yellow lichen *Xanthoria parietina* and associated with it the grey *Physcia adscendens* and *P. tenella*.

One really important group of trees are the ancient or veteran ones. We have hinted at old trees in the landscape, but the really old ones are special. These are the old, gnarled, knobbly, huge, often bent trees compared with other specimens of the same species. Thus, trees that are normally short-lived like birches may be ancient at 150 years old whereas an ancient Oak might be closer to 400 years. Ancient trees tend to have retrenched their canopy and are usually squat, with a wide, often partly hollow trunk. Veteran trees are similar, but might be of any age so long as they show features typical of ancient trees, but these could be due to damage or their environment rather than old age. Ancient trees are thus all veterans, but not the other way round. Ancient trees are renowned for their wildlife, supporting many rare species dependent on old growth. They hold natural pools within their gnarled trunks, contain dead wood, holes and wounds with sap runs, many crevices and flaking bark providing nooks and crannies of every description. Amongst the ancient tree specialists are lichens, fungi, ants, beetles and many other invertebrates. These trees also support many more species than their younger versions and are of the utmost importance for conservation.

There are examples of ancient and veteran trees in a number of settings in the Dark and South West Peaks. Some occur along field or lane boundaries, such as in the Dane Valley where they are mostly Pedunculate Oaks, one estimated to be over 400 years old. Some are buried in conifers, as are Beech in the Goyt

Valley, others occur in churchyards or on woodland fringes. Although surveyed in Derbyshire by the Wildlife Trust and elsewhere for particular projects, there is no inventory for the whole area. The total so far is 1,277 individuals, but there will be more. The best known examples of ancient trees are in parkland, especially Chatsworth Estate, where the biggest Oak diameter is 8.68 m. These magnificent trees lie in the deer sanctuary and are therefore mostly not open to the public.

Lichens are richest on isolated trees rather than in shaded woodland and the best are veteran trees in more sheltered situations such as Chatsworth and Haddon Parks' parkland trees, both tucked into more humid river valleys. These trees are noteworthy not just for the number of lichen species supported but also for the significant populations of both northern and southern species. For example *Lecanora sublivescens*, found on several veteran trees in Haddon Park, has its nearest English colony in Herefordshire. It occurs worldwide on less than 300 trees. In Chatsworth Park, *Hertelidea botryosa*, a lichen previously found only in Scottish Highlands on exposed old bark of Pine, was discovered on fallen veteran oaks.

A number of nationally scarce invertebrates have also been recorded associated with the ancient Chatsworth trees, such as the colourful Longhorn Beetle, *Saperda scalaris* breeding in dead wood; the Cobweb Beetle (*Ctesias serra*), whose larvae feed on insect remains caught in spider webs under dry bark; the wood-boring beetle *Dorcatoma flavicornis*, associated with rotten wood but only south of Northern England; and the plaster beetle *Enicmus rugosus*, a minute brown scavenger beetle found in fungi on trees. Many of the beetles found are considered to be indicative of long continuity of mature woodland habitat. The solitary wasp nesting in hard rotten wood, feeding on flies, *Crossocerus binotatus* and the Alder Kitten Moth (*Furcula bicuspis*), which actually feeds on Birch and Alder, are also nationally scarce species. Such parkland or wood pasture habitat is more abundant in Britain than anywhere else in Europe and protection alongside ensuring continuity of trees for the future are critical to their long term conservation.

It is evident that trees and shrubs outside woodlands are highly valued. They add cover and shelter as well as food and breeding sites for birds, they provide flight-lines and individual habitats for bats, corridors for peripatetic mammals and stepping stones linking woodlands together thus ensuring populations like some ground beetles can disperse across the landscape. More invertebrates such as pollinators occupy well connected hedges with more intersections. Leaves falling from trees and shrubs add more carbon to the soil than in areas without these woody contributors. These benefits show that thick, tall hedges, non-woodland trees and shrubs and ancient trees add much value to our habitats and warrant restoration and expansion where possible.

BIRDS

The Dark and South West Peak woods are home to a specialist upland bird assemblage comprising Wood Warbler, Tree Pipit, Common Redstart and Pied Flycatcher. All are migrant summer breeding species, which together make the upland woodlands special, more particularly since this assemblage is absent or scarce now in surrounding lowlands. Each carries a different story. Pied Flycatchers (*Ficedula hypoleuca*) (Fig. 138) first bred here in 1945, when a pair nested unsuccessfully in Padley Gorge and 1961 saw the first breeding in the Goyt Valley. Numbers fluctuated subsequently across our upland woodlands but it was still a rare visitor in 1982. From this kernel though, this delightful little black and white flycatcher has spread throughout our upland woodlands from the Staffordshire Moorlands, Goyt Valley and Wildboarclough in the west, to Longdendale and Shire Wood in the north, Ladybower and Priddock Woods and the rest of the Derwent Valley in the east, and with a few in White Peak woods too. Just in Derbyshire, 80 pairs were recorded at 24 sites in 2011 all in the Peak District, whilst in the Sheffield Region, there are 61 occupied tetrads, nearly all in our area (Frost

FIG 138. Pied Flycatcher, slightly smaller than a House Sparrow. Male perching on low dead branch, whilst female has browner rump and tail and buff underparts. (Kev Dunnington)

& Shaw, 2013, Wood & Hill, 2013). Gosney (2018b) found a 15 per cent increase between 1988–90 and 2017 in the Baslow area (SK27). The bird favours Sessile Oak woods with little shrub cover, so they can flit between trees below the canopy and find large enough holes for nesting – although many use the boxes erected for them. Nationally, we are at the edge of Pied Flycatcher habitat and therefore very vulnerable to the considerable recent 27 per cent decline accompanied by a westerly and northerly shift in distribution plus a population thinning on its range boundaries accounting for its Red List status (Balmer *et al.*, 2013). This may be related to climate change and the mismatch of the timing of migration (which has not changed significantly) with peak food supplies (which is earlier).

Although Common Redstart is a typical upland woodland bird, it also occurs throughout the Peak District – many farmyards framed by groups of trees support them annually. There was a steep population decline nationally in the late 1970s related to major droughts in Africa's Sahel area, where the bird overwinters, but numbers recovered subsequently in core areas, but without recolonising more marginal habitats. Wood Warblers share the Pied Flycatcher's habitat requirements, although they do seem to be colonising some conifer plantations now as well. They are ground nesters, feeding in the canopy, but singing their distinctive trilling song from lower branches. Like Pied Flycatchers, Wood Warblers showed a marked increase up to 1994, after which there was a large 40 per cent decline. Although numbers have fluctuated since, their overall range and population numbers have contracted, accounting for their national Red List status. The Goyt Valley has been a favoured haunt and numbers there reflect national trends, with a maximum of nine pairs in 1985–7, falling to four from 1992 to 1997, recovering to eight in subsequent years. The Upper Derwent is one of their current strongholds along with Wyming Brook, Win Hill and Padley Gorge, but Gosney (2018b) found that singing birds declined by a huge 92 per cent in less than three decades around Baslow.

Tree Pipits are particularly prevalent on the Eastern Moors (there were 92 singing males in the Eastern Moors Partnership holdings in 2010 and 105 in 2015), nearby on Greave's Piece, in the Upper Derwent Valley, on Win Hill and in the Goyt Valley, utilising in particular, the many birch woods, although Gosney (2018b) records a slight decline in the Baslow area in SK27.

As well as this special bird quartet, our upland woodlands support many more catholic and widespread woodland and woodland edge species, the range and numbers depending on the habitat's structure and composition. Dense woods, with many mature specimens attract a different suite to those with a varied structure and age range, with diverse trees and shrubs as well as glades and variable edges. Surveys in a number of our semi-natural woodlands and mixed mature plantations suggest that the commonest breeding birds are the regular garden birds – Bluetit (*Cyanistes caeruleus*), Chaffinch and Robin, with populations dependent on the number of tree holes or nest boxes and the degree of shrub and ground cover. Robins will be rare where there is little ground cover for example. Wrens are more abundant breeders in some woods than others, whilst Blackcap and Willow Warbler (*Phylloscopus trochilus*) are the most frequent warblers. The number of species is around 21 to 30 in an average-sized wood (data from NT surveys). So there is much to see. Nuthatch (*Sitta europaea*), Treecreeper (*Certhia familiaris*), Goldcrest (*Regulus regulus*), Great Spotted Woodpecker (*Dendrocopos major*), other tits and finches, Blackbird and Song Thrush (also Red-listed) will all be familiar here as well. Chiffchaff (*Phylloscopus collybita*) and Garden Warbler (*Sylvia borin*) favour woods with more undergrowth, although the latter prefers glades and more open woodlands. Chiffchaff numbers have been increasing as woodlands have matured, so listen out for their onomatopoeic calls from March onwards. Garden Warblers will also breed in small patches of scrub in large gardens or along rivers without extensive woodland.

Willow Warbler, once our most abundant warbler, is a quintessential sound of spring and utilises any small scrub patch as well as our better structured woods in the uplands and White Peak. Populations have shifted nationally, vacating the south (28 per cent decline) and increasing across Scotland (33 per cent increase), thought to be related to climate change plus changes in their African overwintering habitats (Balmer *et al.*, 2013); losses which justify its inclusion in the Amber list. We are on the boundary of this change, which is being re-enacted on the regional scale, with losses in the adjacent lowland and more urban areas matched by increases within the uplands. Pairs recorded on the Eastern Moors illustrate this with 160 found in 2010 surpassed by 210 in 2015.

One much rarer woodland bird that is seen occasionally is Hawfinch, also Red-listed; a large, magnificent but secretive finch. It favours woodlands with Beech, Lime, Hornbeam and Sycamore but will feed on Hawthorn berries too. The dearth of Lime and Hornbeam in our upland woods might restrict them, along with the effects of predation by Corvids and Grey Squirrels (*Sciurus carolinensis*), but the best place to see them is on the Chatsworth Estate in summer, and in the wider Derwent Valley in winter where double figure parties have been recorded over the last 20 years. They are absent from the Cheshire side of the South West Peak.

Woodcock (*Scolopax rusticola*), another species on the Red List, is difficult to monitor owing to its crepuscular habits, but increasing sightings are welcome in the well-wooded valleys and uplands in the eastern Dark Peak and the Derwent Valley especially, against a backdrop of national decline. Its preferred habitat is a mixture of open woodland and ground cover such as Bracken, plus damp patches and streams for feeding, but it is intolerant of high disturbance levels. If you are lucky, you might see a male roding over the canopy tops at dusk emitting low pitched grunting noises and shrill whistles. Woodcock are generally thinly distributed but are very much a species of the Dark and South West Peaks, with none known breeding in the White Peak. Our birds are supplemented by winter visitors, roosting in woodlands but feeding largely in nearby pastures.

Conifer plantations, with restricted structural variation and few old trees for nest holes, have a much more limited bird assemblage. Amongst the more numerous are Chaffinches and Coal Tits (*Periparus ater*), but you may also see Siskins (*Spinus spinus*), Great Spotted Woodpecker and Goldcrests. Siskins have only appeared relatively recently, having been first recorded in late winter/ early spring in Derbyshire in 1971–2. They have increased nationally as conifer plantations have matured, where they feed on Spruce and Pine seeds, also seeking out Alder and Birch seeds in winter. They are closely associated with conifer plantations as in Macclesfield Forest and Lyme Park, the Upper Derwent Valley and Goyt Valley along with the reservoir-fringing conifers on the east side of the moors. Siskins (Fig. 139) are gregarious and large numbers can be seen sometimes

FIG 139. Male Siskin, smaller than Greenfinch. Female is paler, greyer and more streaky. (Bob Croxton)

on autumn passage from north to south in the country. The highest recorded is 550 in October and December in 2003 at Ramsley Moor (where Birch is abundant). Wintering birds, often with Lesser Redpolls, can appear in flocks up to 50 or so, but in good years this can reach 1000, more particularly in the Derwent Valley.

Firecrest (*Regulus ignicapilla*), our smallest bird, has appeared recently in the Derwent Valley, but in very small numbers, with only two possible breeding pairs. This first bred in Britain only in 1962 (in the New Forest), and did not reach here until 1986 when a singing male was detected in the Goyt Valley. Breeding was found in Derwent Valley in 1995 and this and Matlock Forest remain the place to hear them. Overwintering also occurs rarely, although not necessarily, in conifer plantations. Another conifer specialist is Common Crossbill (*Loxia curvirostra*), which feeds largely on Pine, Spruce and Larch seeds. It uses its unique crossed bill to prise cones apart. Their breeding is stimulated by seed abundance rather than daylength, so they can breed at any time, but more usually from February to April; much earlier than most woodland species. They are regularly recorded in the Win Hill-Ladybower-Upper Derwent Valley area, more rarely in Macclesfield Forest and with scattered records for Arnfield Reservoir (Tintwistle), Chatsworth, Holling Dale Plantation (Bradfield Moor) and Longshaw. There could be 20–40 pairs or more in the area out of a British total of about 500–1000. Occasional irruptions occur outside the breeding period, accompanied rarely by Parrot Crossbills (*Loxia pytyopsittacus*), which are rare vagrants to Britain. More recent irruptions include 350 Crossbills seen in Derwent Valley in 2003 and 100 in Upper Derwent Valley in 2008.

Some of the more ubiquitous birds seeking out the winter berries on the Hollies and Hawthorn in and outside the woods are Fieldfares (*Turdus pilaris*) and Redwings (*Turdus iliacus*) in ubiquitous wintering flocks, with the largest numbers of the former at roost sites and of the latter (over 41,000 in 2006) on migration at night, as recorded over Ramsley Moor. Just rarely, Fieldfares have nested here. Rooks (*Corvus frugilegus*), another bird of the agricultural landscape, are dependent on groups of tall trees for their noisy rookeries which herald spring. Data suggest that Rooks declined from about 1944, although there was some recovery between 1980 and 1998 in Derbyshire as a whole, as well as in Cheshire and the Sheffield region. This is a familiar bird in all our agricultural areas.

Several birds of prey are also associated with our woodlands. Tawny Owls are perhaps the commonest, but usually need trees large enough for nesting holes, although they have been recorded nesting in rabbit holes, on rock ledges, on the ground and even amongst hay bales in a barn. The persecution story of Goshawks was covered in Chapter 9, but it is a woodland bird, agile and expert at ambushing a wide range of birds, with Woodpigeons (*Columba palumbus*) being their favourite, or mammals like Rabbits or Grey Squirrels, either within the woods or on the moorland fringes. They breed in large mature woods with trees far enough apart to weave between, with an untidy nest of twigs built in forks in mature trees. They are elusive though and fleeting glimpses of hunting birds or of their spectacular displays over their nesting sites are the vigilant watcher's rewards. Their stronghold is in woodlands in the Upper Derwent Valley and on the moorland fringe woods in South Yorkshire, although reports also come from Macclesfield Forest. They do not tolerate Sparrowhawks (*Accipiter nisus*) and frequently kill them.

Our maturing plantations, other woodlands and groups of trees also provide nesting sites for Buzzard (also covered in Chapter 9). They are a daily sight now all over the Peak District, their presence revealed by their plaintiff mewing. They

FIG 140. Long-eared Owl, which appears long and thin with long 'ear-tufts', but is smaller than a Woodpigeon. (Guy Badham)

nest in mature trees or rocky ledges, but their success depends on an adequate food supply. You can watch them swoop from trees at prey (such as Rabbits), feed on carrion like road-kills, or walk around, usually on open ground, searching for invertebrates like worms or beetles.

A much rarer bird associated with dense woodlands or scrub, particular of conifers, is the Long-eared Owl (*Asio otus*). Although suffering from a decline nationally, the local population has possibly increased since the 1970s, but being difficult to survey, numbers may be under-recorded. There may be around 20 or more pairs, all in the Dark Peak (Fig. 140). Successful breeding is linked to peaks in Field Vole abundance (as for Short-eared Owls), although local studies have found that they also catch various birds. Numbers are controlled by predation – Goshawks do not tolerate them in the same wood – although their preferred woodland structures differ, which help separate them. The best time to see these magical owls is after sunset during the breeding season, although ensure you avoid disturbing them.

Much commoner is the Sparrowhawk, which is one of the good news stories. After suffering badly from organo-chlorine persistent pesticides in the late 1950s, the species has largely recovered post 1971 once these were banned and now frequents most of our woodlands, favouring mixed and coniferous woods for nesting. This is a case of the bird being regarded as a particular Peak District species before it spread into adjacent lowlands.

MAMMALS

Like many of the birds, some ubiquitous mammals like small rodents, Grey Squirrel, Fox and Common Shrew are common and widespread upland woodland occupants and indeed, in most other woodlands as well. Some like the moorland deer and Brown and Mountain Hares might seek woodland shelter or browsing in winter. Others, though, are characteristic woodland animals and deserve mention here. Badgers, which are widespread across the whole area except the highest moorlands above about 350 m, excavate their setts more often in woodland, but also in scrub patches and hedgerows. Setts may be a single hole or with multiple entrances and their size is indicative more of length of use than to the number of occupants. Analysis of Derbyshire Badger dung samples shows a typical dependency on earth worms, which will be at their most abundant in woodlands (Mallon *et al.* 2012). Our Badgers do, however, also forage for beetles and other invertebrates, fruit and small mammals as well as wasp or bumblebee larvae, seemingly immune to the angry adults when demolishing nests.

Badgers are under threat. The highest death toll is from traffic, more particularly in March and September coinciding with greater mating and territorial activity. However, a bigger threat in the future is the Government's controversial response to bovine tuberculosis (bTB) through culling, despite the scientific evidence suggesting this would make matters worse by removing territorial animals and encouraging more widespread movement, thus spreading bTB among badgers. Badgers have been shown to be responsible for only some 6 per cent (although this varies) of all new bTB breakdowns, with most derived from cattle to cattle contact. The response from the local Wildlife Trusts has been a badger inoculation programme, which is being closely monitored. This can reduce the chance of a badger testing positively by as much as 76 per cent. Despite this, an area of Staffordshire was added to the Government's cull programme in 2018 and threats are hanging over Derbyshire's Badgers in 2020, despite the scale of the immunisation programme there.

If Badgers are removed, this could have knock-on effects on ecosystem dynamics. Preliminary analysis (Food and Environmental Research Agency, 2011) suggested that Foxes could increase, Hedgehogs (*Erinaceus europaeus*) rebound from being predated by Badgers (and Hedgehogs are in decline), and Brown Hares could also reduce owing possibly to increased Fox activity. Ground nesting

FIG 141. Pipistrelle Bat hanging on a house wall below a roost site in daylight.

birds and other potential prey animals might also suffer from increased predator numbers, thus increasing pressure on them and altering the species balance. We wait to see how this story evolves.

Bats are another group of mammals that depend on woody habitats (Mallon et al. 2012). Some nine of the 17 British species live in the Peak District. All species and their roosts are protected against disturbance, damage or destruction under the Wildlife and Countryside Act, 1981 and the UK Habitats Regulations 1994 (as amended). Pipistrelles (Fig. 141), both Common and Soprano (Pipistrellus pipistrellus and P.pygmaeus) are the commonest bats associated with our Dark and South West Peak woodlands and trees. Common Pipistrelle is more widespread and feeds mostly along woodland edges and hedges, and around trees, venturing into woodland glades more in windy weather. Eating at least 3000 midges or equivalent a night, these bats do us a great service, whilst having an essential role in the food web. Common Pipistrelles venture more out of woodlands than their close relative, the Soprano Pipistrelle, feeding over a wider range of habitats, but mostly avoid the higher moorlands.

Brandt's and Whiskered Bats (Myotis brandtii and M. mystacinus) are much less frequently recorded. Most sightings are in the Derwent and Edale Valleys and the Staffordshire Moorlands. These both deploy a rapid, skilful flight to pick off spiders and moths from foliage, often near water. They have been recorded up to 410 m near Stanage and Hollins Cross. Natterer's Bat (Myotis nattereri), another woodland bat, is rather locally distributed with few sightings in the Dark Peak, but recorded in the Dane Valley. It often feeds with Pipistrelles and other Myotis bats and can take spiders from their webs among shrubs and tree canopies, often again near water. Leisler's Bat (Nyctalus leisleri) is a characteristic species, rediscovered in 1985 and well correlated with woods in the Dark Peak and Peak fringe (excluding Staffordshire), which are now regarded as a national stronghold. Although roosting in buildings or trees (the Chatsworth Estate supports a large roost), they typically feed in fast shallow dives, often in large circles, over woodlands and their edges (Mallon et al., 2012). Brown Long-eared Bats (Plecotus auritus) are probably the third most common in the area, but avoid the higher ground. They have a strong association with trees, particularly broad-leaved and mixed woods, wooded gardens, parkland and mature scrub. They typically feed close to the ground, gleaning insects from the leaves, listening for moths that are warming up rather than echolocating for them.

Many of these bats roost and breed in buildings, which do not have to be old, but rather support appropriate nooks and crannies, under weather boarding or tiles, or in loft spaces. Natterer's roosts are known in Sheldon, Hathersage, Chinley Head and Little Hayfield for example, while Soprano Pipistrelles cluster in large roosts of up to 1,105 animals. They also utilise bat boxes, so it is worthwhile erecting them on garden trees. Brown Long-eared roosts are known in mansions, halls, granges, farms, stables and large rural houses. Although roost sites are used traditionally, populations can split and move between buildings or parts of buildings as conditions (like temperature) change. It helps to keep cats indoors overnight as local bat groups find that cats are responsible for most puncture wounds detected.

Roe Deer frequent woodland more than Red and Fallow Deer, and are spreading, being now more frequent and widespread across Dark Peak woodlands than in the past and also appearing sparsely in the South West Peak. They utilise coniferous and broadleaved woodland and adjacent fields, where they are generally solitary or in small groups. Although a native species, they were reported as dying out in Derbyshire by the end of the 18th century (Mallon *et al.* 2012) and were not recorded again until 1974. They are likely to continue expanding their range, particularly in wooded valleys where habitat is suitable. Reeve's Muntjac (*Muntiacus reevesi*), an introduced small reddish-brown deer, has been expanding its range gradually northwards and our first sightings are not far from Bakewell. As a woodland and dense scrub species, they are difficult to see, but have a loud bark reminiscent of a Fox. High numbers in a wood damage the ground flora and check natural regeneration, although they can keep Brambles under control. They are not a welcome addition to our fauna.

One unfortunately ubiquitous mammal found in our woods and hedges is the introduced Grey Squirrel, but it was not long ago that we could also boast a small population of Red Squirrels, mostly in the Upper Derwent Valley inhabiting its conifer plantations and, earlier, in the Mid Derwent Valley around Padley Gorge. Last seen in 2000, they are now sadly extinct in the area, ousted by the Grey Squirrel, as in so many places, but also possibly predated by Goshawk at a time its population was expanding.

OTHER ANIMALS

There are literally thousands of other animals inhabiting our Dark and South West Peaks' woody landscape, mostly invertebrates, but space permits mention only of the more important or conspicuous that you might encounter. Of the insects, flies of various kinds, moths and beetles are probably the largest groups which are essential to the food web and interesting in their own right. The largest numbers of animals you will see readily, though are probably ants, in particular the Hairy Wood Ant. This is the only one of its genera found in the Peak District.

FIG 142. Hairy Wood Ants on their nest, a northern species found on the eastern side of the Dark Peak in and on the edge of Oak woods and some plantations.

They are long-legged, large ants that forage in shrubs and trees and build large nests of plant debris, which are obvious in many plantations and oak woodlands down the east side of the Dark Peak, but more usually in open sunnier spots, even extending outside woodland edges if favourable habitat is available (Fig. 142). Their nests are effectively mini compost heaps and the plant decay keeps the nests snug and insulated, particularly in winter. The species is at the southeast edge of its range and is nationally locally common in limited areas, but regarded as scarce or threatened in Britain as a whole.

Large groups of interconnecting nests may contain hundreds of queens and marching ants criss-cross the habitat following their pheromone trails. You cannot miss them, but recent research in our woods shows their lives are more complex than it first appears. They have a complex social structure whereby ants in some nests are part of the colony expansion around existing nests rather than foragers. New nests founded near a food supply begin foraging and become part of the colony. Some of the more distant nests begin foraging after some time while others are abandoned. This allows expansion into what turns out to be profitable habitat over time, as if testing the local resources before deciding to stay or move elsewhere (Ellis & Robinson, 2015).

The ants learn their environment through a mixture of light, shade and other features and can travel to the tree tops and re-find their nests on the ground. Their well-defined foraging trails are used by thousands of individuals and can remain in the same position for several years. Foraging consists of either collecting honeydew (for sugars) from aphids, which are carefully nurtured and protected, or capturing invertebrate prey (for proteins) in the canopy. Research in the North York Moors showed that although highly predatory, Hairy Wood Ants do not reduce overall invertebrate diversity, only numbers, particularly of spiders, beetles, a group of springtails, larval flies, centipedes and millipedes. There were more ants, wasps and bees, mites and another springtail group, with no changes in numbers of several other groups (Sudd & Lodhi, 1981). Such control could limit outbreaks of certain pests in the forests and demonstrates the importance of functioning food webs maintaining balances in the ecosystem.

The nests support many ant-dependent insects, especially beetles. Termed myrmecophily, it is particularly well developed in Rove Beetles. Some consume waste materials in the nest or remove fungi that invade. One such rove beetle, *Stenichnus bicolor*, a tiny brown beetle of local distribution was found in a Wood Ant's nest in Broomhead Wood in 2000 (Smith, 2000), although it also lives under wood bark. Any animal living with Wood Ants would need to be able to avoid stimulating formic acid production, which is part of the ants' defence behaviour. Holding your hand close to a nest (without disturbing it) while ants spray you readily reveals the strong vinegar smell. As they also bite (harmlessly but irritatingly), do not stay too close, particularly when wearing sandals!

The Sorby Fungus group has also found ten different fungi on Wood Ant nests, although the association is not well known and needs research. The mycelium must be tolerated in the Ant's nest, and could be associated with the composting debris. Species include False Chanterelle (*Hygrophoropsis aurantiaca*) and Shaggy Parasol (*Chlorophyllum rhacodes*), which may be associated with conifer needles combined into the nests, as well as several other smaller fungal species and a slime mould.

Butterflies are visible day-time representatives of our woodland insects, but few are closely associated with woodland. Speckled Wood (*Pararge aegeria*), a pretty brown and spotted cream butterfly (Fig. 143) flits around sunny glades, their larvae feeding on grasses growing more vigorously in the better-lit conditions. This species has moved north with climate change over the last two decades and has colonised the Peak District only recently, now being regularly seen. This expansion is evident when comparing the current distribution (now quite widespread) with the Butterfly Atlas produced by the Sorby Natural History Society in 1992, when it was absent at least in the eastern half of the Peak District. A much rarer species, Purple Hairstreak (*Neozephyrus quercus*), has been recorded in Oak woods, or even associated with a single Oak tree (Fig. 143). You might see it above several woodlands in the Derwent Valley such as Ladybower Wood or Stainery Clough, on the eastern edge of the Peak District beside Agden Bog, at Kinder Bank Wood Hayfield, along Kerridge and in the Dane Valley woodlands (Machin 2018). Chatsworth is also now a nucleus for it. The larvae feed on different Oak species at night and the adults dance high in the tree canopy seeking aphid honeydew; best seen by scanning Oak crowns with binoculars in late afternoon/evening sunshine in mid-summer. This is another new species that has colonised the Peak District from a gradual westwards expansion since its first sighting in 1995.

Moths are much more numerous than butterflies, but not so often seen or recognised perhaps, but woods are a prime habitat for them. The majority have larvae that feed on trees, rather than the ground flora, although this will vary

FIG 143. Woodland butterflies – the common and the rare: (above right)Speckled Wood butterfly in woodland glade (Steve Orridge); (right) Purple Hairstreak associated with Oak trees, flying from mid-July to late-August. (Alwyn Timms)

depending on woodland grazing levels. Very few are associated with conifers. Oak, Birch, Willow and Hawthorn are the most important food plants and the large populations of insectivorous birds in the woods depend on these caterpillars. Some moths prefer more open woodland, whilst others seek a denser canopy of varied structure or more luxurious ground cover. Each has its ecological niche. The woodlands also support a wide variety of common as well as scarce or rare species.

Moth recording requires access for light traps and other forms of trapping (which do not affect populations as specimens are set free after identification), but access is not straightforward. Knowledge of our moths therefore tends to be patchy. Some rarities have been noted though, exemplified by the records of Lunar Hornet (*Sesia bembeciformis*), which feeds on Willows and is a hornet mimic in Clough Wood SSSI, where at least 250 species of moth have been recorded. This SSSI includes several woodlands which are partly on the gritstones and shales and partly on limestone near Winster. Other rarities are the Beautiful Snout (*Hypena crassalis*), who's larvae feed on Bilberry in woodland or moorland, and Glaucous Shears (*Papestra biren*) with Heather or Willow-feeding larvae. The latter moth is a scarce species in the west and north of Britain exemplifying range edge again here.

Molluscs are inconspicuous animals in our woods, but play a major role in recycling dead plant material. Woodlands on the grits and shales in the Dark and South West Peaks hold very variable assemblages. Out of 17 woods in one sample, only two species, Hedgehog Slug (*Arion intermedius*) and Garlic snail (*Oxychilus alliarius*), a small, flat, rounded, glossy-shelled snail, were found in them all. The considerable heterogeneity in soil type and pH as well as wetness accounted for this. There were generally fewer species with increasing altitude and in the driest conditions. One species, *Zonitoides excavatus* (another small, coiled-shelled snail) was a clear calcifuge, but there was a larger group of moisture-loving species like *Vertigo substriata, Zonitoides nitidus* and *Succinea putris*, all of which also occur in marshes and flushed areas within the moorland (Tattersfield, 1990). Thus, upland woods with oozes and wet areas as well as stream edges are richer in molluscs than dry ones.

Four of the woodland specialists, *Leiostyla anglica* (a chestnut snail with slightly barrel-shaped shell) *Vertigo substriata* (a minute whorl snail), *Zenobiela subrufescens* (a small snail) and *Limax tennelus* (a Nationally Notable pale coloured slug) are reputed to be indicators of ancient woodland, but in our area, Tattersfield found the first two species in unwooded marshes and flushes as well. The other two are better woodland indicators and occur in Abney Cough and Shinning Cliff Woods (Derwent Valley south of Matlock). It could be that they all indicate long-established, stable habitats rather than just ancient woodland here.

Several species are local in their distribution, sometimes at their southern or southeastern range edges like *Leiostyla anglica* and *Zenobiella subrufescens*.

Woodlands are perhaps a surprisingly important habitat for hoverflies, mostly because many breed in them. In more dense woods, having flower-rich habitat on the edges provides a good food source for adults. Some hoverflies are excellent mimics, mostly of bees and wasps, thus deceiving potential predators, but minus the sting. The adults mostly feed on nectar and pollen so will be seen on flowers and are important pollinators, whilst others feed on honeydew excreted by aphids on tree leaves. Larvae are diverse, feeding on, for example, aphids, rotting wood, fungi, bulbs, dung or parasitizing other insects.

Records of hoverflies collated in 2019 for the Dark Peak woodlands and associated habitats on the eastern side of the area (R. Foster, pers. com.) show a remarkable overall diversity – 156 out of the British total of 282. Of 132 for which there are enough data, just over half are woodland species, although some are more catholic and frequent grasslands, scrub or a mixture of habitats as well, including gardens. The rest are more readily associated with wet habitats or grasslands. Of the woodland species, about a third are more regularly associated with woodland edges and glades, and six are regarded as ancient woodland indicators. Examples of common woodland species are the Stripe-winged Dronefly (*Eristalis horticola*) and Hairy-eyed Syrphus (*Syrphus torvus*), both of which have striking yellow and black markings appearing slightly wasp-like. The Orange-belted Hoverfly (*Xylota segnis*), sleek bodied with a black and orange abdomen, and the Blotched-winged Hoverfly (*Leucozona lucorum*), with a black and white abdomen, are also fairly common examples. Of the six ancient woodland indicator species, all are scarce or rare here or on a wider geographical scale. For example, the Forest Hoverfly (*Brachypalpus laphriformis*), is a rarely recorded strong associate of ancient woodland and veteran trees, while *Parasyrphus annulatus*, one of the Forest Syrphus group, another yellow and black-barred species, is a very local, mostly southern species.

The abundance of woodland edge species is perhaps a reflection of the mosaics many of our Oak woods form with grasslands, wetlands, heathland and moorland enabling better survival of woodland edge species. Most are common and widespread and some are strikingly marked. The Large Pied Hoverfly (*Volucella pelluscens*), with an ivory-white band across its middle and large dark spots on its wings, feeds on bramble flowers whilst its larvae develop in wasps' nests. The Pale Saddled Leucozona (*Leucozona glaucia*), with blue-grey abdominal bands, frequents Umbellifer flowers like Hogweed and Angelica. Its larvae, in common with those of many other hoverflies, feed on aphids on the woodland ground flora. The Bumblebee Hoverfly (*Volucella bombylans* var. *plumata*) is much more local and a convincing mimic of the White-tailed Bumblebee, which is common here. The scarce Big-thighed Pipiza (*Pipiza austriaca*), a blackish hoverfly with distinctive delta wings at rest, also frequents woodland margins. The Grey-backed Snout Hoverfly (*Rhingis rostrata*), which has a pronounced rostrum or snout, is a rare woodland relative of the Common Snout Hoverfly (*Rhingia campestris*). It is speculated that its larvae once fed on the dung of wild boar and deer when the High Peak was a hunting forest and that it now breeds on dung in badger latrines.

Woodland is rich in food sources for the development of hoverfly larvae. Narsissus Bulb Fly (*Merodon equestris*) larvae consumes Bluebell bulbs and the grey spotted Ramsons Hoverfly (*Portevinia maculata*) feeds on Wild Garlic. Woodland fungi, particularly the rotting fruit bodies of Boletes, are used by the closely related Truffle Cheilosias (*Cheilosia scutellata*) and *Cheilosia longula*. Tree stumps and their rotting roots provide breeding opportunities for several woodland hoverflies, notably *Xylota* species. The soft wet-rot which develops in the heartwood of live trees, particularly Beech, Elm and Ash, is favoured by rare *Criorhina* hoverfly mimics. Rain-water filled rot holes are commonly exploited by aquatic, rat-tailed hoverfly larvae, particularly those of the Batman Hoverfly (*Myathropa florea*), with pale thoracic markings reflecting a Batman mask. The extendable breathing tube of rat-tailed larvae probes the surface to access oxygen in anaerobic pools.

Sap runs from wounds in tree bark are a rich source of sugars and proteins exploited by some hoverflies and similarly, the sap and cambium layer under the bark of recently felled trees is consumed by the larvae of specialist species: the Dark-shouldered Brachyopa (*Brachyopa pilosa*) being a relatively rare one sometimes found near felled Beech in the Upper Derwent Valley, Chatsworth and Dove Stones. Similarly, the rotting cambium under the bark of fallen branches and even twigs on the woodland floor is exploited by tiny *Sphegina* Club-tailed Hoverflies.

Wet wood or other wet habitats attract some hoverflies, several of which are uncommon and local like the Small Forest Hoverfly (*Chalcosyrphus nemorum*) breeding on the rotten sap under the bark of fallen trees. Marsh-marigold (*Caltha palustris*) flowers attract hoverflies like the White-barred Peat Hoverfly (*Sericomyia lappona*), a Palearctic species that also has rat-tailed larvae. Bearing in mind how limited our wet woodland is, this emphasises its importance for different invertebrates.

Seven hoverflies have been recorded in conifer woodlands, several of which again are rare in the area like *Scaeva selenitica*, a large distinctive fly and *Eriozona syrphoides* a large bumblebee mimic. Both are recent introductions to the UK. They occur in Norway Spruce plantations and may have been introduced from

FIG 144. The very rare Furry Pine Hoverfly found recently in conifer plantations at Dove Stone and at Longshaw. (John Leach)

Northern Europe with imported saplings. Recent discoveries of the Furry Pine Hoverfly (*Callicera rufa*) in conifer plantations on the Dark Peak both on Dove Stone and Longshaw Estate (Foster & Leach, 2019) were also unexpected, being the first records for their counties (Fig. 144). The species was, until recently, thought to be confined to remote Scottish conifer forests. They could be 20th-century introductions or unobtrusive residents that have passed unnoticed for years. Conifer stumps were gouged out at Dove Stones to create artificial rot holes simulating its breeding sites, which the hoverfly then obligingly colonised (Gartside, 2017).

This expanded account demonstrates the importance of just one insect group. Equivalent data are lacking on most others, but the hoverfly story is bound to be repeated for many others over time. There is much still to be discovered.

CHAPTER 11

Upland Grasslands and Wetlands

W E EXPLORE HERE THE ACID grasslands, rush pastures, reservoirs, rivers and other wetland habitats in the Dark and South West Peaks.

ACID GRASSLANDS

The general character of these were described in Chapter 9. The Mat-grass dominance and more diverse acid grassland with Bents, Fescues and herbs like Tormentil and Heath Bedstraw also occur in many of the enclosed pastures but here there are more of the richer versions, often on steeper slopes that have escaped agricultural improvement (Fig. 145). Species like Autumn Hawkbit provide a late summer splash of yellow dandelion-like flowers on drier sites, whilst Devil's-bit Scabious and occasionally Betony (*Betonica officinalis*) tempt butterflies to feed. Sheep's Sorrel (*Rumex acetosella*) provides food plants for Small Copper Butterflies (*Lycaena phlaeas*). More colour may be provided by Selfheal, Harebell or Cat's-ear (*Hypochaeris radicata*) and finding Mountain Pansy (*Viola lutea*), in either its indigo or yellow colour form, is a special delight. There are local community variants with more Yorkshire-fog (*Holcus lanatus*) and White Clover in some more base-rich soil patches contrasting with greater proportions of Wavy Hair-grass and Bilberry on drier, more acidic soils. Wood Anemone may appear out of place (being a woodland plant mostly) in some grasslands, whilst finding the single, lobed leaf of the little fern Moonwort is a treat.

The value of the Bent-Fescue community in the Dark and South West Peaks is greatly enhanced by supporting important grassland specialist fungi. Such fields may not be flower-rich and they may have been dismissed previously as of little nature conservation value. But those not agriculturally improved through

FIG 145. A flower-rich pasture below Gradbach Hill, featuring Selfheal, Bird's-foot-trefoil, White Clover, Common Cat's-ear, Eyebright, Tormentil and Ribwort Plantain amongst others.

liming or fertilising and which are mossy and generally short in autumn through grazing (after hay cropping or as permanent pastures) can support the largest assemblage of waxcaps and their associated fungi. Waxcaps produce often brightly coloured, generally quite small Toadstools. They have waxy or slippery-looking caps, thick waxy gills and are thought to be largely saprophytic (feeding on dead or decaying organic matter in the turf), although some may be mycorrhizal possibly associated with mosses such as Springy Turf-moss *Rhytidiadelphus squarrosus* and Ribwort Plantain. They are part of a grassland fungal assemblage that includes pinkgills (*Entoloma*), earthtongues (*Geoglossum* and relatives), spindles, club and coral fungi (Clavarioids), and crazed caps (*Dermoloma* and relatives). These groups are collectively labelled as CHEGD (the initials represent the different genera), although recent DNA investigations split the *Hygrocybe* genus.

Waxcap grasslands have been widely lost owing to agricultural improvements and to nitrogen pollution, and nearly 90 per cent of all waxcap species are on one or more European national Red Lists for threatened fungi. This emphasises their British importance and the very significant contribution our sites make to their conservation. The Peak District is not unique: new sites are being found for

FIG 146. Parrot Waxcap showing typical green colouring and sticky surface, growing in acid grassland on a steep slope that has escaped agricultural intensification.

example in the Northumberland National Park and across the Welsh uplands, but significantly they nearly all focus on unimproved upland acid grasslands. The threatened state of these and their CHEGD fungal assemblages was only realised in the 1980s. Subsequently, surveys showed that Britain has more valuable waxcap grasslands than any other European Country and the Dark and South West Peaks have more than their fair share, with new sites being found annually and plenty more to be surveyed.

Diverse waxcap assemblages occur on farms in the Upper Derwent, the Edale Valley, the Ashop and tributary valleys (like Alport Dale), at Longshaw, around Abney, Great Hucklow Gliding Club grasslands, Lyme Park, several sites on the Cheshire side of the Peak District, at Tegg's Nose, Macclesfield Forest and Kerridge Hill, and in the South West Peak in various places including near Morridge and Hollinsclough. They also occur on old lawns as at Chatsworth and in a few cemeteries as in Ashbourne. Even vestiges of former better quality waxcap grasslands on steeper slope fragments between agriculturally improved fields can be locally important.

A warm summer and wet autumn will herald a good Waxcap year like 2020. One of the commonest is the orangey-brown Meadow Waxcap (*Cuphophyllus pratensis*), growing in small troops amongst the grasses. Parrot Waxcap (*Gliophorus psittacinus*) is smaller, sticky, glistening and always showing some greenish colouring (Fig. 146). The Scarlet Waxcap (*Hygrocybe coccinea*) has a moist, domed cap and red or yellow gills, whilst the Golden Waxcap (*Hygrocybe chlorophana*) reflects its name. Snowy Waxcap (*Cuphophyllus virgineus*) is another widespread species with a white (usually), moist cap and stem. Some species, such as the Heath (*Gliophorus laetus*) and Splendid Waxcaps (*Hygrocybe splendidissima*), thrive better in acid sandy soils overlying upper gritstones slopes, whilst others like the Pink or Ballerina (*Porpolomopsis calyptriformis*) (Fig. 147) and the Oily Waxcaps (*Hygrocybe quieta*) prefer more neutral areas with deeper soils, often towards the bottom of shaley slopes. Citrine Waxcap (*Hygrocybe citrinovirens*) prefers wetter conditions, whilst the Egg-yolk Waxcap (*Gloioxanthomyces vitallinus*) favours peaty soil at moorland edges.

FIG 147. Pink or Ballerina Waxcap with conical juvenile fungus and older specimen with 'skirt' opening out growing on acid grassland on steep slope in South West Peak.

FIG 148. Young Crimson Waxcap, one of the larger species, indicative often of a good waxcap site, growing in mildly acidic pasture on steep slope.

The NPA has produced a list of about 20 rare or endangered waxcaps of semi-natural grassland, some of which are regarded as indicators of high value sites. These include the Crimson Waxcap (*Hygrocybe punicea*), which is generally much larger than those already described (Fig. 148), the brown-capped Dingy Waxcap (*Hygrocybe ingrata*) and the Nitrous Waxcap (*Hygrocybe nitrata*), marked by the spent gunpowder smell of fireworks.

We have some particularly exciting new records (Rob Foster, pers. com) such as the Jubilee Waxcap (*Gliophorus reginae*) and *Cuphophyllus lepidopus*. The former is a dark, plum-pinkish fungus named to commemorate the Queen's jubilee and 60th anniversary of her coronation. It was discovered in the Edale and Ashop valleys, but a search in Kew's fungal herbarium revealed earlier collections – the first being from the Staffordshire Peak District, where it was re-found in 2020. Despite surveys, it has seldom been found elsewhere although new Peak District sites are being found in 2020. The second, similar to Earthy Waxcap (*Cuphophylus fornicata*), was discovered in Alport Dale by Neil Barden in 2004. The species had been missing from the British list for some

FIG 149. Meadow Coral (*Clavulinopsis corniculata*) in more acidic grassland on steep slope, South West Peak.

years. It is very rare (18 known records according to a Kew study), only found worldwide in Great Britain and Northern Ireland currently, and is on the Global Red List.

Other members of the CHEGD fungal assemblage are equally important. Common Club and Coral Fungi include the White and Golden Spindles (*Clavaria fragilis* and *Clavulinopsis fusiformis*), the Yellow and Apricot Clubs (*Clavulinopsis helvola* and *C. luteoalba*) and the Meadow Coral (*Clavulinopsis corniculata*). There are at least 25 grassland clubs and corals of *Clavaria Clavulinopsis* or *Ramariopsis* genera to look out for (Fig. 149), varying in colour from rose, to violet, smoky or apricot. Amongst the rarer species are Violet Coral (*Clavaria zollingeri*), Rose Spindles (*Clavaria rosea*) and Beige Coral (*Clavulinopsis umbrinella*). The Violet Coral, on the UK and European Red List, is found in the Edale and Alport Valleys and Longshaw. The Straw Club (*Clavaria straminea*) is a nationally restricted spindle found on Chatsworth's lawns and parkland and recently in the South West Peak. Again, to date, Longshaw and fields in the Alport Valley stand out as the prime sites for this group.

FIG 150. *Microglossum olivaceum*, the rare Olive Earthtongue. Neutral to mildly acidic grassland. Occurs in Longshaw Estate and White Peak.

Some pinkgills (a large group) have some bluish, lilac, violet or bluish-grey colouring, whilst others are more dull-coloured. They have pale, crowded gills in their mushroom-like cap and are difficult to identify. A considerable number have been recorded in the Dark Peak, particularly on the Longshaw Estate. The Earthtongues (*Geoglossoid* fungi) are simple tongue or club-shaped structures which are blackish, green, purplish or even dark red. *Geoglossum* and *Microglossum* have smooth fruiting bodies whilst the *Trichoglossums* are covered with tiny bristles (visible through a lens). All the Microglossum species are rare here (Fig. 150). Crazed caps (*Dermoloma* and similar) are dry-capped mushrooms with cuticles that crack in a crazy pattern when flexed. *Dermoloma cuneifolium* is quite common in the Dark Peak, although others also occur, like *Dermoloma magicum*, which blackens when bruised.

Date Waxcap (*Hygrocybe spadicea*), Big Blue Pinkgill (*Entoloma bloxamii* s.l.) and Olive Earthtongue (*Microglossum olivaceum*) are sufficiently rare and threatened to have their own Biodiversity Action Plans. Date Waxcap, with striking brown cap and yellow gills, prefers dry, warm, south-facing slopes and has been found in the Ashop and Upper Derwent Valleys (R. Foster). Big Blue Pinkgill and Olive Earthtongue status has been muddied by recent DNA sequencing splitting them into more species, all of which occur in the Dark Peak. Other rarer grassland fungi include more Pinkgills and Orange, Citrine, Yellow Foot, Dingy, and Fibrous Waxcaps (*H. citrinopallida, H. flavipes, H. ingrata* and *H. intermedia*). They are mostly found in the Upper and Middle Derwent Valley grasslands, which have been intensively surveyed. New, rich waxcap grassland sites were discovered recently in the South West Peak as part of the HLF Landscape Partnership programme, A Landscape at a Crossroads.

Waxcap sites are ranked by the total present, preferably from more than one visit, although the scoring systems have recently been updated (JNCC, 2018). Under the old system, 22+ taxa of *Hygrocybe +Dermoloma* was the threshold for internationally important sites and 17–21 for national importance. Longshaw Estate was ranked the best in the country (Evans, 2004), with 33 taxa, although further surveys increased this to 35 of the approximately 60 UK taxa. This easily qualifies it as Internationally Important along with Alport Valley grasslands with 30 Waxcap taxa (H+D) and a site near Harpur Hill bridging the acid/calcareous grassland divide with 30 taxa also important. The new scoring system includes all CHEGD groups, rather than just waxcaps and crazed caps, and Longshaw and Ashop Valley fields along with the Hucklow gliding club grasslands still qualify as internationally important. New nationally important sites are being discovered in the Staffordshire Moorlands in 2019 and 2020 too, with many more of regional value.

This account emphasises the significance of our high value, acid grassland fungal assemblages and how proud we can be of farmers and land owners working to protect them long term. The waxcap assemblages are beautiful and cheerful additions to any visit but are best left for others to see as they are not worthwhile eating and some may indeed be poisonous. It is increasingly urgent that the value of good waxcap sites is recognised and that they are surveyed, evaluated and adequately protected to avoid pressure, for example, to plant trees on them.

There are other grassland fungi that live on dung (for example Dung Roundhead, *Stropharia semiglobata*, with sticky, hemispherical caps), decompose litter, or are mycorrhizal with different grassland plants. There are a number of *Galerina* species, fragile little Toadstools called bells, and mottlegills (*Mycena* and *Panaeolina* spp.) for example. Scarlet Caterpillarclub (*Cordyceps militaris*), is a remarkable, not uncommon species, with a 1–3 cm red, spindle-shaped fruiting body apparently rising from the soil, but which actually erupts from an underground mummified caterpillar or leather jacket.

RUSH PASTURES AND MARSHES

Unlike the acid grasslands, rush pastures do not support significant fungal assemblages. Their value lies more in sustaining some of our declining wading birds, but they have an interesting and generally little known history. Rush pastures typically support abundant Soft Rush and occupy ill-drained, clayey soils mostly overlying shales, especially coal shales, but also shale beds in the gritstone series. Rush pasture vegetation is adapted to the high moisture levels in soils that frequently exhibit gleying (see Chapter 7), where winter waterlogging occurs, and they are particularly widespread in the South West Peak (Fig. 151).

Grazing animals, especially sheep, avoid Soft Rush, so it can be prominent and sometimes develops into fairly impenetrable masses. Farmers dislike it as it displaces more palatable forage. So why is it so abundant? We should distinguish between more generally limited marshy areas where Soft and other rushes hold

FIG 151. Rushy patches (mostly Soft Rush) in fields on the shales and grits on the southwest side of the Upper Dove Valley contrasting with the limestone grasslands of the hills in the background northeast of the river where there are no rushes.

a key place in the sward and large areas of pasture which are not completely wet but where Soft Rush is still conspicuous. Soft Rush and some of its cousins produce large quantities of viable seed that are readily dispersed by animals or feet, when they are wet and mucilaginous, and can then remain dormant until colonisation gaps appear. If rushes are avoided by grazing animals but grazing levels are high, the surrounding shorter, well grazed swards will provide more opportunities for rush establishment.

There will have been some rushy pastures for centuries, but these could have expanded significantly after disturbance. Agricultural drainage schemes for example could have been a major factor. It was not until the 17th century that underdrainage was widely established using trenches filled with stone or heather. Clay pipes and tiles became available towards the end of the 18th century and underdrainage became widespread after the drainage tax was removed in 1826. Farey (1815) lists many places where he noted recent drainage, including Great Hucklow, Fenny Bentley, Pilsbury, Rushup Edge, Sheen and Stanton; some combined with enclosure. Farey's only rule was 'to lay the land permanently dry at the least expense'. Think how this destroyed the flower-rich grasslands and marshy places along the way. Field drainage continues to this day, with frequent repair of the old stone soughs and drains.

More opportunities for rushes appear when you add pasture re-seeding involving soil disturbance. This started in the 18th century as agriculture was changing from a subsistence to a capitalist economy to feed an increasing population. Manuring and growing clovers to add nitrogen to soils started at about the same time, thus increasing productivity. The Dig for Victory campaign during the Second World War would not have helped either. Virtue (1976), farming below Rushup Edge, complained that he was forced to plough eight acres (3.24 ha) at 1,200 feet altitude (360 m) contrary to local knowledge and accepted practice. The heavy ploughs provided got bogged down and little would be harvestable at that height in any case – previously only oats could be grown. Grants were available for drainage as well (Gregory & Hines, 2019). Applied to the Dark and South West Peaks especially, this policy, albeit with national merit, ignored local circumstances and probably resulted in still more rushes. Ill-drained, clay-rich soils are very vulnerable to trampling damage and are easily poached, resulting in more rushes invading so recent increases in stocking levels and heavier breeds would not help. Farmers are generally aware of these problems and often avoid cultivating wet clay soils.

There is evidence from the West Pennines, probably applicable here, that much more diverse grasslands typically supporting wet meadow communities based around Crested Dog's-tail (*Cynosurus cristatus*) and Marsh-marigold (Fig. 152), but

FIG 152. A wet pasture with masses of Marsh-marigold and Lady's Smock in the South West Peak.

which are quite rich in other plants like Meadowsweet, different buttercups and clovers, have been converted to rush pastures through agricultural improvement. All the factors listed combine to deliver successful invasions especially of Soft Rush on our ill-drained soils. Once established, they are difficult to eradicate. Grazing does not reduce them, they are quite resistant to herbicides, and only complete drainage, reseeding and fertilising along with destruction of the original sward can convert the site into a Rye-grass (*Lolium perenne*) pasture leaving little of wildlife value.

That rush abundance is an historic feature is reflected in their past use and local traditions, some still persisting. Rushlights used Soft Rush, cut in summer before the leaves turn brown, with the outer epidermis stripped off carefully leaving a narrow vertical strip to give rigidity. This bared pith was then dipped into any available household fat or grease (bacon grease or mutton fats) after drying. Small amounts of bees wax would help the rushlight burn longer.

These were regarded as quick and cheap to make, and William Cobbett wrote that the rushlight 'was believed to give a better light than some poorly dipped candles'. The lights were held in a metal clasp at a 45 degree angle – held at right angles, they burnt more quickly, and vertically, they were dimmer. Each lasted only minutes, or at the most, an hour, so many rushes were needed for darker evenings. They would still have been used at the end of the 19th century in rural England and had a revival during the Second World War.

Strewing was a second use for rushes, which we know was adopted locally. Rushes, plus various fragrant or sweet smelling plants like Meadowsweet (try smelling their crushed leaves), were cut and used as an early 'carpet' on stone or compacted earth floors; replaceable when soiled. Gerard in *The Herball*, 1633, enthusiastically supported strewing herbs 'for the smell thereof makes the heart merrie, delighteth the senses'. Not only houses, but churches and other buildings would have been strewn. Rushbearing is an old English ecclesiastical festival involving collecting and carrying rushes for strewing in the Parish Church. The festival was widespread in the Middle Ages but had declined by the beginning of the 19th century. Descriptions of the ceremonies at this time in Chapel-en-le-Frith and Glossop tell of cutting and drying rushes before piling them high on a rush cart decorated with wreaths of flowers surmounted by a garland, streamers and flags. Villagers accompanied the cart with much whip cracking, dancing and music before everyone carries the bundles into the church for spreading. Afterwards, the various parties retire to the village inn, where they 'spend the rest of the day in joyous festivity' (Glover 1829).

This old practice is still celebrated annually in Jenkins Chapel (sometimes called Forest Chapel, Fig. 153) a pretty little gritstone church. Although its history is obscure, an early chapel probably graced this site in 1673, replaced by the current listed building in 1834. It lies on an old salter's route across the Peak District (see Chapter 6), at Saltersford. The rushbearing service, on the first Sunday after 12 August, is a symbolic spiritual renewal, with plaited rushes interwoven with flowers, plus rushes strewn over the Chapel floor. You can see a similar event in Saddleworth, where the local Morris group has revived the highly decorated, 3 m high towering rushcart that is dragged around the local villages by two-man teams with a third Morris man perched on top before strewing at St Chad's Church, Uppermill, along with much gurning, dancing, wrestling and singing!

It is clear that rushes were cut widely for different uses, including hay stack thatching (Virtue, 1976), which would have helped control their vigour and spread. These uses have largely disappeared, but rush patches are being cut again under agri-environment agreements to improve pasture structure for breeding waders, although cut rushes no longer have any value.

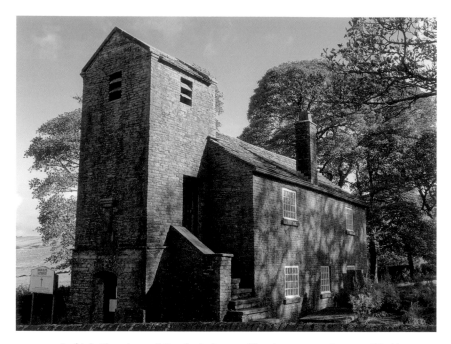

FIG 153. Jenkin's Chapel near Saltersford where rushbearing ceremonies are still held annually in August.

Rush pastures' character varies mostly related to wetness and the degree of agricultural improvement. Drier fields hold more Yorkshire-fog along with mostly Soft Rush, and these may appear in a mosaic with more agriculturally improved grassland marked by increased Crested Dog's-tail, Sweet Vernal-grass and Rye-grass. The rush mixture frequently also supports other grasses like Creeping Bent (*Agrostis stolonifera*) as well as Creeping Buttercup (which prefers wetter conditions). These communities change to wetter marshes where water tables rise either related to springs and oozes (similar to the variation in flushes in the cloughs) or proximity to water bodies. The heavier the grazing, in general, the lower the abundance of additional species, but with lower density cattle grazing more can spread and flower. Some reflect the moorland flushes like Marsh Bedstraw, Lady's Smock, Common Sorrel, Marsh Violet and Greater Bird's-foot-trefoil, but some are more frequent in these lower altitude marshy patches such as Angelica, Common Valerian (*Valeriana officinalis*), Sneezewort, Marsh Arrowgrass and Ragged-robin.

Some of these have interesting histories. Many wild plants were fundamental parts of herbal remedies or household resources. Common Valerian, for example, with sweetly scented, pale pink flowers heads that attract many hoverflies and other flies, has been a medicinal herb since ancient Greek and Roman times. Its esteem was so high in medieval times it was called 'All Heal' and is still used to treat mild nervous tension and aid sleep. It was grown in fields near us around Chesterfield for commercial use at least from the end of the 18th to the mid-20th centuries (Farey, 1815, Grieve, 1976). Farey described how the tops were removed to prevent flowering and the roots (recognisable by their offensive smell) carefully dug up, cleaned, slit down their thickest part, dried in a kiln and packed tight for selling by the hundredweight. Angelica stems were used to make crystalline angelica with the rest of the plant also highly valued for herbal use.

Where more base-rich water oozes through the soil, Sharp-flowered Rush (*Juncus acutiflorus*) appears, which is much more palatable to sheep and cattle. Running fingers down the articulated leaves reveals thin septa that bridge the otherwise hollow leaf-tube. Hollow stems are a common marsh plant adaptation enabling oxygen to reach roots in waterlogged soils. Marsh Horsetail (*Equisetum palustre*), and Greater Bird's-foot-trefoil, which both grow in many of these wetter areas, have similar hollow central tubes. Rarely, Wood Horsetail (*Equisetum sylvaticum*) (Fig. 154) also appears, identifiable by its finer branching compared with other Horsetails, and may accompany Marsh Valerian, a smaller version of its larger cousin. Much more abundant are the tall stems and purple flowers of Marsh Thistle; a major nectar source in these wet rush pastures.

FIG 154. Wood Horsetail, a plant of wet flushes and marshes, often not in woodland. Note the whorled branches which are also branched. It grows in Jubilee Pasture on the Longshaw Estate amongst other wet places.

Look out for other plants where these wet areas merge into true marshes like Common Spearwort (*Ranunculus flammula*), or Water Mint, with its familiar pungent smell and lilac flowers attracting many butterflies. It has long been used in herbal medicine as an emetic, stimulant and astringent (Grieve, 1976). Lady's Smock (Fig. 152), another common occupant of rushy pastures and wet grassland, provides a food source for Orange-tip Butterfly larvae (*Anthocharis cardamines*). Owing to its early appearance and then dieback, Lady's Smock, often along with Lesser Celandine, still appear abundantly in many grasslands that are otherwise agriculturally improved, having avoided late season herbicide application.

A dramatic, uncommon plant to look out for is Greater Tussock-sedge (*Carex paniculata*). The leaves droop down from the pouffe-sized tussocks in winter, making it conspicuous. Look out for them along the north side of the Barmoor Clough road to Sparrowpit, in the ancient landslips northwest of Mam Tor or from the Macclesfield Road just west of Buxton. Alongside streams and where there is less grazing, some rush pastures also contain scattered willow bushes and Alders. These add considerably to the structure of the habitat, particularly suited to birds and invertebrates. There are many more species in the best rushy grasslands, so are worth exploring for any aspiring botanist. The Longshaw Estate rushy pastures are good places to start, but multiple different footpaths in the South West Peak cross this habitat too.

A very special habitat that looks rather like a marsh from a distance is a flood plain mire – part fen/part bog. There are only two examples, neither with public access. One is Moss Carr, Hollinsclough, an SSSI visible from nearby minor roads. It lies on peat developed over mudstones in the upper reaches of a Manifold River tributary. Apart from a thin Sallow and Hairy Birch scrub, there is a range of bog plants like *Sphagnum*, cottongrasses and Cranberry typical of the peaty soil. More base-rich water passes through thus supporting a wide range of typical marsh plants like Water Horsetail and Sharp-flowered Rush, but there are also rarer species like Heath Spotted Orchid and Bogbean.

The South West Peak wet pastures are the breeding wader stronghold, with Snipe, Curlew and Lapwing the principle species, although they all occur too on open moorlands. However, all have declined sharply in recent decades. See Chapter 9 for the Curlew story. The acrobatic, tumbling display ornamented by the 'peewit' cry of Lapwing (*Vanellus vanellus*) heralds spring (Fig. 155), but this is another species of high concern in the Peak District, reflecting its nationwide status as a Red Listed species owing to the enormity of national losses. These are being repeated on a wider scale in Europe too. On the Sheffield side, Lapwings are still on the moorland fringes, but the declines are seen in the post-breeding flocks which regularly approached 700 and now rarely reach 100, for example at Redmires Reservoirs (Wood & Hill 2013). An RSPB survey in 2002 revealed a population of 1,213 pairs in the whole Peak District. Particularly high concentrations were found across the Rowarth-Cown Edge-Chisworth area in

FIG 155. Lapwing or Peewit at Redmires. It favours short vegetation, fields with scrapes or pools and wet fields. (Kev Dunnington)

the northwest with some 97 pairs, and the Harewood Moor area in the east with 59 pairs. Unimproved pasture was the most important habitat followed by rush pasture. Further surveys in 2007 suggested a halving of the total numbers in just five years – an alarming disappearance of a much-loved bird.

Research has shown that low productivity with few young fledged per pair contributes to the population decline, the reasons for which relate to various factors like predation and high grazing levels that result in trampled nests. A study in Glossop in 2000 showed that only 12 young fledged from 30 Lapwing pairs owing to predation by Carrion Crows (*Corvus corone*) and domestic cats (Frost & Shaw, 2013). It is also possible that soil compaction, already mentioned, is having an effect on invertebrate populations and diversity (see also Chapter 12 on the effects of slurry). The Peak District NPA and Natural England's joint Wader Recovery Project, operational up to 2015, worked successfully with farmers to enhance management for Lapwing and other waders and additional projects are furthering the results through wader scrape

FIG 156. Skylark at its nest near Ringinglow. They prefer grassy vegetation on moorland and in fields, where they nest amongst more tussocky vegetation to provide cover. (John Lintin Smith, Sorby Natural History Society)

creation and rewetting as part of natural flood management measures. Predator control seems to be essential where bird densities are so low.

The charismatic eerie fluting or drumming of breeding males early or late in the day reveals Snipe in the wetter rush pastures. They need damp organic soils where their long bills can probe for insects and molluscs, but pasture soil compaction and drainage could be an issue as Snipe numbers have been declining nationally and regionally. They rarely breed now in the Derbyshire lowlands, making the Peak District an important refuge (Frost & Shaw, 2013), but declines on the Cheshire moors (Norman, 2008) and on the Sheffield upland pastures and valleys are concerning (Wood & Hill, 2013).

The rest of the rush and acid grasslands breeding bird assemblage is rather limited. Meadow Pipits and Skylarks (Fig. 156) may be frequent and the Peak District is important for both compared with recent losses in the surrounding lowlands. Reed Bunting is likely to be the other mainstay of rush pastures, although they have been spreading into drier habitats such as Bracken beds and even up to about 490 m altitude (Frost & Shaw, 2013). Other more sporadic breeding birds include Grasshopper Warbler (another Red List species), which requires thick cover for nesting, and Whitethroat (*Curruca communis*) where scrub is denser.

There are hundreds of invertebrates associated with these wet grasslands, but very little understanding of their ecology. A few stand out apart from those already mentioned. Cranefly larvae are a key occupant and provide food for many of the breeding waders. The Hoverfly surveys described in Chapter 10 also reveal their importance in wet grasslands. Out of the 132 species recorded in the eastern side of the Dark Peak, 22 were associated with grasslands, often wet or damp ones, and a further 16 linked to various kinds of wetland habitats. Not all these will occur throughout these habitats, but the numbers indicate the high value of grasslands for hoverflies at least. Some of the species are widespread and common like the Broad-banded Epistrophe (*Epistrophe grossulariae*), often found on Angelica, and the ubiquitous relatively small, orange and black-banded Marmalade Hoverfly (*Episyrphus balteatus*). Others such as the White-spotted Big-handed Hoverfly (*Platycheirus rosarum*) (Fig. 157) and the small, shiny metallic *Lejogaster metallina* are associated with rushes and marshy land. Some of these like the rather small Red-horned Chrysogaster (*Chrysogaster virescens*) are much rarer and may indicate high-value wetland sites.

FIG 157. The White-spotted Big-handed Hoverfly is associated with rushy and wet grasslands. (Rob Foster)

RESERVOIRS AND LAKES

Lakes are rare, with only four in the Dark Peak and none in the South West Peak, and are mostly in large parks (Lyme, Chatsworth). There are also some millponds, but reservoirs are frequent. Most lie in the Dark Peak (63), with only 14 in the South West Peak. Altogether, the reservoirs cover 1,466.14 ha, which is significantly more than some of our scarcer habitats like ash woodland and hay meadows, thus placing these into context (Anderson, 2016). Chapter 6 describes our reservoirs' history. Few are renowned for their animal or plant life, mostly owing to the acidic, nutrient-poor water and sediments, although wintering and breeding birds are locally important. Reservoir edges are often steep and rocky, with drawdown zones linked to water usage and management. Many of the reservoirs are also quite large and therefore generate wave action, which can erode the edges, preventing plant establishment.

The main reservoir vegetation tends to be at the shallower inlet end where sediments accumulate. Soft Rush is ubiquitous, although there are high quality marshes at the inlets of some sites (as in Torside in Longdendale). Some of the most important flora is easily missed as it appears during prolonged drawdown levels in droughted summers as in 1976 and 2018. Todbrook and Combs Reservoirs, near Whaley Bridge and Chapel-en-le-Frith respectively, are SSSIs for the minute, ephemeral Dwarf Bladder-moss (*Physcomitrium sphaericum*) that carpets exposed muds. It was first recorded in 1893 and is rare, with its national distribution very much clustered around our area. It forms part of a community with other short-lived, diminutive mosses like Delicate Earth-moss (*Pseudephemerum nitidum*) and Spreading Earth-moss (*Aphanorrhegma patens*), with its spherical capsules, and the liverworts Common Crystalwort (*Riccia sorocarpa*), which grows in tiny irregular rosettes, and Common Kettlewort (*Blasia pusilla*).

Larger colonisers of exposed reservoir edges are Water-pepper (*Persicaria hydropiper*), its close relative Knotgrass (*Polygonum aviculare*) and Marsh Cudweed (*Gnaphalium uliginosum*), none of which are common in the Dark Peak. Much rarer are Mudwort (*Limosella aquatica*), a Nationally Scarce species, and the little annual Water-purslane (*Lythrum portula*). Mudwort frequents the margins of Combs Reservoir (Fig. 158) – one of only four sites in the Peak District. Water-purslane, in contrast, fringes Combs as well as some of the Upper Derwent, Dale Dike and Damflask Reservoirs (Middleton & Middleton, 2017). More typical wetland species growing in shallow reservoir edges include Common Water-starwort (*Callitriche stagnalis*). There are similar, less common Starwort species

FIG 158. Mudwort, Water purslane and Marsh Cudweed in a low-growing community with the tiny mosses in a period of drawdown in 2018 on the edge of Combs Reservoir, Chapel-en-le-Frith.

which are sometimes difficult to separate. The Flote Grasses, are similarly confusing, with Floating Sweet-grass (*Glyceria fluitans*) the commonest, often producing floating mats of pale green, longitudinally folded leaves. Alder and Willows may fringe the shallower inlets.

One very important invasive plant that constitutes a biosecurity issue is variously called Australian Stonecrop, or New Zealand Pigmyweed (*Crassula helmsii*) (Fig. 159). This was originally sold for garden ponds but has escaped into the wild. Although rare in the Peak District, it has, unfortunately, colonised the edges of Combs and Errwood Reservoirs amongst other water bodies. It quickly forms dense masses and can blanket out native species, especially where they are small and uncompetitive. It is very difficult to control without damaging the native flora. Its sale has been banned since 2013 in an effort to contain it. Since only tiny fragments can form new colonies, avoiding contact with it is important, and thorough cleaning of footwear and other equipment to avoid spreading it is imperative.

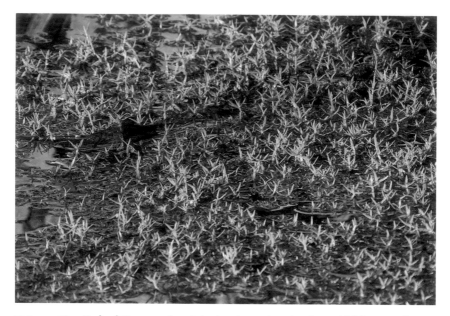

FIG 159. New Zealand Pigmyweed, an introduced, very invasive plant, which has spread into the edges of some reservoirs, meres and ponds.

Some reservoirs are managed as fisheries, such as Tittesworth, Errwood in the Goyt Valley and Ladybower in the Derwent Valley reservoir chain. These focus on Brown Trout (*Salmo trutta*) but may also support Rainbow Trout (*Oncorhynchus mykiss*), which will be dependent on the natural aquatic invertebrate populations. More on reservoir bird life follows the section below on rivers since many of the species share these habitats.

RIVERS AND OTHER WETLANDS

The Dark and South West Peaks are blessed with a wide range of rivers of differing size, force and length. These flow variously through the gritstones and shales or the more superficial boulder clay deposits in the lower valleys (Fig. 160). These will differentially influence the pH, calcium content and other chemical signatures and thus the aquatic animal life. How fast the water flows is also critical in determining its oxygen levels and the ability of plants and animals to withstand its force. The extent of agriculturally improved pasture in the catchments affects nutrient levels that leach or runoff into the water, giving a boost to algae and more vigorous plants, usually at the expense of the less competitive ones and high oxygen-demanders.

As well as rivers, the Peak Forest and Cromford Canals (Chapter 6 gives their history) hold much more slowly flowing waters providing a quite different habitat from the faster flowing streams and rivers. Finally there are ponds and millponds, although not that many ponds in the Dark and South West Peaks compared with the White Peak's dew ponds (Chapter 14). Ponds may be anywhere, as in the river valleys like near Stoke Brook, beside the Goyt, or in fields, or where reservoirs have been decommissioned as in the Lightwood valley near Buxton, plus some in gardens and school grounds. They all provide a network supporting wetland species, many of which can disperse between water bodies quite effectively.

Rivers are not a favoured habitat for plants, so most are confined to the edges or in small back waters. Look out for mats perhaps of Flote Grass, Common Starwort, Brooklime (*Veronica beccabunga*) with its pretty blue speedwell-type flowers, or one of the Marsh Forget-me-nots (*Myosotis scorpioides* or *M. laxa*). Bankside marshy

FIG 160. The Manifold River running through the shales and grits upstream of the White Peak, with planted Beech along one bank and a show of Bluebells and other flowers more abundant where protected from grazing.

depressions may accommodate Soft Rush, or, more rarely, dense swathes of Lesser Pond-sedge (*Carex acutiformis*) intermingled with the striking, tall stems of Great Willowherb (*Epilobium hirsutum*), Meadowsweet or the tall Reed Canary-grass (*Phalaris arundinacea*). The Willowherb is also called codlins and cream, reflecting possibly its pink and cream large flowers or the delicate cool fragrance of slightly bruised leaves reminiscent of scalded codlins (an old country name for apples).

The canals tend to support plants of slower flows like Reed Sweet-grass (*Glyceria maxima*), a large-leaved, luscious grass forming dense stands along canal and sometimes pond edges. This is a highly palatable species rarely found within grazed fields. Branched Bur-reed (*Sparganium erectum*) and Arrowhead (*Sagittaria sagittifolia*) grow in similar places but are commoner in the lowlands around us than in the Peak District. Ponds may occasionally support Broad-leaved Pondweed (*Potamogeton natans*) forming floating mats of elliptical leaves. Common Duckweed (*Lemna minor*), or more rarely, Ivy-leaved Duckweed (*L. trisulca*), can occlude the water with free-floating rafts of tiny plants. Each thallus or leaf, with its single visible root beneath it, can spread amazingly rapidly through budding daughter-plants, although they do produce inconspicuous flowers and seeds. Other invasive species include Bulrush (*Typha latifolia*), which can form dense stands in ponds and other shallow, still waters. It has dense, cylindrical dark-brown seed heads and thick rhizomes that are full of starch (to allow the plant to overwinter) and quite nutritious, boasting a protein content comparable to that of corn-on-the-cob or rice.

More noteworthy is the uncommon Ivy-leaved Crowfoot (*Ranunculus hederaceus*), with its little white, buttercup-like flowers and small leaves, creeping over marginal bare mud. Water-cress (*Nasturtium officinale*) prefers marshy areas, ditches and slow-moving small streams, but is not widespread either. Much too common though is Himalayan Balsam (*Impatiens glandulifera*). This large annual has pink, hooded flowers (likened in shape to a Policeman's Helmet – hence another of is names) and sticky glands, which smell strongly of balsam. It was introduced into Britain's gardens in 1839 (and then much of Europe) from the Himalayas and is highly invasive, owing to its sickle-shaped pods, which, on drying, explode, copiously dispersing seeds. It marches along stream edges and damp places by hedges, roads and fields, shading out the native flora, leaving bare ground all winter after it has died down that can then be susceptible to river-side erosion. It is particularly prolific around Chapel-en-le-Frith northwest to New Mills. Much effort is being expended clearing it systematically from river systems, starting at the top of catchments. Cutting or pulling (or strimming) patches is effective before they seed, although this also removes a valued nectar source. Biological control using a fungal rust is being trialled.

Many of our streams and rivers have been altered historically, straightening meandering rivers and over-deepening them to discharge excess water more quickly, exacerbated by field drainage. The Manifold below Longnor is a good example (Fig. 160), where the remnants of once substantial meanders are visible marooned well above current river level. Reputed to have been altered around 1770, this would have shortened the river's course, discharging water more rapidly, which then eroded a deeper channel. Where this sort of treatment occurs widely, the outcome is a greater potential for flooding downstream as well as damage to the river, its ecology and geomorphology.

Measures are now being introduced to hold water in safe places for longer in upper catchments, enhancing their wildlife value simultaneously, whilst reducing downstream flood risks. The methods are based on natural processes termed Natural Flood Management. You might wonder at leaky dams or woody debris in river channels. Leaky dams allow water underneath in normal conditions, but holds it back in higher flows, not in a watertight fashion, but slowing it down to delay downstream peaks (Fig. 161). Introducing large woody debris is another method. Slowing the Flow – one of the South West Peak HLF

FIG 161. Leaky dams constructed of wooden poles, woven willow branches and stakes over a small stream to help slow the flow, reducing peak flows downstream in the Black Brook catchment. Part of the South West Peak HLF Landscape Partnership programme.

Landscape Partnership flagship programmes is placing woody material across different headwater streams, manoeuvring or felling trees to add large woody debris to water, re-instating meanders where streams have been straightened and excavating new pools on line to hold water back. Staffordshire Wildlife Trust, amongst others, has also been adding large woody debris to the River Churnet near Tittesworth Reservoir, replacing gabions (steel mesh rock-filled baskets) with an engineered log-jam to protect banks and successfully improve river function. One special associated species recently found is the Logjammer Hoverfly (*Chalcosyrphus eunotus*), a Nationally Scarce species, which perches on and breeds in the woody debris. It is mostly recorded from eastern Wales and is rare here.

Not only are these measures cost-effective compared with engineered flood protection measures downstream, but they also enhance wildlife habitats, allowing streams to develop considerable structural improvements. Large woody debris might consist of root plates, trunks or branches and increase the physical structure of the stream forming more pools and riffles. Previously, any fallen trees would have been removed to secure rapid flows through the system, but although critical points like bridges have to be kept clear, there is great scope, as attitudes and understanding change, for allowing rivers to revert to a more naturalised form rather than just being conduits for water flow. If we had Beavers in the Peak District still, they would be engineering rivers for us in a very similar way, creating new and excellent habitats, good for fish and invertebrates.

ANIMALS OF WETLANDS

Many of the reservoirs and rivers animals are the same as those mentioned in Chapter 9, but with some important additions. Common Sandpiper (*Actitis hypoleucos*, Fig. 162) is one of the more important. Thought to be in decline throughout Europe in the 1990s, numbers have reduced too on some of our rivers but mostly remain along reservoir margins. Easily identified by their striking snowy-white underparts and constant bobbing motion, they favour extensive shingle beds and undercut banks produced alongside rivers or reservoir edges after winter spates and wave action. Drawdown zones provide suitable gravelly surfaces and rocks, behind which chicks can shelter. The Peak District is at the southeast edge of Common Sandpipers' main national distribution. Early research on their breeding biology in the Peak District informed the national knowledge-base through a valuable 20 year-long study observing and colour-ringing birds (Holland et al., 1982, Holland & Yalden, 2002). The majority of adults (especially males) were found to return to the same territorial area (a

FIG 162. Common Sandpiper at the water's edge proclaiming its territory.

200m strip along the water's edge) annually. Returning juveniles tended to join the breeding population in adjacent areas (four or more territories away) or in different valleys and did not necessarily establish territories in the same sort of habitat in which they were born (switching between river or reservoir edges). Survival was greater in mild April weather without late-spring snow storms. Hatching dates were earlier when it was warmer and growth rates appeared higher too. Production of fledglings was higher when there were more territories, although why was not clear – possibly related to increased predator-detection effectiveness. Storms in April 1981 and 1989 both affected breeding bird numbers and subsequent levels of recovery. Rain storm incidence during the breeding season affected river-side birds more than those along reservoirs, but the water supply needs and water levels in the reservoirs also determined breeding success, which varied annually. In 1998 and 1999 for example, there was very little water drawdown in the reservoirs, whilst in 1995 and 1996, after very dry periods, reservoir water levels were too low. All these factors affect Common Sandpiper population dynamics in the Dark and South West Peaks.

We have some idea of the range of invertebrates along the reservoir and stream edges from Common Sandpiper feeding studies (Yalden 1985). These found ground beetles, weevils and adult flies to be the most regularly occurring dietary items. Earthworms, click beetles, beetle larvae, stonefly nymphs taken from the water, spiders and rove beetles all appeared in more than 25 per cent of the samples, showing a broadly catholic diet. The main click beetles were the small species *Zorochros minimus* and *Hypnoidus riparius*. The predominant ground beetles were the shiny Black Clock Beetle *Pterostichus madidus*, *P. strenuus* and a range of others typical of upland riparian habitats. The principle spiders were one of the long-jawed orb-weavers *Pachygnatha degeeri*, wolf spiders like *Pardosa pullata* and money spiders.

Interestingly, the proportion of aquatic prey was low despite the birds frequenting water edges, although some of the flies would be breeding in water (mayflies, stoneflies, caddisflies and midges for example). The main aquatic larvae taken were Stoneflies *Plecoptera*, and Mayflies *Ephemeroptera*, the latter peaking after the former, both providing Sandpiper food at the breeding season peak. Of the other terrestrial animals, red ants (*Myrmica* spp.) were particularly abundant in the gravelly edges of the reservoirs and much pursued by chicks in particular. However, the abundance of the different groups varied between years, possibly associated with spates and floods that regularly washed out the habitat, but at different times of year.

In the 1977–80 period, 136 pairs of Common Sandpipers were recorded, with 77 pairs around reservoirs (which collectively have a longer perimeter) and 59 along rivers, mostly in the Goyt, Upper Longdendale and Derwent Valley complex, but also beside several others. More recent maxima show only 85 territories in the Derwent Valley in 2007 and ten pairs in the Goyt Valley in 2006 (Frost & Shaw, 2013), but none on the River Noe (Edale Valley) or Derwent rivers by 2013. In 2005, two pairs were displaying along the edge of Lamaload Reservoir in the South West Peak. The Ashop/Alsop rivers group also declined from 21 pairs in 1977–79 to 11 pairs in 2006 and only three pairs in 2009 – the lowest ever recorded. These changes are of great concern, but the causes generally unknown. One worry is recreational disturbance. Yalden (1992) found disturbance by anglers and other visitors forced Common Sandpipers to take flight within 75 m of an approaching human when with chicks, although only 27 m at other times, which resulted in birds avoiding the most favoured angling shores. Overwintering issues in Africa could also play a part. The longest distance ringed birds have flown is one controlled in Morocco having been ringed 2000 km away in Glossop in 1977. The longest lived bird so far from the study group is a male ringed as a chick in 1992 and re-sighted every year up to 2007 on the western arm of Ladybower reservoir.

Little Ringed Plover (*Charadrius dubius*), opportunists favouring bare ground for breeding, are much rarer than Common Sandpiper. They utilise bared

FIG 163. Female Goosander nests in holes in trees or broken off tree stumps although this one was photographed in November in a garden sycamore.

reservoir banks during draw-down and Longdendale is a favourite location but they have also appeared sporadically at Combs, Errwood and Ramsley Reservoirs. Combs Reservoir is also the main breeding ground for Great Crested Grebe (*Podiceps cristatus*) with occasional birds also in the Derwent Reservoirs. This species is more abundant in adjacent lowlands.

If you are lucky enough to see a handsome diving duck with a clearly serrated bill, this might be a Red-breasted Merganser (*Mergus serrator*). Before 1971 they were rare, but have spread from Western Europe, first colonising England in 1950 and breeding in the Goyt Valley in 1973. Since then, the species developed a stronghold in the Upper and Mid-Derwent Valley – shared here with another newly breeding sawbill duck, Goosander (*Mergus merganser*) (Fig. 163), with only rare sightings in the Goyt Valley. Their peak was in 1990–1994 when 19 broods were located. Since 2007, numbers have fallen with no breeding recorded along the Derwent in 2008. Reasons might relate to perceived competition with fishing interests (they are fish feeders) or to other factors such as climate change or predation (Frost & Shaw, 2013; Wood & Hill, 2013).

The Black-headed Gull (*Chroicocephalus ridibundus*) shares a similar story of shifting fortunes. First recorded breeding on Leash Fen in 1918, substantial colonies built up on Big Moor, on Howden and Broomhead Moors and at Redmires Reservoirs with over 200 pairs at several sites. In the 1950s and 60s numbers declined drastically until a colony re-established on Woodhead Reservoir in Longdendale from 1982 onwards. At its maximum of 600 pairs in 1997, the site was deserted by 2004 after several poor years, possibly affected by Mink predation. Numbers then rebuilt to 226 nests in 2011 (Frost & Shaw 2013, Wood & Hill 2013).

Other breeding birds are more widespread. Pied Wagtails (*Motacilla alba*) and Mallard are typical reservoir species. The Wagtail is found in a wide variety of habitats across the Peak District except the high moors, whilst Mallard can turn up in any water body, even frequenting large garden ponds sometimes. One of the largest heronries in the Peak District lies in Macclesfield Forest near Trentabank Reservoir, with over 20 pairs and birds feed widely including from garden ponds. We have already mentioned the introduced Canada Geese breeding on open moorland (Chapter 8). This ubiquitous species was first brought to Britain in the 17th century and large flocks were found at Chatsworth as early as 1820–30. Derbyshire had the dubious distinction of being one of the centres of the British population between 1967 and 1969. Since the late 1970s there has been a seemingly relentless expansion in range both here and in the country as a whole, with breeding first recorded in the Peak District at Combs Reservoir in 1975, Goyt Valley in 1979 and Longdendale in 1982. Numbers have expanded along the Derwent reservoir chain reaching a combined total of 731 birds by June 2004, even in areas that were regularly disturbed by recreation. Numbers have been partly controlled since, but successful breeding could be linked to predator control on the surrounding grouse moors. Large post breeding flocks appear on most of the reservoirs now with one of the largest winter flocks of 300 recorded at Combs Reservoir (Frost & Shaw 2013, Wood & Hill 2013).

Another introduced species, but less invasive and more striking, is Mandarin Duck (*Aix galericulata*) (Fig. 164). They hail originally from the Far East where they symbolise wedding bliss and fidelity and are frequently featured in Chinese art. A feral population established in the last century in Great Britain and numbers now some 7000 individuals in winter. The Peak District is one of its strongholds. Although longer established on the River Dove in the White Peak, they are now breeding on a number of other rivers where the habitat is suitable such as Stoke Brook (Calver), the upper Manifold River and River Derwent. Winter flocks can be seen on reservoirs like Agden and Lineacre as well as along the River Derwent.

FIG 164. Mandarin Duck, an introduced species that is breeding successfully on many of our rivers and appears in good numbers on several reservoirs.

Various waders drop into reservoirs on passage in autumn and spring, but numbers and diversity are not usually notable compared with coastal and lowland sites that favour more exposed, nutrient-rich mud and associated invertebrates. Ducks include Tufted Duck (*Aythya fuligula*), Mallard and Goldeneye (*Bucephala clangula*) and wading birds are exemplified by Little Ringed Plover and Green Sandpiper (*Tringa ochropus*) at, for example, Trentabank Reservoir in Macclesfield Forest.

Turning to mammals, the importance of moorland Water Voles is described in Chapter 9, but there are still river-side and pond populations. The River Derwent's middle reaches still witness the characteristic 'plop' as the animal takes to water, but it is rare now on the Goyt-Etherow catchment and still declining in Staffordshire, although present along the River Dane. American Mink predation is a major factor, which also take fish and nestlings along the water's edge. 1965 saw the first record for this fur farm escapee in Derbyshire beside the Derwent Reservoirs. They are now widespread here, especially along the River Derwent and its tributaries. Recognisable by its dark chocolate colour (appearing black from a distance), they are usually seen swimming and are notably smaller than otters. Where they are controlled by water bailiffs or angling clubs, water vole and bird breeding success is much better (Mallon *et al.* 2012).

An increase in Otters would help reduce Mink as they outcompete them but, unfortunately, Otter sightings are rare within the Dark and South West Peaks compared with the early 1800s when they were frequent on the Derwent and its tributaries (Glover, 1829). They are increasing in adjacent lowland areas and have lived on the River Don along the northeast fringes for the last 15 years, so we hope for more in the future. The Water Shrew is another wetland specialist, but one which is very scarce in the Peak District, associated with rivers or ponds. This large shrew lives on freshwater invertebrates, so it swims and dives regularly. It digs burrows in the water-side banks or utilises existing holes. Look out for them courtesy of domestic cat victims.

One last mammal particularly associated with water courses is the Daubenton's or Water Bat (*Myotis daubentonii*). It favours smooth water surfaces for feeding, swopping low to catch insects over millponds, rivers, lakes and ponds or reservoir edges such as at Ladybower. The Rivers Derwent and its tributaries and the Manifold provide suitable habitat and a breeding colony is known at Chatsworth. It is worth you helping to clear all discarded fishing tackle from alongside water edges to help this bat avoid being trapped or hurt (Mallon *et al.*, 2012).

Brown Trout and Grayling (*Thymallus thymallus*) are the key fish species along with Bullhead (*Cottus gobio*) in the rivers of the Dark and South West Peak, although Grayling are more restricted in their distribution than the others. Brown Trout are at home even in the higher reaches of the Derwent above the reservoirs and occur in the Noe to Edale as well. These fish are considered further in Chapter 14. The upland character of our streams and rivers do not provide the more nutrient-rich conditions favoured by most coarse fish, so these are more often found in fisheries in pools or lakes such as Turner's Pool near Swythamley, which specialises in Carp, or in the Canals where species like Roach (*Rutilus rutilus*), Rudd (*Scardinius erythrophthalmus*) and Gudgeon (*Gobio gobio*) live. The canals are also a favoured haunt of the Kingfisher (*Alcedo atthis*), but you can spy that magical flash of blue whizzing past on the Derwent and Goyt/Etherow river systems as well as other streams.

Ponds and millponds are the breeding habitat for amphibians. Common Frogs are ubiquitous in all types and sizes of ponds, although preferring some shallow water, whereas Common Toads favour deeper ponds or millponds. Garden and school ponds are also valuable for these amphibians, supporting quite large populations with several species together. In general terms, surveys show that there are about twice as many Frog as Toad breeding ponds in the Peak

District, mostly in the Dark Peak. Toad patrols help avoid them being squashed whilst crossing roads and rescue large numbers heading for ponds, canals and reservoirs as in the Buxworth Basin, Combs Reservoir, Hassop, Watford Lodge and Whaley Bridge (Derbyshire Reptiles and Amphibian Group). In the Staffordshire Moorlands, Frogs and Toads are not widespread and seem to be under-recorded. Although Yalden (1986) found Palmate Newt (*Lissotriton helveticus*) in only 15 sites (out of 142 surveyed), larger surveys reveal that it is the commoner newt in the uplands outside the White Peak (Fig. 165), and is more tolerant of acidic conditions than Smooth Newt (*Lissotriton vulgaris*), which is scarce in the South West and Dark Peaks (Whiteley, 1997). Great Crested Newt (*Triturus cristatus*) is more typical of the White Peak (see Chapter 14), although there are a few records in the uplands.

The invertebrates of these more slowly-moving rivers differ in detail from those in the tumbling brooks typical of the cloughs (Chapter 9) and those in ponds generally favour still water. A number of sample surveys (Allan, 2004, Smith, 2000, plus Sorby Natural History Society data) show that Dark and South West Peaks streams are largely acidic (although they vary) and base-poor in nature, which strongly controls the faunal assemblages. The most acidic streams

FIG 165. Palmate Newt male – the black, webbed back feet and projecting filament from the tip of the tail are clear. (Steve Orridge)

have a compromised assemblage of macroinvertebrates, whilst further down the catchments, higher base-richness associated with the shales helps combat acidification from air pollution. In the faster flowing streams the range of micro-habitats is important in determining the full species assemblage. Riffles of faster flowing water, rocks in the water, tree roots and other vegetation, bottom mud, shingle or sand and slow-moving backwaters all contribute to this. There are a range of riverflies with aquatic larvae such as mayflies, caddisflies and stoneflies that are typical of running water. Larval May and Stoneflies have three or two tail bristles respectively helping to differentiate them. Caddisfly larvae often build cases of tiny stones, twigs or leaves around them to protect from predators, each case type helping identify the species.

The Large Dark Olive Mayfly (*Baetis rhodani*) and Blue-winged Olive Mayfly (*Serratella ignita*) are very common in swift running water (Smith, 2002) although the latter seems to be more restricted on the Sheffield side of the area, whilst the former is absent from the tops of the clough streams (Zasada, 1981, Allan, 2004). Stoneflies are important members of the freshwater assemblages as they need highly oxygenated running water over a stony substratum. This kind of habitat is lost once the rivers become larger and more silt-dominated, often being polluted too resulting in lower oxygen levels. Common Stoneflies in our smaller rivers and streams include the predatory Yellow Sally Stonefly (*Isoperla grammatica*). The stonefly *Leuctra hippopus* was fairly consistently recorded across the Dark and South West Peaks (Allan, 2004) and reported by Zasada (1981a) in the upper reaches of Sheffield's streams in the Dark Peak, whereas *Leuctra moselyi*, an upland species, rare nationally, was more abundant in the Staffordshire Moorland streams Smith sampled.

Caddisflies were generally much less abundant in these streams. The most frequently recorded in the Staffordshire Moorland streams was *Rhyacophila dorsalis*; a widespread species found in small streams to larger rivers. Water beetle numbers were low, but some streams support riffle beetles, *Elmis*, *Limnius* and *Oulimnius* species (some of which are very local or rare, or at the edge of their range) and predatory diving beetles (for example, *Oreodytes sanmarkii* and *Agabus guttatus*, which is most common in the Dark Peak). Much more universal are various midge larvae that occur throughout all the streams sampled such as Chironomid midges and Blackfly larvae (*Simulium* spp.). Chironomid larvae construct tubes in or attached to the substrate, while Blackfly larvae attach themselves to rocks using tiny hooks at the ends of their abdomens and live under stones in small streams where they catch passing debris for food.

Crustaceans, like Freshwater Shrimp (*Gammarus pulex*) and Hog-louse (*Asellus aquaticus*), and molluscs are not abundant in the most acidic water, needing higher calcium levels to form their carapaces or shells. Freshwater Shrimps are only frequent in some lower reaches as on Burbage Moor, Dick Clough, Agden Bog and Black Brook Staffordshire Wildlife Trust Reserve. Both crustaceans occur in Hogshaw Brook in the Lightwood Valley, Buxton, where a few snails are companions – Wandering Pond Snail (*Radix balthica*), Bladder Snail (*Physa fontinalis*) and Pea Mussels (*Pisidium* spp.), all of which frequent ponds as well as streams.

Pond freshwater invertebrates often differ. Commoner mayfly larvae are represented by *Cloeon dipterum* and caddisflies by *Limnephilus lunatus* and *Phryganea bipunctata*. Ponds, especially new ones, support several highly mobile water boatmen, including over a dozen species of lesser water boatman in the Dark Peak alone. Examples are *Sigara lateralis*, a common and widespread species, and *Callicorixa wollastoni*, a species of peaty pools, ponds, reservoirs and stony streams (Merritt, 2006). Greater water boatmen such as *Notonecta glauca* and *N. obliqua* are predatory and swim upside down, hence labelled Backswimmers. *N. obliqua* tends to be an upland species. Hog-louse can tolerate quite de-oxygenated water and can be abundant in some of the less acidic ponds.

Different largely predatory Water Beetles are abundant in ponds and include some small black *Agabus* species. Watch out for Great Diving Beetles, large beetles that collect air from the pond surface storing it beneath its wing cases which produces a silvery sheen. *Dysticus marginalis* is the commonest species, but *D. circumflexus* is known from Ramsley Moor and Ewden Valley, although this group is difficult to separate. Species specialising in using the surface tension are Pond Skaters (*Gerris costae* is the upland species) and Whirligig Beetles (Gyrinidae) that whizz around the water's surface in dizzying circles. The Nationally Local Screech Beetle (*Hygrobia hermanni*) has been recorded in the ponds developed in Barbrook and Ramsley reservoir basins.

Brightly coloured dragonflies and the smaller damselflies are delightful to see and their larvae are all aquatic and predatory. The Dark Peak fauna is enhanced by nine species expanding their ranges from the south and east in recent decades. In 1997 the largest dragonfly, the Emperor (*Anax imperator*) reached Big Moor and is now quite widespread across the Peak. Broad-bodied Chaser (*Libellula depressa*), Black-tailed Skimmer (*Orthetrum cancellatum*), Four-spotted Chaser (*Libellula quadrimaculata*), Common Darter (*Sympetrum striolatum*), Emerald Damselfly (*Lestes sponsa*), Migrant Hawker (*Aeshna mixta*) and Banded Demoiselle (*Calopteryx splendens*, Fig. 166) have followed a similar pattern. The latter prefers clean running water and has spread rapidly; by 2019 reaching the upper parts of the Derwent beyond Slippery Stones. All are now established but still quite localised as Dark Peak breeding species. Of these newcomers, only Emperor and Banded Demoiselle have reached the South West Peak, but dragonflies may be under-

FIG 166. Banded Demoiselle dragonfly, a relatively recent coloniser of the Peak District, prefers clean running water. (Kev Dunnington)

recorded here (Staffordshire Ecological Record on line). In addition, Frost (2007) was particularly excited to locate the first Peak District record for Keeled Skimmer (*Orthetrum coerulescens*) at a Barbrook pond in 2006. The nearest known population is at least 50 km away for this heathland pond specialist. In 2018, another was seen at ponds on Ramsley Moor, so its future establishment is hopeful.

Long-standing residents are Blue-tailed Damselfly (*Ischnura elegans*), Common Blue Damselfly (*Enallagma cyathigerum*) and Large Red damselfly (*Pyrrhosoma nymphula*) all quite widespread, with the latter often making a welcome early appearance in April. The Azure Damselfly (*Coenagrion puella*) is more restricted, but slowly extending its distribution. The typical Dark Peak hawker is the Common Hawker (*Aeshna juncea*), breeding in a range of acidic waters and flitting around the cloughs and sunlit rides in broadleaved and conifer woods from July to September. In autumn it is replaced by the Migrant Hawker, less common, but in a good year flying well into October and early November, although not yet recorded in the Staffordshire Moorlands. Adults are well known wanderers, often flying far from breeding sites. Other large dragonflies are less frequent. The Brown Hawker (*Aeshna grandis*), easily recognised by its brown body and bronze wings, is more of a lowland species, and the Southern Hawker (*Aeshna cyanea*) is more at home in the White Peak, but both occur at a few sites in the Dark Peak. Finally, the Golden-ringed Dragonfly, detailed in Chapter 9, brings the Dark Peak total to 18 species, double the number recorded in the 1970s (McClean *et al.*, 2020).

Although this can only be an introduction to the animal life of the habitats described in this chapter, it hopefully wets the appetite to know more and encourages excursions to see for yourselves. It is more than was included in the first Peak District New Naturalists book where limited information on animals was hidden in Appendices. The Peak District animals deserve more than this.

CHAPTER 12

Neutral Grasslands

Where have all the flowers gone?
Pete Seeger, 1955

NEUTRAL GRASSLANDS FORM THE LINK between the last chapter and the White Peak. By neutral, we mean neither particularly acidic nor alkaline, but buffered owing to the nature of the soils or through past liming. We will cover a range of meadows and pastures, but also road-side verges, across the whole of the Peak District. Meadows usually refer to grasslands that are cut for hay with aftermath grazing, whilst pastures are more often just grazed. We will start with the most widespread types (which are also of least wildlife interest) and then explore the magnificence of flower-rich grasslands.

AGRICULTURALLY IMPROVED GRASSLANDS

The majority of neutral grasslands lie on flatter land and have been effectively drained and reseeded. This usually entails removing the existing swards, cultivation and subsequent re-seeding with agricultural grass seeds, sometimes with clover. Modern seeding favours a Perennial Rye-grass monoculture with fast growth under nutrient-rich conditions dependent on inorganic fertiliser plus slurry and/or farmyard manure. Slurry is largely produced by dairy farming and will be reducing as some farmers have changed to beef or sheep enterprises in response to financial pressures. These applications add enormously to the soil's nutrient status and result in more productive grass swards (albeit still constrained by our climate and altitude) that produce two or three silage crops annually but causes the reduction or elimination of less competitive plant species

FIG 167. General view of farmland on the limestone plateau showing the large number of bright green fields that have been agriculturally improved, where few different wild flowers remain. The Perennial Rye-grass glints in the sun.

in such a strongly competitive environment. Silage is grass that is cut and stored without significant drying, but compressed to remove oxygen and wrapped in plastic in which it then ferments. It is the main crop for most Peak District farms now. The advantage to farmers is less dependency on good weather compared with hay making.

However, the implications for wildlife are manifold. The near monoculture means a dearth of plants to brighten the landscape; insect diversity and abundance strongly decline with increasing nitrogen inputs and decreasing floristic variation; small mammal populations are compromised by regular cutting or high grazing levels; fungal interest declines in favour of a bacterial-dominated soil system and the lack of diversity makes the sward very vulnerable to extreme events like the 2018 drought where longer-rooted plants fared better. The first silage crop is taken in May usually, cut before flowering or seeding so there is little for birds or other animals to glean afterwards; the soil seed bank reduces; and the early cutting prevents successful nesting in the grassland swards (Fig. 167).

The soil fauna is also affected; the most important being earthworms: the soil ecosystem engineers in the web of life. Charles Darwin called them 'nature's ploughs' because of their mixing of soil and organic matter. Indeed, it is calculated that a healthy earthworm population takes only some 11 years to turn over a field's soil. There are different groups of earthworms living in different soil layers. Most cannot tolerate acidic soils below about pH 5, so are especially important in neutral grasslands. There are leaf litter dwellers, shallow burrowers and deep burrowers. The familiar large earthworm with the pink saddle, called a Lob Worm (*Lumbricus terrestris*) is one of the deep burrowers, living in permanent tubes that can extend down the whole soil profile. They pull plant material into their tubes on which to feed. The shallow burrowers excavate temporary, mostly horizontal burrows in the top horizon, also feeding on plant material, incorporating it into the soil. The surface dwellers include the smaller Redworms (*Eisenia fetida*) that you have in your compost heaps or wormery. All these have important ecological niches, helping to maintain healthy soils by decomposing dead plants; concentrating plant nutrients; aerating and draining soil profiles; increasing water infiltration and reducing runoff; promoting root growth and sequestrating carbon. But they are sensitive to ammonium-nitrogen in inorganic fertilisers and slurry. Applying these at high rates reduces the surface and shallow burrowers for some weeks, although the deep burrowers may be less affected. Surface dead worms probably accounts for the gull flocks that often follow slurry spreaders. The shallow burrowing worms are also sensitive to winter cold, so fluctuate in numbers naturally.

Research has shown that slurry addition results in 2.3 times fewer detrivore earthworms (feeding on organic matter), compared with farmyard manure application which supports increasing worm numbers. These are further reduced by low soil organic matter; more likely with intensive silage cutting and grazing. Less organic matter also reduces soil moisture in drying conditions, further restricting worm movement and activity near the surface (Onrust, 2017). The surface or subsurface worms are most affected and are a key food resource for breeding birds. Nematodes, one of the most numerous soil animals, increase under slurry treatments. These are tiny and no substitute for predators dependent on much larger earthworms. Soil fauna impacted by agricultural intensification therefore tends to become smaller-sized. With reduced worm abundance, their critical services to soil condition will be concomitantly compromised. This has knock-on effects on birds like Lapwing which use visual cues to find prey. Fewer earthworms and changes in the soil fauna therefore also add stress to declining farmland birds.

At the same time, where stock are treated with anthelminthics (to control parasitic worms) based on avermectins, damaging effects on pasture fauna can occur – particularly to dung beetles and dung flies which decline significantly and which are vital in the decomposition process (Pope, 2009). The extent of use of these parasiticides in the Peak District is unknown, but they are widely available.

Thus, the combination of slurry spreading; the reduction in earthworms and their maintenance of soil health; the application of inorganic fertilisers; the use of some pesticides; the potential for compaction in clay-based soils; reseeding with monocultures and intensive management for silage or grazing shows that the ability to support any significant wildlife is severely compromised. Moreover, these intensively managed grasslands dominate in the Peak District, as they do in adjacent lowlands wherever pasture systems operate. They are equally widespread on the more gentle slopes of the South West Peak and on the White Peak plateaux, but more restricted in the Dark Peak simply through lack of suitable land. These highly improved fields glare in a rich dark-green hue with bluish overtones reflecting the fertiliser additions and dark shiny-green Rye-grass leaves, standing out in any view or on aerial photography (Fig. 167).

What is particularly troubling is that the National Parks' first purpose is to conserve and enhance natural beauty, wildlife and cultural heritage. That agricultural improvements of pasture and hay meadows are nearly as extensive in our National Park as they are beyond its boundaries (steep slopes excepting) shows the dearth of tools available to meet these obligations. Many of the now Rye-grass dominated fields would have exhibited much greater diversity of plants and animals in 1949 when the National Parks and Access to the Countryside Act was passed under which National Parks were established. However, our area was not wall to wall flowery grasslands in the 1950s as Moss (1913) mentions growing oats and even minor wheat production on shale-based soils. He also describes reclamation of moorland to pasture whereby the Heather and Mat-grass were first burned, before stones were cleared. The land was then ploughed, limed and sown with oats, possibly for several years, before adding grasses to make permanent pasture, usually accompanied by drainage. This process was also described by Farey 100 years previously and more was ploughed between these dates. Moss describes a rotation with oats grown for several years followed by a root crop like Turnips. The list of plants associated with arable land that Moss gives does not suggest we have lost a rich arable flora, although it did include Shepherd's-needle (*Scandix pecten-veneris*), which is now Red-listed owing to widespread herbicide use and no longer occurs in the Peak District. The commonest of the arable flora were those regularly seen now in gardens like Annual Meadow Grass (*Poa annua*), Hairy Bittercress (*Cardamine hirsuta*) and Goosegrass (*Galium aparine*). We will meet some of these again on the bare limestone ledges and rock outcrops in Chapter 13.

Moss also lists plants associated with meadows and pastures, including 12 different grasses. Farmers at that time regarded most broadleaved plants as weeds, but they included those now associated with unimproved grasslands like Bird's-foot-trefoil, Tormentil, Yellow Rattle (*Rhinanthus minor*), Mouse-ear-hawkweed (*Pilosella officinarum*), Mountain Pansy, Harebell, Oxeye Daisy (*Leucanthemum vulgare*) and Common Cat's-ear. Thus the earlier grasslands probably contained numerous additional plants besides those sown owing to vestigial seed banks, surrounding seed sources for colonisation and the practice, as Farey describes, of using hay barn sweepings for seeding new pastures. We know that there were certainly many more flower-rich grasslands 70 years ago than now, but this short history suggests some of our richest grasslands may not necessarily be that old (not equivalent to an ancient woodland for example), but have evolved with time since seeding. The best examples, though, will not have been disturbed through agricultural reclamation measures for many years.

SEMI-IMPROVED GRASSLANDS

The extreme type of agriculturally improved grassland described lies at one end of the spectrum. There are many variants on this theme grading into the truly diverse fields depending mostly on intensity of management and time since re-seeding. There are also many fields with old lead rakes, quarries, or other uneven surfaces or short steep slopes which have escaped agricultural improvement and still harbour vestiges of a once richer flora. The semi-improved grasslands are often paler green and spangled with a few other plants like Dandelions (*Taraxacum officinale*), which reflect higher levels of soil potassium derived from inorganic fertilisers, and Buttercups, mostly the taller flowering Meadow Buttercup on drier soils and spreading Creeping Buttercup on wetter soils. Dandelion-covered fields are widespread (Fig. 168), sparkling in the spring sun, adding sources of nectar for early flies and bees, seed later for Goldfinches and Greenfinches and a glitziness to the view. Dandelions have a long history of use. Young leaves embellish a spring salad; they are boiled as a vegetable like spinach; used in Dandelion beer (which was a North Midlands speciality); the roots made into Dandelion coffee – a useful alternative in the last World War when coffee was rationed; and the plant has a number of historic medicinal uses.

Damper fields, pastures or meadows, may be painted yellow with low growing Lesser Celandines in early spring. Equally at home in woodlands, this was William Wordsworth's favourite flower for which he wrote three poems. Generally common pasture herbs include Yarrow (*Achillea millefolium*), White

FIG 168. Largely agriculturally improved fields full of Dandelions sparkle in the sun. European Gorse covers Eyam Edge in the background.

Clover, Daisy (*Bellis perennis*) especially on tightly grazed grasslands, Common Mouse-ear (*Cerastium fontanum*) and, increasingly with higher nutrient levels, the tall umbellifers Hogweed (*Heracleum sphondylium*) and Cow Parsley. Farey (1813) describes Yarrow as being tied in bunches and dried for dyers giving a luminous golden-brown colour. Common Sorrel is a speciality in many less improved meadows, producing a distinctive rusty-red haze when flowering. The leaves are astringent and quite refreshing, can be used in salads, soups and sauces and are reputed to allay thirst (Grieve, 1976).

Many fields also contain patches of Common Nettle, and a variety of thistles. Spear Thistle and Creeping Thistle (*Cirsium vulgare* and *C. arvense*) are the commonest, but Marsh Thistle is also widespread on damper soils. These provide welcome nectar for different butterflies, bumblebees and other insects in mid-summer and seeds later for Goldfinch, but tend to be controlled by farmers as a perceived sign of poor management and invasive 'weeds'.

Some pastures support a wide variety of grasses but few broadleaved plants, suggesting broad-leaved herbicide applications in the past, but no recent reseeding. Crested Dog's-tail may be abundant here, along with species like Common Bent. On wetter soils, the coarse tussocks of Tufted Hair-grass (*Deschampsia cespitosa*) stand out, being generally avoided by grazing animals. Carefully running fingers towards the inner end of a leaf will reveal the reason – the saw-like edges. In tightly grazed sheep pasture, mosses like Springy Turf-moss *Rhytidiadelphus squarrosus* can be abundant. It is not uncommon to find a few Bluebells hugging the side of walls in the South West Peak in particular in what are otherwise agriculturally improved fields. These seem unlikely to be woodland relicts, and some of the field boundaries are medieval, suggesting long establishment. The walls may be providing shade which reduces grass competition, but their early leafing and flowering helps avoid summer herbicide effects and competition. There may be other hedge or woodland-edge plants like Greater Stitchwort and Foxglove along these boundaries.

Although not as important as the waxcap grasslands, neutral meadows and pastures, not always those with high floral diversity, can also support some interesting fungi. Most are generally less sensitive to soil conditions and occur in other grasslands too. Troops of the tall Parasol Mushroom (*Macrolepiota procera*) are distinctive. The impressively large, related Horse and Field Mushrooms

FIG 169. Mosaic Puffball, a fresh and an old specimen adjacent to each other in flower-rich grassland. A large fungus, fruiting body 6–15 cm across with a coarse net-like surface when young.

(*Agaricus arvensis*, *A. pratensis*) can form fairly large fairy rings over a long period. St George's Mushroom (*Calocybe gambosa*) does the same but fruits in spring. The Giant Funnel (*Leucopaxillus giganteus*) similarly forms metre-wide crescent-shaped scars on pastures. Such features, created by the slow progress of the fungal mycelium through the grass roots, which can take decades or even hundreds of years to develop, are rapidly disappearing following extensive field improvements. A range of blewits (*Lepista*) similarly occur in rings. The Field Blewit or Blue-leg (*Lepitsa saeva*) is more common in neutral grasslands. Unexpectedly, the Wood Blewit (*Lepista nuda*) also appears where the soil is especially enriched with dung and spilt forage, for example around stock feeding sites. Puffballs are common, a large one being the Mosaic Puffball (*Lycoperdon utriforme*) (Fig. 169). The even larger Giant Puffball (*Calvatia gigantea*) occasionally appears in areas which are also dung-enriched.

Even with these small-scale variations, none of these fields are of high nature conservation value, although many have the potential for reversion to a more diverse sward. There may be up to 20 or 30 different flowering plants in these fields, but with the agricultural grasses dominating, they are not prime habitats.

FLOWER-RICH GRASSLAND

Turning now to the glory of the flower-rich grasslands and first the hay meadows. Imagine a probably small field boasting some 50 to 80 different plant species, most well represented and spread out, largely bordered by stone walls, or perhaps fringed with shrubs and trees (Fig. 170). Then envisage seasonal colour changes: a swathe of gaudy yellow as Bulbous Buttercups (*Ranunculus bulbosus*) with their reflexed sepals and neat little leaf rosettes appear in spring, perhaps with patches of the much less common white-flowered Meadow Saxifrage (*Saxifraga granulata*). Then the pinks of Red Clover punctuated by brown spikes and cream anthers of Ribwort Plantain, the eggs and bacon colours of Bird's-foot-trefoil and the bright white and yellow Oxeye Daisies take their place. Keep these in mind whilst they are joined by Common Spotted Orchid (Fig. 176) spikes with white to pink flowers and variable darker pink markings, or by yellow dandelion-like flowers of Common Cat's-ear or Rough Hawkbit (*Leontodon hispidus*). The papery seed-cases of Yellow rattle are conspicuous rattling in the breeze after the small yellow flowers are fertilised, the yellow pea-like flowers of Yellow Vetchling (*Lathyrus pratensis*) followed by purple heads of Common Knapweed (*Centaurea nigra*) or if you are lucky, Betony (*Betonica officinalis*) and Devil's-bit Scabious are the last to flower and may not reach seeding stage before cutting time.

FIG 170. A glorious, flower-rich hay meadow on the National Trust's Longshaw Estate with an abundance of Red Clover, Oxeye Daisy, Yellow Rattle, Eyebright, Meadow Buttercup and Bird's-foot-trefoil visible.

This is the splendour and wonder of our best, most colourful hay meadows. The key to their survival is low nutrient levels, especially of phosphorus. The legumes – the Clovers, Trefoils, Vetches and Vetchlings – all provide nitrogen by fixing it from the air in root nodules – like your garden peas and beans, which naturally fertilises the soils. Phosphorus is an essential nutrient in the formation of proteins and the very core of cells, the DNA, and comes from natural decay of rocks and of litter in the soil.

There can be up to a dozen different grasses in the best hay meadows. The most abundant relate to soil type, with Common Bent more conspicuous on the slightly acidic soils and the dainty flowered Quaking-grass replacing it on more lime-rich soils. Universal species are usually the very fine-leaved Red Fescue (*Festuca rubra*) and Crested Dog's-tail (with flowers all on one side of the stem). Sweet Vernal-grass is another stalwart, growing early with a dense head of flowers turning straw-coloured in summer. Its high levels of coumarin gives cut grassland its sweet smell. There is further variation related largely to soils and hydrology. Because of past liming most of the best hay meadows in the Peak District are on neutral soils, although there are more acidic and calcareous variants. Some are also wetter than others with a heavier clay-based soil and fringing marshes or oozes introducing greater variety. Lady's Smock and Ragged-robin are more regular in damper meadows and striking dark pink spikes of Southern Marsh Orchids (*Dactylorhiza praetermissa*) might stand proud in damper soils or adjacent marshy spots. These also hybridise with Common Spotted Orchids (*Dactylorhiza fuchsii*), often displaying hybrid vigour in their taller spikes and more lush growth (Fig. 171).

The main hay meadow community is linked to the Crested Dog's-tail/ Common Knapweed type described in the National Vegetation Classification (Rodwell, 1992) as the Mesotrophic Grassland no. 5 (MG5). This is the typical hay meadow type across lowland Britain, but which is becoming increasingly rare nationally. It can also occur in churchyards and on roadside verges. The acidic and calcareous variants of this hay meadow type are typical in the Peak District. Our calcareous version may support Downy Oat-grass (*Avenula pubescens*) and

FIG 171. Common Spotted Orchid/Southern Marsh Orchid hybrids showing hybrid vigour in a flower-rich hay meadow above Cressbrook Dale. Abundant Yellow Rattle, Meadow Buttercup and Ribwort Plantain mixed with Sweet Vernal-grass and Red Fescue behind the orchids.

FIG 172. The day-flying Chimney Sweeper moth frequents meadows and pastures supporting Pignut. (Steve Orridge)

even Hoary Plantain (*Plantago media*), which are much more typical of limestone grasslands in the Dales. In contrast, in slightly more acidic fields on gritstones or limestone plateau loess soils, Pignut (*Conopodium majus*) and Tormentil can be abundant, sometimes with plenty of Autumn Hawkbit. Abundant Pignut usually also means a good population of the lovely little day-flying black moth, Chimney Sweeper (*Odezia atrata*), which has white fringes to the upper wing tips (Fig. 172). Its larvae feed on Pignut flowers and seeds. This plant, which is an Umbellifer (along with Carrots and Parsnips), has a highly nutritious tuber which would have been sought by foraging pigs (hence its name) but was coveted during the last War as a substitute for potatoes. Small holes and plant remains in fields now usually signal badger rootling for tubers to augment their omnivorous diet.

There is also a group of hay meadows where Meadow Foxtail (*Alopecurus pratensis*) is the essential grass species, often accompanied by Common Sorrel, Meadow Buttercup and a variable range of other species. This is not typical of the MG5 community but has more affinities with the lowland alluvial hay meadow, labelled MG4, one based on the constant appearance of Meadow Foxtail and Great Burnet (*Sanguisorba officinalis*). Our versions of these meadows are clearly not alluvial grasslands (they do not grow on river floodplains). Great Burnet is regarded as a long-established meadow indicator in lowland Britain, but can behave differently in the Peak District, colonising sites like railway embankments. It is still a good indicator of a high quality grasslands though, with excellent examples in the South West Peak on generally damper soils. There are also wet meadows better aligned to a different grassland community MG8, which features Crested Dog's-tail and Marsh-marigold on gleyed soils, either where there are springs and seepage lines, or where seasonal flooding occurs near to rivers. This type is rare in our area, and mostly in the South West Peak where they are largely used as pasture rather than meadows (Fig. 152).

Finally, we have meadows with affinities to the upland meadow MG3 community, otherwise confined to the upper river valleys in Durham, North Yorkshire and parts of Cumbria. This is characterised as the Sweet Vernal-grass/Wood Crane's-bill (*Geranium sylvaticum*) meadow, and although Wood Crane's-bill is absent from our area, many other typical species are here and meadows with more abundant Great Burnet and Lady's-mantles (*Alchemilla* spp.) may well be variants on this theme. This would not be unreasonable since we have lost so many flower-rich hay meadows nationwide there will be types that have now disappeared. This is illustrated by my finding a tiny patch some years ago of Great Burnet, Melancholy Thistle and Water Avens (*Geum rivale*) growing intimately together on alluvial soils. If these had been more widespread once, they would have been more typical of the upland hay meadow than the lowland MG5. What other communities we have lost, we do not know. We do not think of Melancholy Thistle or Water Avens as meadow species in our area now (they have been found in only one and nine meadows respectively, Buckingham & Chapman, 1997), but perhaps they were once a different community now lost.

Further evidence for lost grassland types comes from Farey (1813). He considered Dyer's Greenweed (*Genista tinctoria*) to greatly infest old pastures and thought that liming these would destroy it despite its importance for Manchester's dyers, yielding a yellow dye. This is certainly a rare plant here now reflecting its English Red List status, occurring in pasture near Glossop in Long Clough and among lead waste heaps at Gang Mine, both Derbyshire Wildlife Trust Reserves. Yellow Rattle (*Rhinanthus minor*) was said to abound in meadows in Peak Forest and Cowdale in Farey's time and, interestingly, Wild Garlic or Ramsons (*Allium ursinum*) greatly infested some limestone pastures, particularly under shade, flavouring the butter of cows grazing them with garlic. Farey complained about this especially in the Matlock Bath area. Wild Garlic is not a meadow species now, but might also represent a different lost community or was it a woodland vestige after clearance? Common Knapweed was also described as

FIG 173. Greater Butterfly Orchids in a flower-rich meadow in the South West Peak.

an unsightly weed which much disfigured dairy pastures and Spiny Restharrow (*Ononis spinosa*), now very rare here, was a 'disagreeable prickly weed in some pastures, destroyed by increasing the sheep'. Cowslips (*Primula veris*) Farey deemed, were also too often seen in great numbers on cold rather poor pastures where draining and liming would often remove them. It seems such advice was taken as for the most part these descriptions no longer apply.

Apart from some individual species, it is the flower-rich meadow as a whole that is rare, and the greater the abundance of the best indicator species and the more diverse the assemblage, the higher the value of the community. But there are also a few plants that are seldom found which uplift the spirits when discovered, such as Bee Orchid and Northern Marsh Orchid; not typical meadow species but which are scarce in some limestone plateau meadows. Greater Butterfly Orchid (*Platanthera chlorantha*) is another (Fig. 173), which also grows in limestone grassland. It has been found on both gritstone (limed) and limestone meadows. The little ferns, Adder's Tongue (*Ophioglossum vulgatum*) (Fig. 174) and Moonwort are easily overlooked and rare in meadows. The first was found in only

16 meadows in Buckingham and Chapman (1997) survey. Both have a single stem bearing spores clustered near the top. Moonwort attracted much superstition in the past. Culpeper describes it in the 17th century as being able to 'open locks and unshoe horses that tread upon it...', which seems rather fantastical.

Flower-filled meadows are important cultural and landscape features, celebrated in folklore, customs and literature and representing an historic display of traditional rural life now lost from many areas. They may be the oldest link with the past that a village or parish still has. Think of these beautiful habitats as works of art (and there is much they have in common). You would not destroy a gorgeous painting, so why let the best hay meadows disappear? They are an example of a truly sustainable approach towards environmental management (Buckingham & Chapman, 1997).

There are other factors that increase flower-rich meadow value that are not widely recognised. Many farms retained a flower-rich or 'hospital' field where sickly animals could self-medicate by selecting natural herbal remedies amongst the flowers. These fields are now rare and expensive veterinary alternatives

FIG 174. Adder's Tongue, a little fern with a slightly fleshy, single oval leaf and simple spike of spore-bearing sporangia in two fused opposite ranks at the top, growing in a flower-rich hay field on the Longshaw Estate.

have to be used instead. Flower-rich fields, including various trees and shrubs within the grazing area, also benefit stock health and nutrition. Plants exhibit different methods to deter grazers (stock or other animals) through, for example, distasteful chemicals, prickles or hairs. On the other hand, most plants also provide different beneficial nutrients that the animal seeks. Thus, stock given choice will take some of one plant, turning then to another to counteract the first's effects, and so on. What is taken changes during the year as maturing leaves produce more tannins for example, deterring animals more latter in the season. The dietary needs of animals change too. The ability to choose successfully without being affected by plant chemicals is ultimately dependent on a good variety of plants in the forage. In this scenario, selectivity is a fundamental trait of grazing animals.

Broad-leaved plants (herbs in ecological rather than cookery-speak) can contain higher crude protein, energy, minerals and trace element levels than grasses and for longer in the season. Investigations into Ribwort Plantain recently showed its protein levels are highest in July and August compared with improved pasture grasses which peak in June – hence silage is cut then, but late cut hay will preserve the higher levels in the herbs (a fact now lost from some farming memories). Anecdotal evidence of stock pulling out the Ribwort Plantain from wildflower hay bales in preference to anything else is clear. Testing some flower-rich hay revealed more than twice as much protein and a third more energy compared with grass-only samples in one investigation (Grayson, 2017). It also contained higher concentrations of all the major minerals and trace elements, showing that cattle selecting this hay over the grass-only sample were choosing what best supported their nutrition and health. We need similar analysis of other species that form the bulk of a wildflower grassland to provide the full story.

Giving animals choice therefore allows them to gain all they need for a healthy and balanced diet on their own terms reacting to natural requirements. Contrast this with a Rye-grass dominated field plus supplements which give minimal choice. Moreover, it has been found that stock with a good level of choice contain higher levels of the beneficial polyunsaturated fats that are better for humans. Remember, 'you are what you eat has been eating' if you eat meat. The major drawback for modern farmers in the scenario described here is that the wildflower meadows have a lower productivity, so produce less forage. In addition, traditional stock breeds bred to cope with native plant material do better compared with more modern breeds which dominate on Peak District farms.

There are other important factors elevating the value of wildflower-rich grasslands if we consider whole ecosystem benefits. In these times of a climate emergency, wildflower-filled grasslands are more resilient to change owing to their more complex and diverse food webs and functioning. They also capture and store more carbon than monocultures of Rye-grass. There are more legumes and slow-growing plants that decay more slowly and the wider range of plants has more diverse root depths, which trap carbon deeper in the soil. Mosses are more abundant than in highly fertile monocultures and these have slower respiration rates and higher carbon:nitrogen ratios, enabling greater carbon capture and retention. Flower-rich fields depend on a fungal rather than a bacterial-based soil system that also ensures more carbon sequestration. Flower-rich grasslands can sequester around 30 to 44 $gC/m^2/yr$ in the top 0–30 cm soil horizon, and with 60 per cent of the carbon more than 30 cm deep (dependent on management above ground), carbon totals can be high. This contrasts with much lower levels in highly improved grasslands (Alonso *et al.* 2012, Thompson, 2008, Ward *et al.*, 2016).

Add to these carbon benefits the impact on human health and wellbeing. Significant health benefits have been demonstrated from contact with greenspace, but this is enhanced where colourful wildflowers and the sounds of nature are all embracing as well – really important attributes in a National Park. The same flower mass supports bees, thus producing rich honey and providing pollination services for any other surrounding habitat; crop or garden. Without additions of artificial fertilisers and pesticides, wildflower-filled grasslands also help reduce potential pollutants in nearby water courses and water supplies. Issues at Tittesworth Reservoir, for example, include herbicide residues from pastures upstream.

Although much of the account so far here has focused on hay meadows, much is also relevant to pastures. Species-rich pastures are not as floriferous as hay meadows as they are grazed, but can be equally diverse. Some species will differ since annuals like Yellow Rattle and Eyebrights in particular need gaps in the sward to establish and are less frequent or absent from pastures. On the other hand, a few pastures also support plants like Primroses and Bluebells, particularly along banks,

Flower-rich grasslands are also vital for a variety of invertebrates as well as breeding birds like Skylarks and Meadow Pipits. Hay meadows are not the best invertebrate habitat since cutting disrupts their life cycles. Nevertheless, the meadows can provide nectar, pollen and many different plant species as food. The flowers attract insects from flower beetles and plant bugs to bumblebees, and from aphids to ladybirds. Crab and orb-web spiders might ambush prey in flower heads while moth and butterfly caterpillars munch through leaves. Orange-tip Butterflies will frequent those fields supporting plenty of Lady's Smock, whilst the common Meadow Brown Butterfly (*Maniola jurtina*) larvae are grass feeding, but adults need nectar supplies. Bumblebees like the Red-tailed, White-tailed, Buff-tailed and Common Carder are the commonest. These all have different tongue lengths, metabolic requirements and abilities to reach into flowers so tend to specialise on different species – making the whole array of floristic diversity even more important in supporting a wide variety of invertebrates along with their predators and parasites.

The Cinnabar Moth (*Tyria jacobaeae*) is a common species associated with Ragwort (*Jacobaea vulgaris*), which may be more abundant in less improved fields but will be rigorously controlled where silage or hay is cut owing to its toxic properties. In grazed pastures, most animals avoid it, making it more conspicuous. The moth caterpillar, with its familiar yellow and black stripes, uses the alkaloids to render it poisonous to predators and is just one of 61 Ragwort-feeding invertebrate species, particularly micromoths and bugs (Buglife website). Other insects like butterflies and hoverflies regularly visit Ragwort flowers, highlighting is wildlife value when in the right place.

Breeding Skylarks and Meadow Pipits need good sources of invertebrates for their chicks and are therefore more frequent in the semi-natural or unimproved grasslands. Indeed, there are large areas of improved fields where soaring Skylarks and fluttering dives of Meadow Pipits are now absent. Curlew breed in many semi-improved or improved fields across the Peak District, although agricultural improvements will limit the numbers of earthworms on which adults depend. The chicks have a wider diet, but are often led by the parents to better feeding sites once hatched.

Common in many fields are flocks of feeding Jackdaws, often intermingled with Rooks and Carrion Crows. Locally, there are large Jackdaw populations, with breeding centred on villages or quarries. They utilise rock ledges and holes in trees, defending them against intruders (from other Jackdaws to Grey Squirrels), and are fairly sedentary, often roosting en masse in local woodlands or trees. They usually feed on open ground on a wide variety of invertebrates but will also forage for fruits and can take nestlings. In winter, roving flocks of Redwing, Starlings (*Sturnus vulgaris*) and Fieldfares, visitors mostly from Scandinavia and Eastern Europe, scour the fields for food, utilising the Hawthorn and other berries along the boundaries as well as soil invertebrates. Middleton Moor above Stoney Middleton around the old Cavendish Mill lagoons is where to see murmurations of Starlings swooping into roost in the reed beds. Up to 100,000 were recorded there in November 2019 and the awe-inspiring ballet of fluid, sweeping movements are essential viewing. Kestrels (*Falco tinnunculus*) and Buzzards frequently hunt over grasslands, although, again, they will find more potential prey from worms or beetles to small mammals in the less improved fields. Indeed, Kestrel numbers are feared to have declined in recent years (Frost & Shaw, 2013), although artificial nest boxes have helped support them.

Barn Owl (*Tyto alba*) is benefiting from new nest sites too, largely in restored barns. It suffered severely from pesticide poisoning in the 1950s and 1960s, with some recovery since. It is still quite rare in the Peak District, possibly related to its vulnerability to cold or wet winters. It is largely sedentary and most adults remain paired all year (Frost & Shaw, 2013). Its breeding success also varies with peaks in Field Vole abundance, which are higher in rough grass than in short-grazed vegetation and agriculturally improved fields (Mallon *et al.* 2012). Some better habitats are larger road-side verges, but hunting over them increases risks of traffic accidents.

Mammals that frequent neutral grasslands more perhaps than other types include Brown Hare (Fig. 175) and Moles. Brown Hare can be seen throughout the Peak District grasslands, but more particularly in the White and South West Peaks. They need some cover in scrub, rank grass or a hedge to shelter by day in a shallow form, but in which they can be incredibly well camouflaged. With a

FIG 175. Brown Hare, not uncommon throughout the Peak District neutral grasslands. (Kev Dunnington)

preferred diet of herbs in summer and more grasses in winter, the loss of flower-rich grassland would have affected numbers and local declines have been noted, such as in the Wardlow basin (Mallon *et al.* 2012, Cheshire Mammal Group, 2008). Molehills indicate adequate worms in soils, although Moles will also eat beetles, fly larvae and molluscs. They are one of our most widely distributed mammals, but avoid stony, waterlogged, sandy and shallow soils which restrict construction of their burrow systems. It is an eerie experience watching a molehill erupt beside you, although the perpetrator rarely appears, but these are the temporary hunting galleries only. Moles also have an underground, more permanent tunnel system that, being cleared, is rarely marked by soil debris in the form of molehills. This network can extend 100–200 m and will include a nest (Cheshire Mammal Group, 2008). Moles are essentially solitary and defend their tunnel systems aggressively. These can be very abundant in some flower-rich pastures and meadows, no doubt benefiting from the higher worm numbers, and are generally scarcer in Rye-grass fields regularly spread with slurry. Unfortunately Moles are still generally regarded as pests and gibbets of dead moles are not unusual, although more frequent in the past (Mallon *et al.*, 2012). Tawny Owls and Buzzards are their main predators, particularly in summer when juveniles are more abundant. Herons are perhaps an unexpected predator too (Mellanby, 1971). Most predatory mammals find them distasteful.

The total amount of neutral flower-filled grassland in the Peak District is unknown. The data available are variable and not all up-to-date. The NPA (Buckingham & Chapman, 1997) has surveyed and assessed hay meadows in the National Park, but there has been no systematic surveys of flower-rich pastures. The best estimates (made in 2016) for the whole Peak District are shown in Table 8.

A and A/B grades are from the Hay meadow surveys, favourable condition from other sources. Buckingham and Chapman (1997) developed a meadow evaluation protocol using botanical content whereby where only common and widespread species occur, the field is graded C. With increasing frequency and abundance of other species, fields reach Grade B or A value. The highest value depends on at least three species from the most restrictive plant list occurring occasionally or one being more frequent. This list includes species like orchids, Adder's Tongue Fern, Cowslip, Devil's Bit-scabious, Great Burnet, Lady's Bedstraw (*Galium verum*) and Rough Hawkbit. Of the lowland grassland priority BAP habitat mapped by the National Park, there is a rather limited total area of 2,511 ha, a minute proportion of all the Peak District. Those noted specifically as hay meadows (Table 8) constitute about 47 per cent of this area and only 47 per cent of these (by area) are of the highest value. The total of 554.5 ha is tiny and one of our

TABLE 8 Lowland Grassland area

	Whole (ha)	Dark Peak (ha)	White Peak (ha)	South West Peak (ha)
Lowland priority BAP grassland (% of area)	2,511 (1.6%)	286 (<1%)	1,792 (3%)	433 (1%)
Hay meadows (% of lowland grassland)	1,178.5 (47%)	149 (52%)	705.5 (39%)	324 (79%)
Hay meadows Grade A, A/B or Favourable condition (% of hay meadows)	554.5 (47%)	85.7 (57.5%)	267 (38%)	201 (62%)

most threatened habitats. There may be some more as not all hay meadows have been identified and new sites surveyed since 2016 are not included here.

In the NPA's hay meadow survey in the National Park, 959 meadows were assessed between 1995 and 1997. Only 125 (15 per cent) qualified as Grade A. A further 285 (30 per cent) met Grade B thresholds (usually exhibiting some agricultural improvement and not as rich or diverse as Grade A fields). 366 meadows were classed as Grade C, but were considered to have enhancement potential. Concentrations of better meadows were found around Bonsall, Bradwell and Sheldon. More occur within the South West Peak, particularly on the Warslow Estate, which is owned and managed by the NPA through their tenants. Others occur on the gritstones and shales in the Woodlands and Edale Valleys and on the Longshaw Estate, many owned by the National Trust.

Analysis of available data shows a massive 50 per cent loss of flower-rich meadows between the mid-1980s and 1995 in the National Park. Taking 56 meadows with adequate earlier data and comparing these with observations from the mid-1990s showed a decline from around 10.4 old meadow indicators to just 1.7 – a dramatic loss illustrating well the outcome of increased agricultural intensification. Even between the 1995 and 1997 surveys, 24 flower-rich meadows were known to have been lost, which is substantial in minimum time illustrating this habitat's acute vulnerability. A further analysis of selected species mirrors these losses. Taking just Lady's Bedstraw, Cowslip and Hoary Plantain, the number of meadows supporting these declined sharply in around 10 years from 5–17 per cent down to just 1 per cent of 436 meadows. These species are typical of calcareous meadows which would appear to have lost more than some others. Other specialist plant species show similar declines, like Field Scabious (*Knautia arvensis*), Great Burnet and Rough Hawkbit being lost from 70 per cent of the 1980s' sites. Even the commoner Oxeye Daisy and Common Knapweed disappeared from 54 per cent and 62 per cent respectively of meadows, suggesting wide scale agricultural intensification and earlier silage cutting dates which prevent later flowering plants from seeding successfully (Buckingham & Chapman, 1997). More will have been lost since these surveys were conducted, mostly to agricultural improvement, but also to abandonment as rich grassland can lose its floristic interest within a year or two with no regular cutting and inadequate grazing.

These losses are not unique here, but occurring, often at a higher rate, nationally. The total amount of lowland species-rich grassland (meadows and pastures) declined by a huge 97 per cent between the 1930s and 1984 (an astounding 3 million ha) according to the national State of Nature Report (2013). Closer to home, the Derbyshire Wildlife Trust's survey of semi-natural grassland in lowland Derbyshire between 1983 and 1999 demonstrated another massive 91 per cent loss from an already low base, with more still found in the White Peak and Peak fringe outside the National Park than in the rest of the area. A similar decline in valuable grasslands has occurred across the Sheffield and Rotherham Wildlife Trust area over 15–20 years. The losses are also of the animals associated with the fields – invertebrates like bumblebees (already under threat from loss of habitat nationwide), butterflies and a wide range of other insects and breeding birds such as Meadow Pipits and Skylarks that depend on diverse invertebrates. Their wider environmental benefits have also gone.

To counter these losses, efforts are being made to secure the future of as many flower-rich grasslands as possible. Many of the best fields are owned by organisations that seek to protect them – including the NPA, the National Trust and the County Wildlife Trusts. Agri-environment schemes support farmers in managing and safeguarding many of these sites and some have been designated SSSIs which also offers good protection. There are multiple fields much loved by their owners who are proud of their beauty and diversity. However, there are many more, possibly more than half, without such care or protection. Bearing in mind that a single application of artificial fertiliser or one dose of broad leaved herbicide are all that is needed to destroy diversity, their vulnerability is clear.

There are also several meadow restoration projects. In total over 100 ha have been set on the path to restoration over some 20 years, with a further 50 ha targeted by the South West Peak HLF Landscape Partnership programme by 2022. For example, Natural England restored 5 largely improved fields adjacent to Lathkill Dale from 1999 trialling different methods. Monitoring shows massive increases in average plant numbers from 6–8 per metre square to 17–19 after 20 years, matched by expansions of high value indicator species and concomitant declines in more

FIG 176. Meadows on the edge of Lathkill Dale NNR which have improved dramatically following restoration. Plenty of Common Spotted Orchids visible amongst the Oxeye Daisies, Ribwort Plantain, Rough Hawkbit, Yellow Rattle and Common Knapweed.

competitive species, especially of grasses (Fig. 176). Similarly welcome uplifts of species richness and diversity are visible elsewhere as on the National Trust's Longshaw Estate where better management combined with species addition in several hay fields are showing up to 90 per cent herb cover increases from a low base over only seven years. All the measures used in the detailed recording by NT volunteers in the Yarncliffe meadows, for example, show very positive responses overall (Fig. 177), although the 2018 drought depressed them.

It can take 8–15 years before special species like Greater Butterfly (Fig. 173) and Frog Orchid (*Coeloglossum viride*) appear, for example in the Lathkill fields. Common Spotted Orchids emerge sooner (Fig. 176), and have expanded into a substantial population here and are also frequent in some National Trust meadows. This is matched on a small meadow I developed, where this orchid first appeared three years after sowing and then spread extensively before decimation by the 2018 drought, albeit on very skeletal soils (Fig. 231). The Lathkill Dale hay meadows have also developed an excellent waxcap assemblage, including the Ballerina waxcap (Fig. 147) (Abrahams, 2019). Do check out whether the self-guided

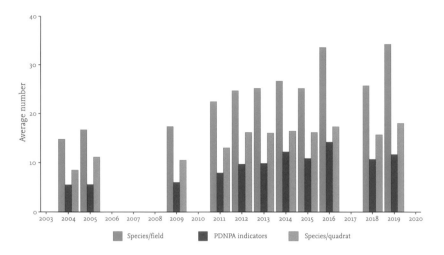

FIG 177. Floristic data for a group of 6–9 hay meadows at Yarncliffe on the National Trust' Longshaw Estate. The graph shows the average number of species per field across all the quadrats (teal), the average number of species per quadrat (green) and the number of Peak District NPA indicators for better-quality meadows (red). Data collected by NT volunteers and staff.

trail (starting just east of Monyash) is open in June/July to enjoy these meadows. Other hay meadow restoration has been undertaken by the Derbyshire Wildlife Trust in their beautiful Hartington Meadows by removing invasive 'weed' species. Eight meadows on the NPA's Warslow Estate have received wildflower seed, mostly by cutting and strewing green hay, and the NPA has also been promoting restoration on other sites. These are all vital for creating more vibrant, colourful and valuable flower-rich grasslands. There is much more to do to though to match targets set locally and nationally in the Government's 25 year Environment Plan and to reverse in any significant way the recent losses of this habitat.

VERGES

Many road and track verges share the characteristics of some agricultural grasslands, although there are also differences. Road verge character is largely related to their soils, management and history. In general they are cut, often annually in mid-summer which removes all flowers, prevents seeding and leaves the cuttings as a thick thatch that blankets out smaller, less competitive species. Some verges are cut later by agreement in order to allow flowers to develop better. Road verge management is the responsibility of the County Council Highways Departments, not the NPA, thus making management for wildlife more difficult to achieve. The back edge of many verges and those alongside tracks rather than roads may not be cut at all or cut only occasionally. It is unfortunate that a few land managers, probably fearing invasion into their adjacent fields, apply broad-leaved herbicide to verges, thus depriving animals like butterflies and bumblebees of even road-side food plants. Contorted dying plants are a depressing sight.

The majority of our verges support few plants or animals, but mostly display characteristic tall-herb vegetation in which False-oat Grass (*Arrhenatherum elatius*) predominates, being an indicator of little or no management. Other tall, vigorous grasses like Cock's-foot (*Dactylis glomerata*) and Yorkshire Fog vie for space, along with Rough Meadow-grass (*Poa trivialis*) – a familiar understory species supported by stronger neighbours. Goosegrass (also called Cleavers) scrambles up this framework using its backward-pointing stiff hairs to find purchase. Equally hooked hairs on its round fruits get entangled in any fur or feathers (and clothes) that try to pass, thus spreading it far and wide.

These tall herb verges are graced by a suite of white umbellifers during the growing season. First is the introduced Sweet Cicely (see Chapter 6), a tall clumped umbellifer with soft green leaves that were used to sweeten stewed fruits (Mabey, 1972) and dense, foamy flower heads in late spring (Fig. 51). Its crown of upward pointing large, dark-brown, seed cases are visible even if passing at speed. It is native in central and southern European mountains. Sweet Cicely is scarce further south (including immediately south of the Peak District in Derbyshire and Staffordshire) and more often found in the northern and western half of the country. It occurs regularly on our verges, more often on the limestone, but only infrequently elsewhere.

Cow Parsley is next, with a more open, taller, lacier appearance (an alternative name is Queen Anne's Lace). Hogweed, a much sturdier, rough plant, replaces this in mid-summer. Both these seem to have increased in recent years, along with Goosegrass – possibly related to accumulative deposition of nitrogen (mostly from vehicle pollution), fertilisers, or/and litter accumulation in unmanaged areas. Even within high altitude verges as along the A53 from Leek to Buxton, Cow Parsley has spread recently where it was absent for decades, although it has not reached all the higher Dark Peak verges yet.

Commoner later flowering plants on these tall herb road verges include Creeping and Spear Thistles, Common Nettle and Rosebay Willowherb. These provide colour, as well as nectar and pollen for Bumblebees and Butterflies, and the

FIG 178. Elephant Hawk Moth caterpillar which can inflate its head to flash the enormous eye spots to alarm predators (and us). (Steve Orridge)

Willowherb is a key food plant for the Elephant Hawkmoth (*Deilephila elpenor*). The caterpillar of this beautiful, large, pink-marked moth always surprises you (Fig. 178). Fully grown, it is very large with black markings along the side and large eye spots which expand dramatically to repel predators: a very impressive response.

Mid and late summer roadside colour is provided by Common Knapweed, Meadow Crane's-bill (*Geranium pratense*) and Field Scabious (*Knautia arvensis*). These are all magnets for bumblebees and butterflies amongst other invertebrates. The last two plants are more frequent on limestone verges. Scrambling plants like Hedge Bindweed (*Calystegia sepium*) take advantage of the tall herb flora and hedgerows, but occurs more at lower altitudes.

More restricted are Perennial Sowthistle (*Sonchus arvensis*) and Chicory (*Cichorium intybus*). Both are characteristic verge-species and rare or absent elsewhere. The tall, yellow-flowered Sowthistle avoids the higher moorland roads and is commoner outside the Peak District. Chicory, a Red List species with sky-blue flowers, is much rarer here and declining (Fig. 179). Also called Succory, it is related to Endive (*Cichorium endivia*): both having long histories. Chicory is nearly universal in every European language and its use dates at least from the Ancient Greeks and Romans who used it in salads and as a vegetable. Melancholy Thistle is a quite rare but majestic plant of roadside and trail edges, mostly on limestone. Look out for the white-backs to the large oval leaves and the typical purple, 'Scotch' thistle-type flower head (Fig. 75).

Where verges are wetter or near rivers, swathes of the large, roughly hairy-leaved Butterbur (*Petasites hybridus*), perhaps accompanied by Meadowsweet, occur. In contrast, on many drier, gritstone verges, vestiges of moorland communities of Bilberry, Heather, Common Cat's-ear and Autumn Hawkbit, often punctuated by tall ferns and even occasionally accompanied by Primroses and Bluebells, show the once greater extent of our heathland vegetation, now fringing mostly improved grassland.

There are some magnificently diverse and colourful verges boasting limestone communities with many similarities to the more calcareous hay meadow with masses of Red Clover, Lady's Bedstraw, Yarrow, Hoary Plantain, Quaking Grass and Field Scabious. Some of these might be graced by Wild

FIG 179. Chicory, an uncommon Peak District plant, growing along a roadside near Foolow with Meadow Crane's-bill and False-oat Grass.

FIG 180. Flower-rich limestone grassland growing alongside Blakemere Lane. Small Scabious (*Scabiosa columbaria*), Burnet Saxifrage, Yellow Rattle, Wild Thyme, Lady's Bedstraw and Ribwort Plantain all visible.

Thyme (*Thymus drucei*) and little annuals like Fairy Flax (*Linum catharticum*) where soils are thin and swards short. A good example fringes Blakemere Lane; a track on the limestone plateau on Bonsall Moor (Fig. 180). It might seem more remarkable to find some of these lime indicators on the moorlands too, but there is a history of using limestone for road surfaces prior to tarmacking. An early 20th Century account describes undertaking road repairs on Rushup Edge as piece work using broken limestone. A machine crushed the stone, which was then moved in carts and steam-rollered into place with water and binding materials, creating much dust when dry. The heaps of limestone sludge swept up in wet weather by the councils were carted off and used by local farmers for their fields, which they regarded as less harsh on the land and slower in action than burnt limestone (Virtue, 1976). The effects of this lime dust and debris have persisted in some verge soils where the lime-indicators may also be mixed with moorland plants. The best examples are near Swallow Moss/Revidge in the South West Peak and Big Moor/Leash Fen on the eastern side. Look out for Common Spotted Orchid, Devil's-bit Scabious, Common Knapweed, Harebell, Meadow

Crane's-bill, Oxeye Daisy, Common Cat's-ear and Eyebrights. But then see how the Heather, Sneezewort and Angelica are intermingled in this moorland/road verge interface (Fig. 181). There are other shorter lengths of flower-rich verge in a variety of locations, more particularly on the less busy road edges. Look out for banks of Oxeye Daisy, Common Knapweed, even with a little Heather or Scabious in them and a variety of other meadow plants including orchids.

Trees and shrubs have colonised many verges, not continuously necessarily, and provide shelter and cover for various animals. As mentioned in Chapter 10, young Ash, Sycamore and many Hawthorn are the principle species, but there are also Roses, Brambles, some Blackthorn forming little thickets and both European and Western Gorse. The Roses are mostly difficult to identify and tend to hybridise. Dog-roses are the commonest, but much rarer are the Northern and Glaucous Dog-roses (*Rosa caesia* and *R. vosagiaca*) and Sherard's Downy-rose (*Rosa sherardia*). The latter has deep rose-pink flowers contrasting with Dog Rose's pale pink ones. Verges are also where garden escapes or purposefully planted non-native species grow. Lines or clumps of garden Daffodils (*Narcissus* spp.) are

FIG 181. Roadside flower-rich grassland beside Swallow Moss in the South West Peak. Common Knapweed and Autumn Hawkbit mixed with Quaking-grass on the verge contrasting with heather moorland behind the fence on the Moss.

everywhere, even far from habitation sometimes – garish yellow briefly before reversion to monotonous green. Garden throw-outs might establish rather than die, so clumps of Montbretia (*Crocosmia* spp.), Spotted Loosestrife (*Lysimachia punctata*), Comfrey (*Symphytum* spp.) and, much more rarely, Winter Heliotrope (*Petasites fragrans*), appear. Snowberry thickets (*Symphoricarpos* spp.) and sprawling Japanese Rose (*Rosa rugosa*) are not uncommon either.

The value of verges is variable. The richer stretches are very important, primarily owing to the dearth of similarly diverse fields. Although we do not know their area, what is left is important as a habitat for invertebrates in particular, as part of local history and as an attractive road-side backdrop. They supplement species-rich grasslands, but in some areas may be the last vestiges of these once splendid meadows. This is a common theme country-wide and Plantlife has highlighted how the total UK rural verge cover matches that of remaining lowland species-rich grassland and supports 45 per cent of our total flora; both being quite remarkable statistics (Bromley *et al.*, 2019). Recent research found that verges more than two metres wide on less busy roads and away from the immediate vicinity of the road are best for pollinating insects, but that these are damaged by summer cutting (Phillips *et al.*, 2019). Plantlife is campaigning to manage road verges as a nationally significant resource to help create more, bigger, better and joined up habitats in keeping with the Government's 25 year Environmental Strategy. This means that away from safety constraints, management needs to foster rather than constrain species-rich grassland.

CHAPTER 13

The Limestone Dales: The Crown Jewels

T HE Limestone Dales are the Peak District's crown jewels. Indeed, in describing chalk and limestone flowers, Lousley (1969) considered our Carboniferous Limestone to be one of the richest in central England for wildflowers, both in terms of the species numbers and their luxuriance. This has not changed. This richness emanates from both intimate mixtures of different habitats and variation in soils, aspects and slopes as explained in Chapter 7. A huge range of plants exploit these multiple opportunities, some with very narrow niches, others more widespread and catholic. The same meeting of northern and southern species also makes the overall character unique within the British context.

The dales are not just a botanists' dream, they form the stage for some literary giants. In Charles Dickens' *The Miner's Daughters – A Tale of the Peak*, for example, children scramble through the dales picking flowers and the fruits of the mountain bramble 'known only to the inhabitants of the hills'. The story is of hard-drinking lead miners, pinched framework-knitters and their children, but starts '…true beauty lies in the valleys that have been created by the rending of the earth in some primeval convulsion and which present a thousand charms to the eyes of the lover of nature.' (Shimwell, 1981).

The most striking features are, as Dickens describes, the steep, sometimes precipitous slopes, craggy outcrops, tors, rock faces and screes (Fig. 182). Woodlands, scrub and grassland often intermingle and grade gently into each other, or scrub decorates the grasslands with woodlands more clearly bounded. This interactive mixing is ideal for a variety of birds and other animals that are featured after the habitats have been explored. We will deal with woodlands, scrub and grasslands here, creeping up onto the plateau when the subject merits

FIG 182. A view down Miller's Dale looking north-west from the top of Priestcliffe Lees showing limestone grassland, both north and south-facing steep slopes, lead waste heaps (hummocks with some bare rubble), woodland, scrub and limestone cliffs. Litton Mill village is just visible in the valley bottom which also accommodates the River Wye.

it, whilst wetlands, heathland and the remains of mineral extraction are covered in Chapter 14, but many of the communities described in both chapters are intermingled or grade into each other in different situations. Best to read the chapters together.

DALE WOODLANDS

In the early descriptions of the dale woodlands, Ash was perceived as the dominating high forest tree. It was not until the 1960s that this was questioned through studies that measured the age of tree cohorts, examined the woodlands' structure and species composition and considered the effect of historical activities. This links with where we left the limestone woodlands composed largely of Oak, Elm, Hazel and Lime with some Ash in Chapters 5 and 6, being

cleared for agriculture. This clearance was probably more pronounced on
the plateau than in the dales and there is hardly any original plateau forest
remaining, even Farey (1813) commenting on the extraordinary paucity of trees on
the limestone plateau then.

We can now identify different kinds of dale woodlands based on their history,
which has implications for the variety of other flowers you can see. Ancient dale
woodlands (defined in Chapter 10) are mostly those with low historic estover rights
(removal of wood for domestic uses) or protected through other means such as in
Royal Forests or merely being very inaccessible on precipitous slopes. Botanically
they are blessed with a greater plant diversity including those that struggle to
colonise new sites. They might exhibit a good structure and diverse tree and shrub
components, or they might be regenerated Ash woods into an original woodland
ground flora after clearance, but maintained as woodland over time.

Ancient woodlands exhibiting the best variety of trees and shrubs and a rich
ground flora are scarce now. The best support Limes – both Small-leaved (Fig. 183)
and Large-leaved (*Tilia platyphyllos*) plus swarms of their hybrids (*T. europaea*, which
are very rare in the wild), Pedunculate and, rarely, Sessile Oaks, Hazel, Wych Elm,

FIG 183. Small-leaved Lime on
very steep slope in the National
Trusts' Hinkley Wood, Ilam.
Note the small, beige, stick-like
plant in front of the tree which is
the very rare Bird's-nest Orchid.

Field Maple, possibly Yew (*Taxus baccata*) along with Rock Whitebeam (*Sorbus rupicola*), Hazel and shrubs like Dogwood, Bird Cherry and Buckthorn (*Rhamnus cathartica*). Ash may be present, but not in dense stands. There is considerable variation on this theme. The Limes grow on cliffs and screes and on dale-side knolls, usually on deeper soils. They may be small and scrubby specimens, bushy from past coppicing, or multiple and single stemmed trees. The biggest known Large-leaved Lime in the Peak District was found recently in the Via Gellia at 7.5 m girth (Wilmott & Moyes, 2015). The natural hybrids are frequent in some of the Via Gellia woodlands and in Matlock Dale. Field Maple is typically a lowland and southern species, more frequent in southern rather than northern White Peak dales. Wych Elm was affected by Dutch Elm disease in the 1960s and 1970s, but has managed to survive in reduced amounts. It was clearly more abundant in dale woodlands prior to the spread of this destructive disease (eg. Merton, 1970). It is an important food plant for White-letter Hairstreak Butterfly larvae (*Satyrium w-album*), which are more frequently seen in the White Peak, even though it tends to frequent the tree canopies (Fig. 184).

FIG 184. White-letter Hairstreak butterfly on creeping thistle, the caterpillars of which feed on Wych Elm. (Steve Orridge)

Yew is not abundant in our woods. Look for its dark-green foliage clinging to tors and cliffs in Chee Dale for example. A hollow branch in an ancient Yew tree named the Betty Kenny Tree in Shining Cliff Woods, was apparently used by a charcoal burner's wife, Betty, as a children's cradle, purportedly giving rise to the rock-a-bye baby nursery rhyme (Wilmott & Moyes, 2015). Like Yew, Rock Whitebeam is quite scarce. Clearly identifiable by the white-felted leaf undersides in spring, this is a Nationally Scarce species which favours rock outcrops and cliffs. It is not confined to ancient woodlands as it occurs on disturbed areas as near Miller's Dale's lime kilns (Chapter 6). We have already met Bird Cherry (Fig. 77), a northern species quite abundant in some dale woodlands and Dogwood, a southern species with limited powers of spread in our cooler climate (Chapter 7).

Close examination of these woods tends to show that there is little regeneration of either the Limes (explained in Chapter 7) or the Oaks. The latter need light to establish and are more often found as young trees outside woodland, but neither the Limes nor Oaks produce viable seed regularly and small mammals eat many. The amount of possibly ancient woodland with well-structured and diverse trees and shrubs is very limited; perhaps only about 5 per cent of the dale woodlands (Piggott, 1969). There are small areas in Lathkill, Monk's, Monsal and Cressbrook Dales, some of those in the Manifold and Hamps Valleys, the Griffe Grange section of the Via Gellia woods, beneath High Tor, Matlock and Hinkley Wood near Ilam.

Woods with a similar although sometimes more depauperate ground cover but dominated by Ash indicate possible ancient woodlands where the tree cover has been removed and then regenerated after a period of grazing or other disturbance. These woods have a much simpler age-structure, usually with Ash in fairly even-aged stands, with sparsely occurring shrubs or small trees like Crab Apple, Rowan, Holly and Guelder-rose (*Viburnum opulus*), although none of these are restricted to woodlands. Sycamore commonly invades, Wych Elm often occurs and Hawthorn and Hazel provide the main understory. These more recent tree eruptions originate within the last 200 years after past widespread disturbance like lead mining or coppicing Ash for puncheons about every 25 years (Farey, 1813), for which demand was high. As the mine depths increased and water ingress became problematic, soughs were driven into hillsides, the first in about 1650 with concomitant disturbance to dale sides. Grazing in the dales would have been greater prior to the reclamation of the plateau rough grazing following 18th and 19th century parliamentary enclosures. Rabbit population increases in the 19th and 20th centuries would have grazed off woody saplings and seedlings, only relieved after the spread of myxomatosis starting in 1954.

As we have seen (Chapter 10), woodland planting became more prevalent in the same period as Enclosures were established, adding Beech, Sycamore and various conifers to many dale-sides. At the same time, scrub was being cleared in the late 1790s to accommodate tree planting, only to spread again as dale-side grazing values declined, a process which accelerated after myxomatosis decimated Rabbits. In addition to these external pressures, some favouring woodland, others reducing it, trees have different powers of establishment and spread, with mast years, competition and predation all playing a part (Merton, 1970). Ash fruits, for example, are vulnerable to Wood Mice and Bank Vole depredations (up to 70 per cent of seeds); their viability is also variable; and infestations by a moth larvae particularly in warm, dry summers reduce seed performance. It also has a long dormancy of 18 to 20 months and needs light to develop.

The interplay of these interacting forces creates the patterns and appearance of the rest of the dale-side woodlands today. Where disturbance was considerable, such as in the Via Gellia, the woods show a complex pattern, whereas with little later disturbance, extensive stands of more uniform woodland developed as in lower Cressbrook Dale. Thus new Ash woodlands can form in different ways. If the dale-side was grazed grassland, Hawthorn is usually the primary coloniser, even under moderate grazing intensities. Ash can grow sparsely through the protective thorny thicket, but if grazing levels are reduced (including Rabbits), it can establish in the open too. It favours bare soils for establishment and will eventually overtop the Hawthorn, which then remain as an understory shrub. Where these are within the same general age range, or the Hawthorn is older, then this is the probable origin of a particular stand.

Hazel can colonise the finer talus generally at the bottom of screes, even persisting despite quite high grazing levels. Ash can colonise even active scree; indeed trees are noticeable on several screes outside woodland such as in Monk's Dale and Deep Dale at Topley Pike, but its colonisation of thick grassland would be slow without bare ground. Woodland developed from grasslands would have a wider Ash age range reflecting this slower establishment rate. In contrast, sites where mining or other activity such as quarrying (as in Miller's Dale) were abandoned, Ash trees quickly invade the bared ground if there is no grazing, sometimes with Sycamore as well once this became widely established in amenity planting in the late 18th Century. Volunteer Sycamore and Ash can also take over from failed planting. Pines and Oaks often grow badly when planted on steep dale-sides for example, visible on south-facing slopes in Lathkill Dale and on the west side of lower Cressbrook Dale.

As an example of change over time, the detailed history of some Lathkill Dale woodlands has been knitted together from historical documents and pollen analysis (Shimwell 1977, Anderson and Shimwell, 1981). Ash was lost from the woodlands in the 1100s during a warmer period. This was when extensive lands were given to several abbeys, which established sheep granges until the dissolution of the Monasteries in 1539. The Lathkill River was one Abbey boundary with populous Over Haddon to the north. The effects on the south side Dale woodlands is not significant in the absence of a large rural population plus carefully controlled sheep grazing. Estover levels were low within the monastic lands, compared with the north side where Over Haddon inhabitants were in frequent conflict with the Grange through pasturing cattle and felling trees on the wrong side of the river. After the Dissolution of Monasteries and during the following Little Ice Age (starting in the mid-1500s), Lime disappears from the pollen diagram from near Lathkill Dale and woodland regeneration failed owing to clearance for lead mining. It was not until after its peak that regeneration of dale-side woodland occurred, although in a piecemeal fashion relating to where extraction was concentrated. This regeneration has been dated to the period 1810 onwards through counting tree rings, and the swarms of Ash, many of which still survive, illustrate colonisation after disturbance when there is plenty of light.

Placing the Ash and Lime woodlands with a richer ground flora in the national context shows these are a sub-community of the widespread Ash-Field Maple-Dog's Mercury woodland type, which is more-or-less unique here. These are the woods with Lime. Other woodlands have affinities with a more westerly sub-community where Ash, Sycamore and Wych Elm are more prominent with a rich ground flora but with what appears to be an even-aged tree canopy. Silver Birch may appear in this community, often picking out less calcareous soils, but Hazel is less common, with Hawthorn and some Elder replacing it (Rodwell, 1991a). This assemblage fits into the category of recently regenerated woods described above.

Semi-natural plateau woodlands, being now so rare, are difficult to describe, but one, now lost, on Coombs Dale's upper edge was Sessile Oak dominated with a ground cover reminiscent of the upland Oak woods of the Dark Peak of Bracken and Creeping Soft-grass accompanied by Bluebells, Pignut, Wood Sorrel and Common Dog-violet (Piggott, 1970). Although ungrazed at the time, this sounds like a woodland missing a shrub and small tree layer, and might have been grazed previously. It was growing on a brown earth soil developed over loess deposits (see Chapter 7). Where planting has been prevalent, mostly since the late 19th century, in both the dales and on the plateau, Beech, Larch, Spruce and Pine survive, sometimes still providing a full canopy, as in Great Shacklow Wood, or as scattered vestiges of former plantations as in Hay and Biggin Dales.

FIG 185. Mountain Currant in Monk's Dale. It has a tumbling habit from rock ledges and outcrops and erect flower spikes.

The dale-side woodland ground flora is as varied as its origins and can be splendidly rich. Additional shrubs are Spindle (*Euonymus europaeus*) and Wild Privet (*Ligustrum vulgare*); the former being more a lowland species and scarce here, with many past uses as spindles for wool spinning and for butchers' skewers. The brightly coloured fruits are poisonous. Another, much more special shrub is Alpine Currant (*Ribes alpinum*), a Nationally Scarce species restricted to North Wales and northern England, but the best place to see it is in the White Peak. Its tangled curtains hang over short rock outcrops as in Monk's Dale and along some roadsides, for example in Dovedale and the lower Manifold Valley (Fig 185). This is not to be confused with upright bushes of wild Red and Black Currants (*Ribes rubrum* and *R. nigrum*), which are more scattered in dale woods but commoner in lowland Britain's woodlands.

Other scarce shrubs are related: Mezereon and Spurge-laurel (*Daphne mezereum* and *D. laureola*). Mezereon is commonly grown in gardens supporting clusters of bright pink flowers before spring leaf emergence. The wild version is generally less showy growing under the tree canopy and struggles to regenerate.

It is Red Listed but grows in the Wye Valley and Via Gellia woods, although even here the number of plants is limited. Miller's Dale is a good place to see it near the lime kilns. Spurge Laurel is much more a southern species reaching its northwestern boundary in our woods where it is rather rare.

The ground layer in the woods is particularly colourful, at its best in spring prior to canopy leaf emergence. Ash leaves are frost sensitive so they unfurl later than many trees in late May, giving plenty of light to the vernal ground flora. The woodland herbs start with Dog's Mercury's spring growth, although the observant will note that male and female flowers grow on separate plants, and as they spread vegetatively, whole banks can be of one sex or the other. Interestingly, our climate seems to favour female plants, identifiable by their swollen ovaries and missing anthers, whilst the opposite pertains in most of the country. This green matrix is spangled by Common Dog-violet, spiked by the hooded flowers of Lords-and-ladies (*Arum maculatum*), and followed by carpets of Wood Anemones – more abundant here usually than in shale-based woods. The Arum has many vernacular names, including Wild Arum, Cuckoo-pint and Friar's Cowl; all of which reflect its unusual shape with inferences of males and females and copulation. The flowers are at the bottom of the central flower stem (spadix), female flowers in a ring below the male ones, sealed by a ring of hairs, with the whole partly enclosed in a cowl-shaped hood (spathe). The spadix produce a faeces-like smell and, remarkably, is up to 15 °C warmer than the surrounding environment. This attracts various insects, particularly owl-midges (*Psychoda* spp.), which are then trapped by the hairs while they are dusted with pollen, before escaping to repeat the process; thus pollinating female flowers on the next plant they visit. Look out for the red berries in autumn but beware they are poisonous.

Bluebells are the next species to push through the ground cover, but patchily rather than the high density found in some Dark Peak and South West Peak sites. Primroses share this cover, perhaps with patches of Woodruff (*Galium odoratum*); a delightful little white-flowered bedstraw with distinctive ruffs of leaves. Sanicle (*Sanicula europaea*), another ancient woodland indicator, is occasionally found, more often in the limestone woods than elsewhere. Early-purple Orchid (*Orchis mascula*) and Common Spotted Orchids are similar in having spotted leaves, but the first flowers much earlier with bright rather than pale pink flowers (Fig. 198). Both are equally at home in the dale woods and grasslands.

There are subtle variations on this general picture. On lower damper slopes, a distinctive garlic smell heralds sheets of white-flowered Ramsons in spring – detectable even at speed passing along the A6 east of Buxton or along the Via Gellia, for example. The fresh green leaves of this onion relative can dominate for the few months it is above ground. Lily-of-the-Valley (*Convallaria majalis*) features on a number of scree patches or on shallow soils within the woodlands. Broad-leaved Helleborine (*Epipactis helleborine*) is another scree specialist, both in the woods, on their edges and in open scrub associated with screes. Later summer-flowering plants of similar spots include Nettle-leaved and Giant Bellflowers (*Campanula trachelium* and *C. latifolia*). These are two more species on their range edges, Nettle-leaved bellflower (Fig. 186) being a southern species and fairly scarce, with Monk's Dale plants at their altitudinal maximum (320 m). The generally taller Giant Bellflower is the more northern and western species, with its flowers opening from the bottom upwards (opposite to Nettle-leaved Bellflower's) and is paler coloured. Narrow-leaved Bitter-cress (*Cardamine impatiens*) is an annual or biennial plant which prefers humid woodlands, riverbanks and moist rock outcrops, often under a tree cover. The pointed ends of the leaves and often petalless flowers make the plant distinctive. It is a Nationally Threatened species which occurs rarely in our area. Look for it in Monk's and Chee Dales.

FIG 186. Nettle-leaved Bellflower in woodland in Cressbrook Dale. The flowers are richer and darker than those of Giant Bellflower which also grows in dale woods.

The grasses demonstrate similar ecological niche preferences. Tufted Hair-grass' rough tussocks grow in the damper soils lower down slopes, whereas the limey-green leaves of False Brome (*Brachypodium sylvaticum*) are more closely associated with rock outcrops on shallow, dry soils and may represent vestiges of invaded grassland in these situations. The soft green leaves and delicate flowering heads of Wood Melick (*Melica uniflora*) are less widespread in the woodlands. The much rarer Mountain Melick (*Melica nutans*) is another northern species distinguished from its cousin by having flowers clustered on one side of the stem. It is only found in Wye Dale and Via Gellia woods in the Peak District.

The woodland herb layer is augmented by several ferns that are commoner in or restricted to dale woodlands compared with those elsewhere. The most recognisable is Hart's-tongue (*Asplenium scolopendrium*) with a robust clump of undivided, strap-like, dark green fronds. Hard Shield-fern (*Polystichum aculeatum*) is similar to Male-fern (abundant in acidic as well as limestone woodlands), but the leaflets' spiny points and leathery fronds are distinctive. Common in all types of dale woodlands are more widespread species like Red Campion, Herb Robert (a Geranium that favours scree or rocky places) and Wood Avens (*Geum urbanum*). Water Avens, also called Granny's Bonnets, is locally common in damper places and is more common on the limestone than elsewhere. It can hybridise with Wood Avens producing a highly fertile Hybrid Avens (*Geum* x *intermedium*). It is not common but is distinguished by its pale yellowy flowers, scarcely notched rounded petals and purple and green stalks. It grows with both parents, for example, in several sites along the Wye Valley, in Grinlow Woods, Buxton and in Lathkill Dale. Ivy is another ubiquitous plant but favours secondary woods and plantations, often climbing up the trees.

Mosses and liverworts add luxuriance to the ground flora, particularly on rocks and ledges with little soil. The most abundant are quite distinctive. Common Tamarisk-moss (*Thuidium tamariscinum*), Fox-tail Feather-moss (*Thamnobryum alopecurum*) and Big Shaggy-moss (*Rhytidiadelphus triquetrus*) produce sprawling carpets amongst the flowering plants (Fig. 187). Tamarisk-moss has very evenly thrice-branched stems with yellowish-green or dark-green shoots, while the Feather-moss has the appearance of a miniature tree. The Shaggy-moss is large and bushy with irregularly branched red stems and large leaves (to 6 mm) that stick out untidily. Greater Featherwort (*Plagiochila asplenioides*) is a profuse, robust liverwort with stems up to 12 cm long and bright or pale green translucent leaves that drools from ledges in shaded or damp locations.

There are many more species for the specialists to pursue. One that has puzzled the experts is *Brachythecium appleyardiae*. This was regarded as not only endemic to Britain and Ireland, but also one of the special features of

FIG 187. Example of some of the many luxurious mosses that grow in limestone woods. Here Fox-tail Feather-moss appears as a miniature tree-form, with the regular branching patterns of Tamarisk-moss mixed with the curled ends of Cypress-leaved Plait-moss.

the Wye Valley SSSI. However, DNA analysis has shown that it is too close to *Scleropodium cespitans* (Tufted Feather-moss) for it to be considered a separate species and it has also now been found in Belgium. Not uncommon in southern England and Wales, *Scleropodium* forms mats on dry limestone rock ledges, often near streams. It is found in Cressbrook Dale but is no longer worthy of special conservation measures as it is not now considered as endemic or rare (Blockeel *et al.*, 2005).

Climbers in the limestone woodlands are scarce or absent as they are mostly southern or lowland species. The epiphytic flora is sparse too. The main moss on the tree boles is Cypress-leaved Plait-moss (*Hypnum cupressiforme*), closely related to the moorland *Hypnum*. Polypody fern consists of three separate species, of which the Intermediate Polypody (*Polypodium interjectum*) is more frequent in limestone woods, both perched on tree branches, where it is damp enough, and on walls and rocks.

There are some really rare species that deserve mention amongst this sample of our rich flora. Herb-Paris (*Paris quadrifolia*), is uniquely shaped, with a whorl of four large broad leaves and a curiously wispy, spidery flower (Fig. 188). It is a good ancient woodland indicator of restricted distribution nationally and very rare in the Peak District, found only in Monk's Dale and Via Gellia woods. More esoteric is the Fingered Sedge (*Carex digitata*), also associated with ancient woodland. It is very local nationally and limited in the White Peak to Cressbrook Dale, Priestcliffe Lees and Coombs Dale. It flowers early often in light shade and is identifiable by the female flower spikes' crimson tinge. Another real rarity is the Daffodil (*Narcissus pseudonarcissus*), which, compared with the many garden daffodils, has two-toned colouring, is shorter and has narrow grey-green leaves. Once more widespread in woods and pastures and more familiar in the west and north, the only Peak District limestone site now is Taddington Woods, although it also appears in the Derbyshire Wildlife Trust's Lea Wood on the southeast edge of our area (which is not a limestone wood). An old name is Lent Lily related to its long association with Easter celebrations.

Other specialities are Toothwort (*Lathraea squamaria*) and Bird's-nest Orchid (*Neottia nidus-avis*) (Fig. 183). Toothwort is a root parasite on Hazel (usually) or

FIG 188. Herb Paris in Monk's Dale. The plant's unique shape and flower form are distinctive.

Poplars (*Populus* spp.) and Beech and is therefore leafless and colourless. In the Peak District it is principally found in dale-side woods as in Cressbrook Dale, along the Manifold Valley and in Shacklow Wood. Bird's-nest Orchid also has no chlorophyll, but is mycotrophic (obtaining all its nutrients from mycorrhizal fungi). It is very rare here and in adjacent counties, usually below Beech as in Shacklow and Hinkley Woods, but also in Ash woods in Miller's Dale. It is classified as Near Threatened in Britain.

Plants dependant on fungi lead neatly into consideration of fungi in these White Peak woodlands. There is far less information on these than for some Dark Peak woods. Ash, although the main tree, is not ectomycorrhizal, so has few associated fungi unlike most trees in the upland Oak woods. It does, though have several conspicuous fungal associates. The commonest is King Alfred's Cakes (*Daldinia concentrica*), also called cramp balls. This is a black, hard, perennial ball which looks like burnt growth, growing in concentric rings on dead wood – it is a saprophyte so does not kill the tree – each ring representing the previous year's growth. Another common fungus, *Nectria galligena*, causes canker in Ash – visible as dead tissues that grow slowly and open out. This is a parasitic fungus, but one that provides opportunities for other organisms such as bats that might roost in the developing crevices or a myriad of other invertebrates that benefit from the damp holes and hollows. Dead wood is the most important element in any wood owing to the number of fungi, invertebrates and other organisms that live in or on it, with different species favouring wood hanging in the trees still or on the ground.

Different fungi will be more abundant in the dale woodlands with a greater variety of trees, but these are limited to some degree too. Limes, for example, despite their long history in the country, host a surprisingly meagre number of fungi. Neither Field Maple nor the more recently introduced Sycamore are ectomycorrhizal and there are no specific agaric relationships. Sycamore, though is renowned for the common pathogen, Tar Spot fungus (*Rhytissma acerinum*), producing black blotches on leaves in summer and early autumn. This has no obvious effect on the tree's long-term health. This is also an Ascomycete fungus like King Alfred's Cakes that produces spores in an ascus; Greek for a sac or wineskin, the microscopic structure in which the spores are formed. There are few Oaks in our dale woodlands, but these are renowned for their ectomycorrhizal and saprophytic fungal complement of the agaric and Boletus types, although less so on calcareous soils. Hazel, though is the commonest undershrub and this has several specialists associated with it. It is ectomycorrhizal and Fiery Milkcap (*Lactarius pyrogalus*) is the commonest associate, with its greyish-fawn caps, pale orange gills and a white acrid fluid exuding if broken open. Completely different in appearance, the common Glue

Crust (*Hymenochaete corrugata*) spreads by producing a mycelium that binds or 'glues' touching branches together, thus permitting it to move between trees. This is a common survival strategy in tropical rainforests but rare in temperate ones. Another fungus found under Hazel, for example in Deep Dale, is the Greenspored Dapperling (*Melanophyllum eyrie*), this time appearing on soils. This quite rare fungus is unusual in having green spores – unique amongst UK fungi. The White Peak is good for Dapperlings (*Lepiota* species mostly) in general. The Common Earthstar (*Geastrum triplex*) is often associated with the ground under Yew, not only in woods but also in churchyards. This lovely little brown, bulb-shaped fungus opens up like petals round a puffball-like central dome. It has been found in quantity in woodland in Cromford and Darley Dale but look out for it along the trails in the White Peak.

The total White Peak woodland cover is around 3,096 ha, including plantations. Of this, only some 1,074 ha is considered to be ancient woodland or with an ancient woodland ground flora (Anderson, 2016). This is a paltry 2 per cent of the White Peak. The woods are generally long and narrow and quite well connected within each dale, but there are very few links across the plateau between them, rendering them rather isolated at the landscape scale. Around 83 per cent of the ash and lime dale woodlands are protected in SSSIs (Fig. 3) together with more in County Wildlife Trust sites or National Trust land. In a European context, the woods are classified as ravine woodland and the Peak District is enormously important for these, supporting the largest area in the UK, even though the richest woods are restricted. These ravine woodlands form a priority element in the European Special Area for Conservation (SAC), which embraces all the main limestone dales (Fig. 3).

This small area is further threatened by ash dieback disease, first confirmed in the UK in 2012 (although it might pre-date that). It originates in Asia where it does little harm to their native ashes. It was introduced to Europe some 30 years ago but our native Ash has no natural defences against it. A fungus *Hymenoscyphus fraxineus* causes the disease, being spread by windborne spores. It kills the most recent growth before spreading down the tree. Symptoms are dark patches on leaves in the summer, before wilting, turning black and falling prematurely. Dieback in the shoots and leaves is then visible and lesions develop where branches meet the trunk, stimulating epicormic growth below the dead wood, which is a common response to stress (Fig. 189).

Unfortunately it is estimated that some 95 per cent of European Ash trees have been infected. Death can be rapid or a slow decline over several years. The disease is affecting all the UK, but it could be particularly devastating in the White Peak owing to our high concentration of ash woodlands. Already Ash are

FIG 189. Road-side Ash showing symptoms of dieback with many of its outer twigs without leaves, pale brownish colouration to the upper branches and a much reduced canopy.

affected throughout the Peak District with mature trees succumbing rapidly in places like Monk's Dale and Dovedale. Smaller trees and saplings (which the disease often preferentially affects) are infected or dying everywhere, not just on the limestone. There is a glimmer of hope. It was found in northeast France that, although the disease spread to virtually all Ash present, many stayed relatively healthy after 10 years, with only about 5 per cent loss for more isolated mature trees rather than those closely growing in woods. However, this was related to the fungus' intolerance of heat-waves that are rarer here (Grosdidier *et al.*, 2020). British research is currently searching for resistant trees from which new Ash could be grown.

To increase ash woodlands' resilience, the National Trust and Natural England, both of which own numerous dale woodlands, have commenced woodland management programmes in partnership with the Forestry Authority, NPA and others. The strategy separates the ancient woodland with the richest tree and shrub flora from those where Ash-dominates, with or without a rich woodland ground flora. Coupes of some 0.1 to 0.2 ha of the dominant Ash are selected, covering up to 15 per cent of any woodland compartment over five years for selective felling. This is followed by planting new trees and shrubs using the rich areas as species templates. This targets reversing the past human intervention and damage and adds to tree diversity, whilst simultaneously increasing woodland resilience to Ash Dieback or future threats such as climate change or pest outbreaks. The aim is to maintain woodland conditions where most beneficial for the richest ground flora, while accepting some woodland thinning and reversal to scrub and grassland. This management has only just started and there will be more in the near future (Barley, 2018).

There is a natural resistance to managing woodlands based on our reverence for such long-growing plants and our love of trees. We have told the story of these woods in terms of their history and shown that many are very much derived from past human activities and that Ash-dominance is not related to the imagined original 'wildwood', even though the richer areas with a range of trees and shrubs, especially with limes, possibly have a much longer history and largely escaped the axe and grazing. Do not therefore be alarmed when you see the felled patches and replanting – but recognise these as the best way forward to ensure these wonderful woodlands remain in the future.

SCRUB

We have seen how scrub might develop either from invasion of grassland with reduced grazing or as woodland vestiges after partial clearance or increased grazing. We have some very distinctive scrub communities in the dales, although their origin might sometimes be rather obscure. A particularly rich hazel scrub is widely distributed, often on north or west-facing slopes, frequently incorporating screes, and surviving in a similar state for decades. It is a botanical gem of high nature conservation value – a feature of the SSSI dales. It shares a mixture of woodland and grassland plants that seem to have found a unique niche. Hazel is a key component but shared with occasional sprawling masses of Dogwood, some Common Privet and Guelder Rose. The largely southern chalk-loving bush, Wayfaring Tree (*Viburnum lantana*) is decidedly rare here. Buckthorn, an important food plant for the delightful yellow butterfly, Brimstone (*Gonepteryx rhamni*), is slightly more frequent. Occasional Whitebeam (*Sorbus aria*) and other commoner species also appear.

This is also home for several roses, mostly difficult to identify as they hybridise freely. Dog Rose, with its drooping branches and pale pink flowers is readily recognisable. Soft Downy-rose (*Rosa mollis*) is distinctive through its

FIG 190. Burnet Rose, with its beautiful, large cream flowers, and leaves with 9–11 small leaflets, which contrast with the larger leaves and fewer leaflets of most other roses. Bloody Crane's-bill's fuchsia-pink flowers in the background.

strong, upright patches of suckering stems, plus straight prickles, deep pink flowers and softly hairy leaves. The much rarer Burnet Rose (*Rosa spinosissima*) forms low suckering thickets, but has small neat leaves divided into 9–11 leaflets (Fig. 190). Most roses have fewer and larger leaflets. Burnet Rose is also fiercely prickly, with reddish acicles (slender prickles without a broadened base) intermingled with true prickles. It is largely restricted to the White Peak here but is more familiar on coastal sand-dunes. Look out too for an unusual gall on its stems causing bright red swellings that become heavily fissured over time. Found in 1999 in Cressbrook Dale, it is caused by a new species of gall midge *Janetiella frankumi*, named after its finders (Higginbottom, 2010). While recording in Cressbrook Dale in 2014, I found a Burnet Rose look-a-like, but with atypical glands and hairs on the black fruits. After specialist checking, it was identified as a Soft Downy-rose and Burnet Rose hybrid, *Rosa sabinii*, which had been declared extinct here and not found since 1911. More was later discovered in Monk's Dale in similar scrub. So look out for some strange roses.

The herbaceous cover in this hazel scrub is equally distinct. The assemblage has affinities with woodland glade and edge species, mixed with grassland specialists. Those more typical of woodland and its edges are Dog's Mercury and Mountain Melick, sometimes with Lily of the Valley and Broad-leaved Helleborine. A profusion of mid-summer colour depends on grassland plants like Bloody Cranes-bill (Fig. 190) and Common Rock-rose (*Helianthemum nummularia*) with Marjoram (*Origanum vulgare*) providing a mecca for nectar-loving insects. There are also some open scrub specialists that are more often found at the foot of rock ledges or in scrubby woodland exemplified by Stone Bramble (*Rubus saxatile*), Woodsage and False Brome. Stone Bramble is one of the distinctive White Peak plants found in a limited range of sites in Wye Dale and its tributaries and Biggin Dale, where it is very much at its southeast limit. Its long-growing runners might trip you up on slopes, but are barely prickly and produce scarlet fruits like bramble but with fewer segments.

The once more widespread, but always rare orchid relative, Dark-red Helleborine (*Epipactis atrorubens*) seems to have declined radically in its well-known haunts in Monk's Dale, Priestcliffe Lees, Cressbrook Dale (Fig. 191) and Biggin Dale, but still persists in Combs Dale. Despite close monitoring and

FIG 191. The declining Dark-red Helleborine in Cressbrook Dale. The leaves are more reddish and the flowers are completely red-purple compared with Broad-leaved Helleborine.

protection from rabbits and from slugs in wet years, numbers are declining. Ongoing research is comparing the situation with the stronger population in the Yorkshire Dales, but the future does not look bright for this beautiful plant.

We have already looked at other scrub types that are moving towards woodland habitats, including the effects of myxomatosis on the rabbit population and its suppression of scrub.

GRASSLANDS

Revel in the luxuriant growth of bushy yellow-flowering Common Rock-rose, savour the pale pink mats of Wild Thyme, delight in the fuchsia-pink petals of Bloody Cranes-bill (*Geranium sanguineum*) or marvel at the exoticism of orchids – just a touch of the beauty, colour, scent, shape and structure of the dale-side grasslands (Fig. 192). Note where you see different plants and patterns start emerging. Every twist and turn of aspect, steepening or reduction of slope,

FIG 192. A colourful limestone grassland bank with abundant Common Rock-rose, Bloody Crane's-bill, Salad Burnet and Cowslips amongst a wide range of plant species.

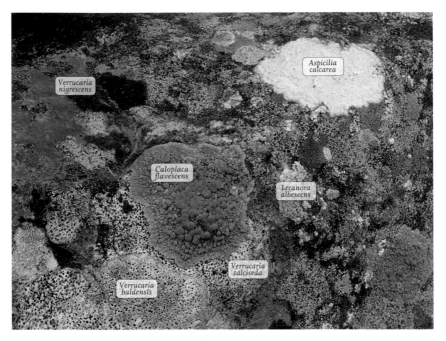

FIG 193. Some of the lichens that grow on exposed limestone rock surfaces. (Steve Price)

emergence of rock buttress or cliff, are all correlated with a change in soil and growing conditions and it is only in the limestone dales that you can see all these at once, within metres sometimes, but certainly all in a day's botanising, and this is unique. We will start with the most open communities that perch on rock ledges and in the most skeletal soils, move onto the steep slopes, crest the top, see what grows on the flatter edges, and explore different aspects. We will clothe the different soils explained in Chapter 7 with plant communities.

Cliffs, outcrops, boulders, scree, crevices and ledges, variously shaded, exposed or sheltered, offer opportunities for colonisation by different lichens giving us an exceptionally rich lichen flora in the White Peak. The crustose white patches of *Aspicilia calcarea* and the orange of *Caloplaca* species are the most obvious ones on exposed surfaces (Fig. 193). Closer inspection reveals several 'black dot' lichens, mostly *Verrucaria* species, often seen in mosaic patterns, the black dots being their fruiting bodies. The thalli of these lichens live in the rock surface and display various shades of black, white and grey, the latter sometimes indistinguishable from the rock itself. More fleshy foliose lichens can also be found in abundance. *Collema* and *Leptogium* species have gelatinous thalli, mostly dark green-brown or red-brown, swelling markedly when wet. One of the larger and most appealing species of the limestone dales is *Solorina saccata*. This is a rather dull and hard-to-spot grey-green colour when dry, but on wetting turns a bright apple-green with contrasting chestnut fruits sunken in the thallus. It grows on south-facing outcrops in moss-covered crevices and ledges.

North-facing cliffs and outcrops, out of direct sunlight, generally exhibit different species. For example, shaded, dry, vertical faces support *Lepraria nivalis*; with fluffy centres appearing to fall away in its pale pinky-grey, circular powdery thallus. Careful searching in shady crevices might reveal *Gyalecta jenensis*, with its pie-crust notched, orange fruits, along with another shade lover *Porina linearis* which has a dark brown thallus.

Rock surfaces, ledges and crevices can all suffer acute water stress, especially if south-facing. Tolerant plants can control this through structural or ecological adaptations. Some more extreme responses are exhibited by mosses like Woolly Fringe-moss (*Racomitrium lanuginosum*), Wall Scalewort (*Porella platyphylla*) and Frizzled Crisp-moss (*Tortella tortuosa*). The upland species *Racomitrium lanuginosum* is one of the more conspicuous and notable mosses that has long whitish leaves often with curved hair points. It is not restricted to calcareous rocks and forms grey-green cushions, although not commonly in the White Peak. The limited limestone pavement at High Edge and nearby Upper Edge are good sites. Incredibly, it can survive water contents as low as 10 per cent of its dry weight, even recovering to grow normally again after 239 days desiccation (Dilks & Proctor, 1974). *Tortella tortuosa*, another western and northern species, commonly forms yellow-green mounds on well-lit rocks. It has long narrow leaves with wavy margins when wet, but when dry shrivels into distinctive contorted spirals to survive long droughted periods, as in 2018, before unravelling and growing again.

Hairy leaves or water storage tissues are other adaptations useful for droughty environments. Biting Stonecrop (*Sedum acre*) occupies many limestone rock ledges or wall tops (Fig. 194). It has a limited root system but succulent leaves that are resistant to drought. Plants like Shining Crane's-bill (*Geranium lucidum*), in contrast, have waxy leaf cuticles to resist water loss. Its reddish green, shiny lobed leaves are conspicuous on or around rock ledges (Fig. 194). Hairy leaves are more frequent on some of the Hawkweed species (*Hieracium*) and the white felty under surface of the creeping rosettes of Mouse-ear-hawkweed are diagnostic.

The Hawkweeds (yellow-flowered cousins of Dandelions) are particularly interesting as many populations are isolated leading to restricted cross-pollination. They are renowned for the ability to produce seeds without this (apomictic reproduction), which results in swarms of identical plants and many

FIG 194. Limestone rock-ledge community with Biting Stonecrop, Shining Crane's-bill, Herb Robert, Maidenhair Spleenwort and dead seed-cases of Hairy Bittercress overlooking Glutton Dale.

micro-species with rather small differences making identification difficult. The Derbyshire Flora shows how different species favour different parts of the area, with some restricted to the White Peak. There are several that favour dry rocky ledges such as Shade Hawkweed (*Hieracium decolor*), a northern species, and Thick-leaved Hawkweed (*H. dentulum*). Both are rare. Dales Hawkweed (*H. dalense*), British Hawkweed (*H. britannicum*) (Fig. 195), and Derby Hawkweed (*H. naviense*) are even more special, unique to the White Peak and not found anywhere else in the world (endemic). They occur variously in Wye Dale, Dovedale, Tideswell Dale, the Winnats and Monk's Dale. Particularly exciting in 2017, two small populations totalling only 62 plants of the endemic Leek-coloured Hawkweed (*H. subprasinifolium*) were found on different banks along the Monsal Trail in Chee Dale. First described in 1863 from the Peak District, it had not been seen for over 60 years. Its leaves are the same chalky-green as those of leeks. Seeds have been collected for long-term storage as part of the Kew Gardens' Millennium Seedbank Partnership (Anon, 2017).

FIG 195. British
Hawkweed growing on
a limestone ledge at the
upper end of Cressbrook
Dale, one of several
Peak District endemic
Hawkweeds.

Rather than withstanding dry conditions, winter annuals employ an
avoidance strategy. These usually establish in autumn, overwinter, flower early
in the year and then survive the desiccating conditions in summer as seed.
Widespread species include Common Whitlowgrass (*Erophila verna*), Hairy
Bittercress, Dove's-foot Cranesbill (*Geranium molle*), Thyme-leaved Sandwort
(*Arenaria serpyllifolia*) and Parsley-piert (*Aphanes arvensis*). All are small. The
Whitlowgrass can be flowering as early as February, and a later spring visit will
witness the explosion of seeds from Bittercress' long narrow pods, growing
amongst the soft green leaves of the Cranesbill.

There are some special species of this community though. Rue-leaved
Saxifrage (*Saxifraga tridactylites*), Early Forget-me-not (*Myosotis ramosissima*) and
Hutchinsia (*Hornungia petraea*) are worth searching out. The Saxifrage (Fig. 196)
is not uncommon on the limestone and even appears on lime-rich mortar in
gritstone walls sometimes, but it is restricted mostly to calcareous soils in similar
habitats in England and Wales. Similarly, the Forget-me-not is typical of dry

open places on sandy or limestone soils, mostly in lowland Britain. Much rarer is Hutchinsia, an inconspicuous little annual with pinnate leaves (Fig. 196) growing mostly in the White Peak, Yorkshire Dales, South and North Wales, with only scattered records in a few other places. It is Nationally Scarce found here on south-facing rocks in Cressbrook Dale, Dovedale, Wye Dale and Monsal Dale. Look for it on thin soils with bare gaps just behind bare rock edges. Its numbers vary annually, possibly owing to the availability of colonisation gaps (which are greater after droughts) but Edwards (1956) hints at possible susceptibility to severe winter weather controlling seedling survival too.

Plants concealed more in crevices include two common little ferns; Wall-rue (*Asplenium ruta-muraria*) and Maidenhair Spleenwort (*Asplenium trichomanes*) (Fig. 194) with contrasting leaf shapes. Much rarer are Green Spleenwort (*Asplenium viride*) and Rusty-back Fern. The former is similar to Maidenhair Fern but with a green rachis (mid-rib) and less vigorous growth. It favours north-facing or sheltered ledges and crevices and is another upland Britain species at

FIG 196. Hutchinsia and Rue-leaved Saxifrage on skeletal soils just behind a rock ledge. Hutchinsia shows seeds developing on the lower stem and inconspicuous white flowers at the top. The saxifrage has small white flowers. Both are only about 4–6 cm high.

its range boundary in the White Peak. There is a substantial population close to the dale-bottom footpath in Back Dale, but also odd plants in a variety of shady crevices in other dales. Rusty-back Fern is commoner in west and southwest of Britain but very rare in the White Peak. It also appears in odd places associated with mortar such as a railway bridge near Combs Reservoir and some garden walls in the Edale Valley.

Perhaps more surprisingly, trees like Ash, Hawthorn, some Purging Buckthorn, Ivy and large herbaceous plants like Greater Knapweed (*Centaurea scabiosa*) and Wall Lettuce (*Mycelis muralis*) frequently cling to rock ledges, although mostly not south-facing ones. Much rarer and Nationally Scarce is Angular Solomon's-seal (*Polygonatum odoratum*) restricted now to just two inaccessible, sheltered ledges in the Wye Valley and Deep Dale. Other perennials of rock ledge and edge communities include Sheep's Fescue. Its fine, rolled leaves help reduce water loss in dry conditions, although the severe damage to plants on south-facing slopes and ledges during the 2018 drought shows inadequate tolerance in extreme conditions.

FIG 197. Limestone Fern on scree in Monk's Dale. The fern spreads by rhizomes producing individual fronds at regular intervals.

Screes are common features in many dales, mostly on the lower slopes. They present similar droughty problems for their occupants, but this depends on the amount of fines below the surface. Limestone Fern (*Gymnocarpium robertianum*) is a specialist here in many dales, although it also grows in rock crevices as along the Monsal Trail in Wye Dale and in Back Dale. This Nationally Scarce little fern is very restricted largely to the White Peak and Yorkshire Dales plus sites in Wales. It forms patches in summer with short, finely divided fronds spreading vegetatively (Fig. 197). Companion plants are often Dog's Mercury or Herb Robert, which reddens in response to lack of nutrients and drought.

Away from bare rock exposures, soils increase in depth (marginally), but are still dry, thin, and constitute difficult growing conditions. The variable depth of underlying rock surface accounts for some of the matching small-scale disparities in soils, plant density and growth. The soils have a high pH, low levels of other nutrients, especially phosphorus, and can be droughted in summer. Such conditions are conducive to very high numbers of plant species able to survive together, mostly in sub-optimum conditions for growth. This huge diversity is celebrated in calcareous grasslands throughout Europe and ours are no exception. Up to forty-five different plants can be found in a metre square – more than in any other habitat and you can easily find well over 100 different species in a day's outing.

The main vegetation type is one in which Sheep's Fescue and Meadow Oat-grass (*Helictochloa pratensis*) are the core grasses plus the delicately pretty Quaking Grass and small wiry, greyish tussocks of Crested Hair-grass (*Koeleria macrantha*). Some of the rock-ledge species persist in small gaps in the more open swards and are joined by two common sedges, Glaucous and Spring-sedges (*Carex flacca* and *C. caryophyllea*). These are easily distinguished as the first has grey-green upper leaf surfaces with striking purple-black fruits in summer, whilst Spring-sedge is earlier flowering with short, yellowish-green inflated flower heads. Interestingly, Glaucous-sedge also frequents some of the moorlands flushes in a completely different habitat.

This colourful community is where Common Rock-rose (Fig. 192), Rough Hawkbit, Salad Burnet (*Poterium sanguisorba*), Harebell, Small Scabious (*Scabiosa columbaria*) and the little annual Fairy Flax all vie for space. Early-purple Orchids (Fig. 198) and Cowslips are usually striking spring additions. Kidney Vetch (*Anthyllis vulneraria*) can be abundant but its legume cousin, Bird's-foot-trefoil is usually more widespread. These yellows are intermingled with Wild Thyme cushions, Bloody Cranes-bill (Figs. 190 & 192) and white flowers of Burnet-saxifrage (*Pimpinella saxifraga*), a small umbellifer.

There are some specialities in this group too. Horseshoe Vetch (*Hippocrepis comosa*), another yellow legume, is rare in our area, being mostly a southern

FIG 198. A cluster of Early-purple Orchids in limestone grassland showing considerable colour variation.

species on the chalk and limestones. Its few localities are in the Manifold Valley, Dovedale and Chee Dale. Nottingham Catchfly (*Silene nutans*) grows more on rock outcrops and scrub edges. It gained its name from Nottingham Castle's cliffs where it was first recorded, but its stronghold nationally is now the White Peak, and it is Nationally Scarce. Look for it in Deep Dale (Fig. 199), Peter Dale, Chee Dale, the Manifold Valley and Dovedale. In flower, this campion looks rather scruffy in daytime, with curled creamy petals, but this hides an interesting flowering strategy geared to pollination. As flowers open, the five stamens project between closed petals in late afternoon. But the petals reflex tidily by sunset, pollen is released and the sweetly-scented flowers attract night-flying moths. By dusk on the second day, a second wave of anthers appear to repeat the process, but on the third day, the anthers wither, the petals reflex, and the female stigmas are held up for moth cross-pollination. This night-time saga avoids the pollen or anthers drying out (Anderson and Shimwell, 1981).

FIG 199. Nottingham Catchfly growing amongst outcrops in Deep Dale, Topley Pike. The flowers are in their daytime curled-up mode.

The White Peak is also a national stronghold for Limestone Bedstraw (*Galium sterneri*), which grows amongst the rock ledges and in the diverse limestone grasslands. It is very restricted nationally to northern and western calcareous rocks. Much less common in the White Peak but equally restricted nationally is Pale St John's-wort (*Hypericum montanum*). The dull, pale green leaves are distinctive but the yellow flowers rather inconspicuous.

The wonder of the dales would be incomplete without mentioning other orchids. Several flower from June to September. Bee Orchids (Fig. 200) are one of nature's wonders. The large genera of *Ophrys* flowers not only resemble a pollinating insect but also produce the pheromones (chemical sexual attractants) produced by the female insect in question. Bee Orchid is unique though in being able to self-pollinate, fortunately, as the bee it resembles is absent from the UK. Bee Orchids tend to colonise newly bared calcareous surfaces so appear in disused quarries or other disturbed sites, with often large population

FIG 200. Bee Orchid, an early coloniser of a worked-out quarries as well as flower-rich grassland.

fluctuations. Fly Orchid (*Ophrys insectifera*) is our only other related species in the White Peak but is extremely rare here (and Red Listed nationally) and more a scrub edge specialist.

Easier to find are the pink (sometimes white) inflorescences of Pyramidal and Fragrant Orchids (*Anacamptis pyramidalis* and *Gymnadenia* spp.) (Fig. 201). The latter has now been separated into three species. Look out for these lovely flowers in the Wye Valley dales and the south-eastern limestone areas although they do not always occur together and neither are common. In comparison, the smaller and inconspicuous Frog Orchids are just as delightful with their little pale yellowish-green sunhat and 'legs' forming their hooded petals and lip. This rather rare orchid tends to favour lightly acidic soils with reduced competition. Wardlow Hay Cop and other sites in the wider Wye Dale and south-eastern limestone areas are its home.

There are many more species, too numerous to list, associated with these flower-rich calcareous grasslands. The community has high affinities to several national vegetation types collectively labelled CG2 (Calcareous Grassland number 2, Rodwell, 1992), which is confined to calcareous soils in England and Wales, mostly on south-facing slopes. This is the Sheep's Fescue-Meadow Oat-grass (*Festuca ovina-Helictochloa pratensis*) community. One variant of this found in the White Peak supports Dropwort (*Filipendula vulgaris*), the diminutive relative of Meadowsweet, usually with Common Rock-rose and Salad Burnet. Dropwort's pretty whitish-flower with finely divided leaves grows extensively in Cressbrook and Tansley Dales for example. Stemless Thistle (see Chapter 7 and Figure 74) is another associate, but only in warmer spots and more often in the southern dales, although it reaches 450 m on High Edge and is abundant on Wardlow Hay Cop above Cressbrook Dale. These communities are unique with their mixing of northern and southern as well as more cosmopolitan species.

Move from a southerly aspect to a more westerly one and this grassland community changes, sometimes quite subtly. Devil's-bit Scabious can be more conspicuous in damper soils away from the strongest insolation effects. This is

FIG 201. Pyramidal Orchids in Cressbrook Dale showing the parched grassland behind them in the 2018 drought.

also a plant of some moorland base-rich flushes, demonstrating wide ecological tolerances. Its domination in parts of Miller's Dale Quarry that can flood over its basalt base in wet weather is not therefore surprising. Named after its abruptly severed root stock attributed to the devil, it has a pincushion of pretty violet-blue flowers that attracts many late butterflies, bees and other insects. Betony is another companion species, preferring more acidic soils (and not just on the limestone). It can cover whole banks, such as in the Dove Valley north of Hartington, and in Monk's and Cressbrook Dales.

Usually westerly or more northerly facing limestone slopes might be graced with a pretty little sedge, Flea Sedge, which can be more abundant here compared with its general scarcity in some clough flushes. Its distinctively downward-pointing fruits help you find it. Joining this is another sedge, Carnation Sedge, with completely glaucous leaves, which is also abundant in short sedge-rich flushes in cloughs. In just a few places other damp place specialities appear: Marsh Valerian, Lousewort (*Pedicularis palustris* – a semi-parasitic, England Red List species), the exquisite little flowers of Grass-of-Parnassus (*Parnassia palustris*, also on the England Red List) and Butterwort are just a few, usually on north-facing limestone slopes as at Solomon's Temple, in Chee Dale, Monk's Dale, Back Dale and Biggin Dale, although not usually found together.

Mossy Saxifrage (*Saxifraga hypnoides*), another species at its eastern limits, generally occupies northerly-facing spots. Also known locally as Dovedale Moss, it is confined to the White Peak in our region but is not common and showed signs of damage from the 2018 drought in places. It is subtly different from the widely grown garden version. Look out for it at the north end of Cressbrook Dale, in Dovedale, Manifold Valley, Deep Dale Taddington and in the Wye Valley. Jacob's Ladder, the native species (Fig. 78) rather than the familiar garden escapee, is much rarer nationally, only occurring in the White Peak and the Yorkshire Dales. Adopted as Derbyshire's County flower by popular vote, it favours sheltered north-facing spots such as the lower slopes in Lathkill Dale and around Hobbs House, where, fenced from grazing, it produces a colourful show in mid-summer.

There is a very distinctive limestone grassland community that occurs usually on north-facing slopes where the limestone is close enough to the surface for the soils still to be alkaline. Many of the species already mentioned are present, but distinction is added by Water Avens, Wood Anemone, Dog' Mercury (all outside their normal woodland environment) and Greater Burnet-saxifrage (*Pimpinella major*) along with Tufted Hair-grass. Rarely, this is where to find Globe Flowers (Fig. 202) and Melancholy Thistle, both northern species (Chapter 7). Look for these in Peter, Monk's and Miller's Dales for example. One of our largest Globe Flower populations is in Cressbrook Dale south of the large Aspen clump. I counted over

FIG 202. Globe Flowers on the north-facing side of Tansley Dale inside a fence to prevent stock grazing.

600 plants here in 2019, but only found 69 at the west end of Miller's Dale Nature Reserve. Accounts of the latter hillside being yellow with them pre-1967 no longer hold and they have declined since the spread of scrub in several places.

You can detect more acidic conditions progressing up to the more gentle brows and on some of the more northerly-facing slopes by the greater abundance of Common Bent, Tormentil, Pignut and sometimes Heath Bedstraw, all of which we have met before on gritstone grasslands. Mountain Pansy is a delight in these slightly acidic swards (Fig. 203), and the faded pink/blue flowers of Bitter-vetch (*Lathyrus linifolius*) are a distinctive companion. Wood Anemone is again not uncommon here.

The upper dale brows and some north-facing slopes are where the calcareous grassland can change to heathland, or have affinities with it. Regularly occurring moorland species are Bilberry, Mat-grass and Wavy Hair-grass, while Cowberry is very rare here and Heather scarce – look for it in an enclosure at Middle Hay (east of Cressbrook Dale) or in Long Dale (near Pikehall). Gorse (European or Western) is another indicator of acidic conditions on limestone slopes – although these

FIG 203. The yellow (and more common) form of Mountain Pansy on slightly acidic grassland on the brow of a limestone slope, growing next to Wild Thyme.

may be derived sometimes from loess materials creeping down from the plateau, as in Biggin Dale. Bracken too might seem a contradiction on limestone soils, but occurs quite widely, such as in Gratton Dale, Coombs Dale and upper Stoney Middleton Dale, beneath which moorland mosses like Red-stemmed Feather-moss (*Pleurozium schreberi*) occur. Some of these heathy edges are the vestiges of the once widespread limestone heath described in the next chapter.

Chapter 7 describes the early research that showed how Wavy Hair-grass can change soils at the expense of the plants preferring mildly acidic conditions like Common Bent and Tormentil. A layer of black undecomposed humus develops which is better at holding water, prevents free drainage and becomes water-logged at the surface compared with surrounding soils. The result is increased acidity, re-distribution of iron and aluminium in the profile and a reduced environment in which manganese toxicity is manifest. These conditions spread over time as the grass grows out of its original patch, eliminating species like Betony quickly and plants like Wood Anemone and Pignut more gradually.

A podsolised soil develops through these interactions. Its expansion is probably also determined by management (Piggott, 1970). Higher levels of grazing will contain Wavy Hair-grass spread; lower levels will encourage it. This illustrates the often complex relationships between plants and their environment which is part of the White Peak limestone grasslands' character, more typical of northern, higher and wetter situations, but a feature of our dales.

Dropping now to the dale bottom, where deeper soils have accumulated over time, gone are the colourful diverse mixtures, to be replaced by more tussocky grassland marked by Cock's-foot and Tufted Hair-grass. Wetland plants like Meadowsweet mark where the water table rises in winter, or a haze of flowering Lesser Celandines might carpet the floor in spring. Meadow Saxifrage prefers deeper soils here or on the dale brows.

Embedded in some of the dale bottoms are small alkaline fens. Those with a short sedge-rich flora, sometimes with Common Butterwort, and those marked by an abundance of Curled Hook-moss (*Palustriella commutatum*) occurring with tufa (explained in Chapter 4) are Habitats Directive Priority Habitats (see Appendix 2). Most occur where springs emerge either on a dale-side or near the foot of slopes. Several vegetation communities are associated with these wet areas, with the best Priority Habitat examples being in Monk's Dale in the NNR and Monsal Dale (on private land). Other examples are all limited in scale. Some support uncommon sedges, the nationally Red Listed Flat-sedge (*Blysmus compressus*) and other wetland species like Marsh Valerian. The Flat-sedge's compact terminal inflorescence and flat, grass-like leaves are distinctive but still difficult to find. It grows by the stream in central Monk's Dale, along with Butterwort (Fig. 204) and even some Common Cottongrass and Bristle Scirpus. Take care not to trample these rare plants here.

Appreciating the subtle changes in the limestone grassland communities and delighting in the rarest plants does need some agility on steep slopes. The best place for the less athletic to see many limestone flowers is Miller's Dale Station where many species, including some of the special ones, have colonised or been established around the former station platforms. The land around Solomon's Temple south of Buxton is also relatively accessible and very rich in species with some surprises for the keen botanist. Many plants can be seen from the different dale bottom footpaths as well, or from the National Park trails on former railway routes, which are essentially flat and highly accessible with a reasonable surface. Numerous car parks service these trails as shown on the OS map.

The importance of waxcap grasslands is expounded in Chapter 11 and there are similarly important sites on the limestone. Not all dales have been surveyed, but Coombs Dale, sites near Longstone Edge, Deep Dale, Hay Dale, parts of

FIG 204. Common Butterwort and Carnation Sedge in a small fen with some Meadowsweet growing beside the stream in Monk's Dale.

Dovedale, a site at Harpur Hill, Thorpe Pastures and Wetton Hills all feature highly for their grassland fungi (some on private land). Many of the waxcaps described in Chapter 11 also occur on the calcareous grasslands, but some specialise in them like Date Waxcap (*Hygrocybe spadicea*), one of the rarer ones with a dry brown cap and yellow gills, mostly found so far in Dovedale. Others are Blushing, Toasted and Persistent Waxcaps (*Neohygrocybe ovina*, *Cuphophyllus colemannianus* and *Hygrocybe autoconica*) all of which are rare in the White Peak and on the Fungal Red List. The rare Olive earthtongue (*Microglossum olivaceum*) has also been recorded from Dovedale and near Tideswell (Fig. 150).

More remarkable perhaps is the discovery of an important fungal assemblage associated with Common Rock-rose. Puzzlement at finding woodland fungal species of Webcaps, *Cortinarius*, *Boletus*, *Amanita* and Knights, *Tricholoma* in limestone grassland led to the discovery that Common Rock-rose is ectomycorrhizal and associated species are still being identified (Barden, 2007). Growing in a stressed environment where nutrient supply is limited, further mycorrhizal associations might be found in the future with other species.

Although the calcareous grasslands total in the White Peak is modest at 2,514 ha, this is only 4.7 per cent of the White Peak area, and therefore very restricted. Although most sites are in the dales, some rocky or unimproved calcareous grasslands occur outside these too. Although agricultural intensification within the dales has been minimal owing to the largely inaccessible slopes, there are more modest slopes where slurry spraying from below or above has effectively reduced the floristic interest, as evidenced by their brighter green appearance. A few dale slopes are overgrazed and scree slopes are developing, but the main threat in most dales has been loss of grassland to scrub and woodland in recent decades, which, when compared with old postcard views, shows how significant this has been in numerous locations.

It is imperative to manage dale-side grasslands through grazing to maintain their richness and diversity and control further spread of scrub and trees. Much scrub clearance has been undertaken, for example in the Wye Valley nature reserves, Hopton Quarry, Deep Dale and Gang Mine, the Derbyshire Dales NNR, NT land in Dovedale, Manifold/Hamps Valley and on Longstone Edge. Most sites are in agri-environment schemes to support this management, whilst the NPA obtained funding in 2011–2 to restore grasslands on the Trails through scrub/tree removal and grassland cutting or grazing for the first time. This was particularly important to maintain some very special species such as Orpine (*Hylotelephium telephium*) on some rock ledges and Common Wintergreen (*Pyrola minor*) under scrub and trees just east of Miller's Dale Station in its Peak District stronghold.

Grazing in nature reserves is undertaken to optimise the wildlife value and the season in which it is applied influences the vegetation. For example, grazing mid-summer onwards may deliver masses of flowers early on, but does not control scrub or bramble invasion, their leaves being less palatable by autumn. In Miller's Dale Quarry, for example, the Derbyshire Wildlife Trust grazes all year every 4 or 5 years, but only later in intervening years to maximise floral display, seed development and dispersal. In Monk's Dale (part of the NNR), cattle are the preferred animal and Natural England is breeding a Belted Galloways herd for Cressbrook Dale, which are lighter, stockier animals better suited to dale-side scrambling. In the past, grazing would be more clearly defined by farming patterns, with perhaps higher levels in spring when hay fields on the plateau were shut up. When the limestone heaths were widespread over two centuries ago (see Chapter 14), the dale-side grasslands were, in comparison, higher value grazing than heathland and would have been more consistently used. When plateau farming changed to more intensive dairy farming, dale-side grasslands were unsuitable for heavier animals and consequently left unmanaged leading to scrub and woodland invasion. Changes in stock types also impacted the habitats.

Sheep farming by the Granges would have been widespread and sheep are very selective grazers which would have controlled some species more than others. The change to cattle would have had an impact as they are less selective, feed on taller vegetation mostly and browse higher up the shrubs. Add to this the fluctuating rabbit populations and you can see how the grasslands have varied over time under these different pressures. Their general resilience is seen in the persistence over centuries of many species, but their detail and abundance could well have fluctuated.

The majority of the Dale grasslands are included in SSSIs and a suite of these also fall in the Peak District Dales SAC (Fig. 3), where the calcareous grasslands, alkaline fens and lead rake communities (Chapter 14) are qualifying features. The very dispersed occurrence of these grasslands and the general lack of other species-rich grasslands between the dales on the plateaux, point to the need for more linking flower-rich grasslands that could be critical for adaptation to climate change in enabling less mobile species to migrate as their climate envelope moves. These are goals for future agri-environment schemes being developed as we leave the EU.

ANIMALS

Having flower-rich grasslands, scrub and woodland intermingled in many dales provides a perfect habitat mosaic for a wide variety of animals. There is ample edge habitat, with scattered trees and shrubs interplaying with open areas or more dense wood. This is perfect for birds like Spotted Flycatchers and Common Redstarts. Although present in the whole Peak District, the dales are a Spotted Flycatcher stronghold and a delight to see darting out from a branch catching unsuspecting insects. This Red List species has suffered a serious national decline as well as one in Derbyshire as a whole, but more outside our area, making our population more important. They regularly return to traditional sites to breed, including gardens and parks. Redstart is a dale species where there are scattered trees, open woodland or rocky outcrops (Fig. 205). Lathkill Dale is a hotspot along with Monsal Dale, Longstone Edge and Via Gellia. It also frequents the plateau top where there is patchy hawthorn cover and drystone walls or derelict buildings in which to nest such as around Brassington and Bonsall Moor (Frost & Shaw, 2013).

Other common birds already met in similar habitats frequent the dale woodlands, the most abundant being Chaffinch, Wren, Robin and Willow Warbler. Blackcaps, Goldfinches, Garden Warblers, Tree Creepers, Coal Tits and Nuthatches are all typical and numerous in the dale woodlands and scrub. In

FIG 205. Male Common Redstart at its nesting hole in a tree. (John Lintin Smith, Sorby Natural History Society)

more open areas, particularly those that are well grazed as in Long Dale (north of Hartington), Wheatear nest in scree and rocks, but they also frequent some of the large quarries. Meadow Pipits and Skylarks are more typical of these less wooded habitats often at the top of the dales as they open out.

High rock ledges and crevices are the domain of breeding Jackdaws, Stock Doves and Kestrels. This is where Peregrine Falcons and Ravens also breed, though in smaller numbers. Many of the huge quarries provide safe havens for these last two with enhanced productivity. First recorded nesting in a White Peak quarry in 1990, they now support the bulk of Derbyshire's Peregrine Falcon population. The quarry sites provide many of the same features as natural cliffs, with clean-cut faces, high visibility, and wide enough ledges to establish a nest scrape. Those working in quarries are generally proud of their Peregrines and keep them safe.

Barn Swallows (*Hirundo rustica*) and House Martins (*Delichon urbicum*) breed throughout the Peak District and, equally, feed through and over many dales. However, a number of House Martins breed on natural cliffs and quarry faces in the White Peak, especially in the Wye Valley and around Stoney Middleton. This would once have been traditional as House Martins only starting nesting on houses and in built-up areas in the early 1900s, where nest sites can be re-used year after year. This species has declined by 47 per cent nationally since 1970 according to RSPB data, with a 10 per cent loss just in the period 1995–2014 through a combination of factors related to weather and conditions in overwintering grounds. This has resulted in the loss of many traditional nesting sites in our villages. They do, however, now nest in quarries such as in Ballidon and Tunstead (Buxton) and quarry nesting seems to have increased at the expense of natural cliffs, possibly owing to greater potential disturbance on the latter where climbing is popular (Frost & Shaw, 2013). Barn Swallows, in comparison, are not traditional cliff nesters, although there have been past records of a few breeding in or around caves and rock faces (Frost & Shaw, 2013).

Most of the common mammals we have already met occur widely in the dales too, but some deserve special mention. Although Rabbits are widely distributed, their role in moulding the limestone grasslands has been critical and Rabbit warrens are still visible for example in Cressbrook Dale and Long Dale near Pikehall. They are an important prey for Stoats, Weasels and Buzzards as well as Badgers and Foxes. Some 10 per cent of Rabbits are black in various White Peak localities like Monsal Dale and Deep Dale (Mallon *et al.*, 2012). Amongst the small mammals, Bank Voles are widely but unevenly distributed. Their chestnut-brown fur, which gave them the old name of 'fox mouse', distinguishes them from Field Vole. Bank Voles are more typical of woodland and hedges than open grassland and, in a live-trapping programme in Lathkill Dale ashwoods, was the most abundant species found associated with a good autumn seed fall, more so than Wood Mouse (Mallon *et al.*, 2013).

The limestone dales have an interesting and diverse invertebrate fauna, including assemblages requiring basic calcareous rocks or specific limestone plants. The White Peak holds some significant populations of national importance and has been fairly well studied, although much remains to be discovered. Only selected species and groups can be mentioned here.

There are close links between some invertebrates and plants. Various plant galls are often conspicuous. These produce abnormal growths caused by organisms like bacteria, virus, fungus, or animals such as mites, midges or wasps. There are often some very intricate relationships and life cycles, which are fascinating to unravel (Redfern, 2011). In the limestone dales, white felty hairs on Marjoram and Wild Thyme, for example are caused by mites in the *Aceria* genus. Another *Aceria* thickens the leaves of Bloody Cranes-bill to produce a loose mop-head, whilst Common Rock-rose might be attacked by a midge, *Contarinia*

helianthemi, which produces thickened shoot tips and a hairy, artichoke-like
gall. Small fleshy swellings between the leaves of Lady's Bedstraw are caused
by the midge *Geocrypta galii*, whilst the hard swellings on the underside of
Mouse-ear Hawkweed midribs are the work of the gall wasp *Auladidea pilosella*
(Higginbottom, 2010). Robin's pincushion will be familiar on roses, caused by a
gall wasp, *Diplolepis rosae*.

The distribution and abundance of land snails is strongly influenced by
calcium availability so they are generally commoner in the White Peak than on
more acidic soils elsewhere. A total of 44 species (31 snails and 13 slugs) were
found in a survey of 21 White Peak woodlands; each woodland supporting 23 to
35 species. This mollusc fauna was rather uniform, with only a few geographically
localised species, including the Ash-Black Slug (*Limax cinereoniger*) that frequents
woodlands along the Hamps and Manifold valley, plus Glossy Glass Snail
(*Oxychilus navarricus* ssp. *helveticus*) and Three-toothed Moss Snail (*Azeca goodalli*)
that are associated more with lower elevation woods, especially along the River
Derwent valley around Matlock. Within the woodlands, snails show preferences
for different microhabitats. For example, the tall-spired Two-toothed (*Clausilia
bidentata*) and Plaited (*Cochlodina laminata*) Door Snails and the Tree Slug
(*Lehmannia marginata*) browse algae and lichen on rock surfaces and tree trunks,
whereas several small species like Slippery Snail (*Cochlicopa lubrica*), Crystal
Snail (*Vitrea crystallina*), Tawny Glass Snail (*Euconulus fulvus*) and Prickly Snail
(*Acanthinula aculeata*) live in leaf litter and moss (Tattersfield, 1990).

Few molluscs are restricted to open habitats, although many woodland
species may occur in unshaded stands of tall herbs and dense grassland on
north-facing dale-sides. Several obligatory open habitat species are dependent on
high levels of soil calcium such as the native Heath Snail (*Helicella itala*), which is
uncommon and declining, found in short, grazed, south-facing grassland in a few
limestone dales (like Lathkill Dale) and disused limestone quarries. Heath snail
populations are sensitive to increases in sward height and density that follow
grazing relaxation or decreases in rabbit populations caused by myxomatosis.
Large Chrysalis Snail (*Abida secale*) is another unshaded limestone grassland
and rock habitats species sensitive to similar habitat changes. It was formerly
recorded in Monsal and Miller's Dales around the turn of the 19th century, but,
despite searching, has not been found since. Other calcicole species associated
with limestone grassland include the subterranean Blind Snail (*Cecilioides acicula*),
the minute Common Whorl Snail (*Vertigo pygmaea*) and Moss Chrysalis Snail
(*Pupilla muscorum*). Limestone scree represents a further specialist mollusc habitat
for species such as the Rock-Cutter or Lapidary Snail (*Helicigona lapicida*) and
the minute Rock Snail (*Pyramidula pusilla*). Drystone limestone walls provide

FIG 206. The Pink-striped Millipede which can erupt in impressive swarms in the limestone dales or on heathlands. (Bob Croxton)

a surrogate habitat for these rock-dwelling species and also for the Nationally Scarce Wall Whorl Snail (*Vertigo pusilla*).

The White Peak millipedes are well studied with 20 species so far found. The colourful Pink-striped Millipede (*Ommatoiulus sabulosus*), a large snake millipede up to 40mm long, with two distinctive pink or salmon coloured stripes (Fig. 206) seems equally happy in limestone dales and sandy heather moorland and erupts in spectacular swarms in summer when thousands suddenly appear. In contrast, the tiny, almost colourless pill millipede *Geoglomeris subterranea* is something of a White Peak celebrity. Only 2–3 mm in size, living under moss-covered limestone, it rolls into a tiny ball the size of a large sand grain. This Common Pill Millipede (*Glomeris marginata*) relative is quite scarce and very difficult to find. Cressbrook Dale and Miller's Dale are known sites, but it is certainly overlooked.

Of our 15 woodlice species found, four relatively unusual species have special affinities for limestone habitats. Limestone Pill Woodlouse (*Armadillidium pulchellum*), an attractive colourful woodlouse, rolls into a ball leaving a slight gap rather than forming a tight sphere. It favours calcareous substrates and is widely found in the dales amongst sparsely vegetated screes, in rock crevices, and under mats of moss and low herbs. The Peak District population is of national and European significance (Richards, 1995). Picture Pill Woodlouse (*Armadillidium pictum*) is equally colourful but even scarcer, associated with deadwood within old woodland in hilly areas. It is recorded from standing dead trees in Miller's Dale, Priestcliffe Lees, Lathkill Dale, Rose End Meadows and in Bretton Clough on the White Peak fringe. It is Nationally Scarce and has a preference for calcareous rocks or base-rich features in other strata (Richards & Thomas, 1998). Rosy Woodlouse (*Androniscus dentiger*) is an attractive limestone-lover, commonly found in dale-side scree and quarries. Rare records in the Dark Peak are always associated with dumped limestone or discarded mortar and bricks. Finally, *Porcellio spinicornis*, a large, elegantly coloured woodlouse, is abundant in the White Peak but rare elsewhere in our area. It favours dry, relatively exposed calcareous substrates such as dry stone walls, quarries, cuttings, cliffs and buildings with loose flakes of stone or mortar.

Butterflies are relatively abundant in the dales. With diligence a tally of 30 species is possible in a year, although some are locally rare. Grizzled Skipper (*Pyrgus malvae*) is only known from one or two sites, and Small Blue (*Cupido minimus*) is almost certainly now extinct in the Peak District. Wall Brown (*Lasiommata megera*) was once a common sight, but in line with a drastic national decline since 2000, it became a rarity by 2015. It persists in the White Peak where basking sites like walls, rock outcrops, quarries and patches of bare ground in sunny locations, combined with short, open, dry grassland provide suitable refugia. Lead rakes are a favoured habitat. In the past few years it seems to be showing some recovery.

On the credit side, four species have expanded from the south and east and are now regarded as commonplace. Ringlet (*Aphantopus hyperantus*, Fig. 232), Speckled Wood, Gatekeeper (*Pyronia tithonus*), and Essex Skipper (*Thymelicus sylvestris*) were rare or absent thirty years ago. Silver-washed Fritillary (*Argynnis paphia*) will soon join this list. It seems to be in the 'invasive phase' as sightings have become increasingly frequent in recent years. There is optimism that it could become a future common resident. Holly Blue, Orange Tip, Brimstone, Small Heath and Comma (*Polygonia c-album*) have all seen significant expansion in numbers and distribution in the past two decades.

Arguably the celebrity amongst White Peak butterflies is the Brown Argus (*Aricia agestis*). For many years it was attributed to a different species, the Northern Brown Argus (*Aricia artaxerxes*), but detailed work by Dr Bill Smyllie over three decades concludes that we have a unique 'Peak District Race' of the more southerly Brown Argus. It is single brooded (very rarely there is a partial second brood), in flight from May to September, on sunny grassland dale-

FIG 207. Dark Green
Fritillary on Red Clover.
Good numbers of this large
butterfly can be seen in
most dales.

sides seeking its main food plant Common Rock-rose. Dark Green Fritillary
(Fig. 207) is another elegant flagship species with a White Peak stronghold. It
is single brooded, flying from late June to late August in good years, in sunny,
open, flower-rich grasslands, lightly grazed with a little scrub. Favoured nectar
plants include Thyme, thistles, knapweeds and Ragwort; but the caterpillars
feed on violets. In July it is possible to count over 200 on open hillside such as
Longstone Edge and some of the dales. Numbers and distribution have increased
dramatically in the past three decades.

   It would be unfair to conclude an account of White Peak butterflies without
special mention of an assemblage of four species that typify a late spring day in
the limestone dales. Green Hairstreaks (Fig. 124) first appear on warm days in
early April, followed by Dingy Skippers (*Erynnis tages*) from late April onwards.
Common Blues (*Polyommatus icarus*) become common in early May and, being
double brooded, can appear in good numbers through to September or even

early October. The Small Heath, often overlooked but a delightful little butterfly, graces dale-sides, quarries, limestone heaths and hay meadows in large numbers throughout the summer.

Amongst the hundreds of moth species recorded in the White Peak, a selection of noteworthy macro-moths includes Cistus Forester (*Adscita geryon*). It is an attractive, metallic-green, day-flying moth favouring sunny south-facing slopes in late May and June with Common Rock-rose feeding caterpillars. In favourable conditions, it can be abundant in places like Longstone Edge. The Peak District has a nationally important population. Speckled Yellow (*Pseudopanthera macularia*) is a pretty day-flying moth of open woodland, scrubby grassland and rural gardens in the White Peak in May and June, whilst Chalk Carpet (*Scotopteryx bipunctaria*) is a Nationally Scarce, declining moth with a White Peak stronghold where its caterpillars feed on trefoils and clovers. It favours open short-turf and quarries with bare patches, and flies readily during the day.

FIG 208. Wood Tiger moth in limestone grassland.

Three more limestone day-flying moths include the poorly-named Wood Tiger (*Parasemia plantaginis*), which actually prefers open close-cropped grassland with a sunny aspect (Fig. 208). On good days at sites like Longstone Edge, dozens fly in June sunshine. Latticed Heath (*Chiasmia clathrata*), not unlike a small butterfly, flies in sunshine and basks with its wings open rather like a Dingy Skipper. Mother Shipton (*Euclidia mi*) with its unique 'hooked nose' markings enjoys flower-rich grassland and hay meadows in May and June too. The Herald (*Scoliopteryx libatrix*) and The Tissue (*Triphosa dubitata*) can hibernate, sometimes in large numbers, in limestone caves and abandoned mines. Old quarries are the principal habitat for moths such as Pimpernel Pug (*Eupithecia expallidata*), Muslin Footman (*Nudaria mundana*), The Shears (*Hada plebeja*) and Gold Spangle (*Autographa bractea*).

Turning to hoverflies, *Neoascia obliqua* and *Cheilosia pubera* are characteristic dale species. *Neoascia*, a small dark species, uses Butterbur leaves for feeding on honeydew, sunning and courtship. Its larvae probably develop in Butterbur's rotting leaves or stems. The White Peak is one of its national strongholds where it is widespread along the River Wye and its tributaries. *Cheilosia pubera* is a blackish hoverfly with a metallic bronze sheen, easy to spot sunning on leaves or visiting Marsh-marigold flowers. Its larvae feed on the roots of Water Avens. Nationally it is fairly scarce but adults are quite common along the River Wye in May.

Limestone woodlands support hoverflies in abundance, some preying on other insects and others exploiting deadwood habitats (saproxylics). Rarities include *Xylota tarda*, a scarce deadwood hoverfly associated with Aspen, which has been found at Miller's Dale Quarry. *Xylota xanthocnema*, breeding in rot-holes of Yew, is known recently from Lathkill Dale. *Xylota segnis*, *Brachypalpoides lentus*, *Xylota sylvarum* and *Myathropa florea* are more widespread deadwood species. Some hoverflies have very specific requirements. *Brachyopa insensilis* breeds in oozing sap runs on tree trunks notably Horse Chestnuts, even in town parks, as in Bakewell. The Ramsons Hoverfly (*Portevinia maculata*) characteristically rests on Ramsons leaves in dappled sunshine, with wings held in delta fashion. It has a slow, sluggish flight always near its food plant, where the larvae feed in Ramsons' bulbs, maturing in winter. Blackthorn blossom in April attracts a good range of early species. The furry bee mimic *Criorhina ranunculi* is quite common in some dales, the tiny *Platycheirus ambiguus*, a nationally uncommon species, is known from Cressbrook Dale and Monk's Dale, whereas *Platycheirus albimanus*, which has silver-grey spots, is very common.

Some invertebrates are closely tied to their host plants. The nationally uncommon Herb Paris Fly (*Parallelomma paridis*) has recently been found in Monk's Dale forming blister mines on its host plant. It is a member of the

FIG 209. Yellow Meadow Ant nest in Biggin Dale. Note the abundance of mosses on the north-facing right-hand side, compared with the Wild Thyme and lack of mosses on the south-facing left-hand side. Sheep's-fescue grows throughout the mound.

Scathophagidae, popularly known as dung flies, although not all species are associated with dung. Another national rarity in the White Peak is the Giant Bellflower Fly (*Platyparea discoidea*), a pretty picture-winged Tephritid fly which develops in the stems, and is recorded from Miller's Dale, Cressbrook Dale and Lathkill Dale.

Finally, in this cruise around the invertebrate fauna of the dales, ants must be mentioned. There are a number of species, but one of the most obvious is the Yellow Meadow Ant (*Lasius flavus*), which builds long-lasting nest mounds in undisturbed grasslands. Over 100 can occur in clusters and they provide their own micro-habitat for plants and animals with, in Cressbrook Dale and Biggin Dale, distinct mossy northern and drier southern ends. Thyme and other dry soil species favour the southern end (Fig. 209). Look out for the rare Wall Whitlowgrass (*Drabella muralis*) on the top and Green Woodpeckers (*Picus viridis*) feasting on the ants.

CHAPTER 14

From Rivers to Dew Ponds: The Other Limestone Habitats

L OGIC SUGGESTS EXAMINING THE REMAINING limestone habitats starting with the rivers in the dale bottoms, then moving up slope to lead rakes and spoil heaps, before venturing onto the plateau tops for the limestone heaths and ponds.

THE DALES RIVERS

Linked to the fens and temporary pooling in the dale bottoms in Chapter 13, the rivers are one of our most important White Peak habitats, whilst also being highly vulnerable to damage through pollution (Fig. 210). They also have a long cultural history tied to their fishing interest. White Peak rivers will be mostly more alkaline than those in the moorland environment, although this will vary with their sources – the moorlands or springs and caves. Water emanating from the latter will have a much higher pH but will be diluted once in the main stream. This generally high alkalinity enables more crustaceans and other invertebrates needing calcium for exoskeleton or shells to survive. These, in turn, supply large fish populations with food and so the White Peak rivers are generally more valuable fisheries than those particularly on the upper moorland river reaches.

Aquatic plants are generally more abundant in alkaline waters too. Layers of waving, submerged, green leaves on long stems undulating in the river flow signal an abundance of Stream Water-crowfoot (*Ranunculus penicillatus*). This locally rare plant is typical of running water with a degree of nutrient enrichment, although more usually found in the lowlands. It predominates in the Wye, Dove and lower Derwent Rivers, where its white flowers sparkle across

FIG 210. The River Wye in Miller's Dale with beds of Stream Water-crowfoot in the water. Note the clarity of the water.

the surface in early summer (Fig. 210). The ubiquitous Greater Water-moss (*Fontinalis antipyretica*), our largest aquatic moss with streaming shoots 15 cm or more, is usually attached to stones or waterside trees and provides a darker green mass. Most of the other stream plants creep in from the edges until pruned by the current, such as Brooklime and floating mats of Creeping Bent.

Smaller streams in some tributary dales may support beds of Water-cress (*Nasturtium officinale*), Water-starwort or Fool's-water-cress (*Helosciadium nodiflorum*) as in Monk's Dale or Deep Dale near Taddington. Cut Water-cress releases phenethyl isothiocyanate, which delivers its peppery taste, as a defence against herbivores, but it affects some invertebrates, including depressing Freshwater Shrimps' reproductive success. Tall wetland herbs growing adjacent might include Meadowsweet, Hemp-agrimony (*Eupatorium cannabinum*) though not commonly, or Great Willowherb. Hemp-agrimony is a mecca for butterflies and other nectar-seeking insects, but it was also an important herbal remedy, recognised for its cathartic, diuretic and anti-scurvy properties and considered as a blood purifier. All stock except goats are reputed to avoid it (Greaves, 1931).

FIG 211. Stands of Butterbur in flower growing alongside a limestone river with flowering Blackthorn on the far bank in spring.

Large Butterbur stands fringe many river edges and dominate seasonally dry sections on the Wye, Manifold and Hamps (Fig. 211). Their large, hairy, pliant leaves made them perfect for wrapping butter to take to market. The hairs keep the leaf cool and the absorbent texture mops up any leaking butter (Mabey, 1996). The leaves are also large enough to be mini-umbrellas or sunshades and indeed, the scientific name *Petasites* is derived from the Greek for a felt hat that shepherds used to wear. The dusky-pink flowers emerge on sturdy spikes in spring attracting early insects before the leaves fully emerge.

The aquatic plants add significantly to the habitat quality for animals and interact with the habitat itself, which is very diverse on a small-scale, exhibiting riffles, slides and pools. Riffles are the faster flowing water through gravel (limestone in our case) beds, whilst the slides reflect shallow water passing over smooth rock, with quieter, often deeper stretches forming pools holding more sediment over the bottom stones. Walk beside the dry Manifold or Hamps beds to see how uneven the bottom is. All these produce a diversity of water flows, sediment and oxygen levels when wet, suited to different stages in animal life cycles.

The abundant invertebrate groups in White Peak rivers are stoneflies, mayflies and caddisflies, along with Freshwater Shrimps, all much more plentiful and varied than in more acidic streams. Mayflies date back over 300 million years; some of the oldest winged insects known, just stretching into the Carboniferous period, which seems fitting here. Mayflies are unique as they moult again after developing functional wings and omit the pupal stage. Eggs hatch into nymphs living in water for one or two years feeding on microscopic organisms and algae. They moult many times to grow, but when this is complete, the nymphal skin splits and a winged form, the sub-imago or dun, emerges at the water surface (Fig. 212). This seeks shelter before re-shedding its skin, emerging as the more brightly coloured imago or spinner (sexually mature adults), ready for mating, usually at the end of May/beginning of June for the mayfly, *Ephemera danica*, with others appearing between March and November. The accumulative temperature measured by degree days suitable for growth determines timing of mass emergence events. Thus, if spring is cold, the hatch is delayed.

The synchronised hatch produces brown clouds of adult mayflies, although in reduced numbers now possibly compared with the past. The classic mating dance occurs in specific areas near rivers, often the same places annually. The males

FIG 212. The male sub-imago of Mayfly (*Ephemera danica*) emerging from its larval case on the river surface. (Don Stazicker)

yo-yo up and down and once mated, the females fly upstream to lay their eggs before dying (and being eaten by fish and other animals from the surface). If the imagos hatch in unsuitable conditions, as when it is windy, they can be blown away from the river, although marginal vegetation helps shelter them. If blown towards a wet road, like the A6 close to the River Wye at the end of Monsal Dale, the mayflies can lay eggs on the road attracted by its reflection (as witnessed by Don Stazicker, pers. com). Mayfly nymphs, as well as many other aquatic invertebrates, can drift downriver to redistribute themselves in a process termed behavioural drift to optimise their environment and food supply.

Data on the range and abundance of some adult stages of aquatic invertebrates in the River Wye have been provided by Cressbrook and Litton Fishing Club (Don Stazicker, pers. com.) and for aquatic invertebrates stages in the Manifold and Dove catchments through surveys for Natural England (Everall, 2010). The English names tend to be those allocated by fishing interests. At least 22 out of a national 51 mayfly species live in these White Peak rivers. The commonest in the Wye, Dove and Manifold catchments are mayfly *Ephemera danica* and Large Dark Olive (*Baetis rhodani*). Several others are widespread, although not as consistently found, including Olive Upright (*Rhithrogena semicolorata*), Angler's Curse (*Caenis rivulorum*) and Small Dark Olive (*Baetis scambus*).

Much rarer mayflies are Medium Olive (*Baetis buceratus*) (Regionally Notable) and Southern Iron Blue (*Nigrobaetis niger*) (a Priority BAP species), both found during the Dove/Manifold surveys. This last species is particularly interesting as historically it is considered part of the southern chalk clean river assemblage where it has recently declined up to 80 per cent. It was only recently discovered in the Staffordshire White Peak, and may have moved north as water temperatures increase with climate change. It is very sensitive to pollution and therefore indicates clean water habitats. This mayfly has two generations annually, one slow-growing in winter and a much faster summer one, so winged adults appear over an extended period. The Northern Iron Blue (*Baetus muticus*) is another potentially useful climate change indicator as it seems already to be moving up river catchments (for example in the Manifold).

Another recent discovery is the Upland Summer Mayfly (*Ameletus inopinatus*) (Everall, pers. com). Found rarely in White Peak and South West Peak rivers, this is the only British arctic-alpine mayfly and is very much on its range-edge here, requiring clean water and slacker flows. It is potentially another useful species to clock climate change effects if it moves further up catchments, into cooler spring waters or we could lose it altogether.

Research into the phenology of the mayfly *Ephemera danica* on two stretches of the River Dove with contrasting temperature regimes shows it emerging after

FIG 213. Stonefly larvae (*Isoperla grammatica*), from the River Wye. (Don Stazicker)

only one year in Beresford Dale's warmer water, but reverting to its normal cycle in a cooler year. In cooler spring water-fed stretches, it maintained its two year cycle. There are risks of a one year cycle both due to poor weather (like high winds and heavy rain) leading to reduced egg laying and hence larval populations, and to one year Mayfly adults being smaller, producing fewer eggs, and thus also negatively affecting populations. A two year cycle is more robust with inbuilt resilience to environmental conditions. Spring water input into White Peak rivers could well provide critical cold-water refuges from warmer water resulting from climate change (Everall *et al.*, 2014), possibly extending some 30 m downstream from each spring, as found in the Wye Valley between Buxton and Bakewell (Smith, 2002).

There are at least 18 species of stoneflies in White Peak rivers out of a national total of 34 (Fig. 213). The most widespread in the Manifold and Dove catchments are the Needle Fly (*Leuctra hippopus*) and the Small Yellow Sally (*Chloroperla torrentium*). Different species occur on the Wye, such as Yellow Sally (*Isoperla grammatica*), and Willow Fly (*Leuctra geniculata*). One Regionally Notable species, Common Early Brown (*Protonemura meyeri*), occurs on all the rivers, but seldom, whilst another, the Needle Fly (*Leuctra moselyi*). is found more in the north and west of Britain than elsewhere, but was only recorded in a few locations in the Dove/Manifold catchments. Stoneflies are another ancient insect group frequenting cooler, usually fast flowing, well-oxygenated waters, which makes them good pollution indicators. Most of the species complete their life cycle annually, although *Dinocras cephalotes*' three-year life cycle is an exception. This is one of our largest stonefly species, reaching 5 cm in length, common amongst stony river beds but largely in the north and west, with the White Peak being on its southeastern boundary. The female is too large to fly, so swims across the water surface after emerging to mate before egg-laying.

Caddis or sedges (Trichoptera) are another very important group of aquatic insects (Wallace, 2004). Their original name seems to have been caddisworm or codworm. Izaak Walton (in *The Compleat Angler*, 1653) refers to the cod-worm or caddis. Adult flies are mostly nocturnal and may be attracted to lights. Their wings are hairy, indeed Trichoptera translates as hairy wings, easily identified by their moth-like appearance with wings folded tent-like over the resting body and long, forward-pointing antennae. Many males swarm, zig-zagging across the water surface in their mating dance. The female drops a little bag of jelly containing her eggs that becomes stuck on river-bottom stones or other material, or she crawls down exposed roots to deposit eggs in the water.

There are two main groups of caddis, one with a unique larval stage which uses silk threads to attach debris, vegetation or woody fragments together to fabricate mobile protective cases and the other that remains caseless. The case materials used and their shape help identification. A caddis is a piece of cloth and cloth-sellers would have worn colourful strips of their wares to advertise them. Shakespeare's *The Winter's Tale* mentions 'caddyssus to sell'. The cased larvae create this same neat patchwork of materials (Marren and Mabey, 2010). All caddis larvae develop a protective, camouflaged shelter for the pupal stage to reduce predation, but exiting this is a more vulnerable step. The adult emerges in its pupal skin and swims to the surface, but a gas secreted to help it surface and shed the pupal skin gives it a silvery appearance which renders it conspicuous to predators like fish, although many will emerge at night to avoid them. Daubenton's Bats feast on them instead whilst foraging low over water courses.

Seventy two caddis species have been recorded in the Dove and Manifold catchment surveys out of 199 British species. The most widespread are Silver Sedge (*Odontocerum albicorne*), Small Silver Sedge (*Lepidostoma hirtum*), *Halesus radiatus* and Welshman's Button (*Sericostoma personatum*), all being true-cased caddis. Sand-fly (*Rhyacophila dorsalis*) and Grey Flag (*Hydropsyche siltalai*) are the most widespread free-living caddis. Several caddis are also much more restricted in the river systems, some occurring in only a few sites, whilst others are rarer nationally. Good examples of the latter are the caseless caddis (*Hydropsyche fulvipes*) and the free-living caddis (*Rhyacophila septentrionis*), both Nationally Notable species found in the Dove and Manifold catchments.

Freshwater shrimps (which are amphipods not true shrimps) are the most abundant Crustacean in White Peak rivers and more than 300 can be caught in a standard sample, although this may indicate some pollution as the best sites on the River Wye produce counts in the thousands. Their side-swimming habit is distinctive. They are important inhabitants, feeding on organic debris, while also being food for a wide range of fish. They are a useful pollution indicator too since they require oxygen-rich water.

Although not caught in every sample, there is a range of water beetles in White Peak rivers. Several common Riffle Beetles *Elmis aenea* and *Limnius volckmari* occur as adults and larvae, for example, and predatory diving beetles, whirligig beetles and smaller beetles all feature in limited locations and numbers, related to the habitat quality and type. *Elmis aenea*, a common species amongst stones, is more tolerant of silty conditions than the *Limnius*. *Esolus parallelepipedus*, another riffle beetle, is locally uncommon and found more in the Dove than other White Peak rivers. *Hydraena gracilis* is a midstream beetle, but one which is also Local, more often recorded in the north and west. *Riolus subviolaceus*, another Nationally Scarce riffle beetle, is a characteristic limestone dales species in the riffle sections of the Rivers Bradford, Dove, Wye and some of their tributaries (Merritt, 2006).

This is just a sample of the diversity of aquatic invertebrates that also include River limpets (*Ancylus fluviatilis*), various blackflies, whose larvae are sometimes abundant, some aquatic snails like Jenkin's Spire-shell (*Potamopyrgus antipodarium*), Pea and Orb mussels (*Pisidium* and *Sphaerium* spp.) and a few leeches such as *Erpodella octoculata*.

We cannot conclude this section without mentioning White-clawed Crayfish (*Austropotamobius pallipes*) (Fig. 214). This, our largest native freshwater invertebrate, was once abundant in White Peak rivers but has been decimated through competition and disease (crayfish plague) from introduced animals, principally the American Signal Crayfish (*Pacifastacus leniusculus*), after establishment in crayfish farms in an attempt to diversify incomes in the 1970s. Possibly predictably, these escaped and invaded catchments. They are much larger than the native species and can also move overland between catchments, thus spreading both competition and disease. Moreover, plague-infected animals, caused by a fungus, can release spores that quickly disperse downstream and, although American Crayfish are relatively immune to the disease unless under stress, the native species is highly susceptible. It is now listed on the IUCN Red List of Endangered species as it is considered to be facing a high risk of extinction in the wild throughout its European range.

There are still very small native populations in a few spots in the White Peak. Surveys in 2011, for example, discovered a healthy population on the

FIG 214. Freshwater Crayfish, our largest freshwater invertebrate at around 10 cm long, but smaller than the American Signal Crayfish (16 cm). (Alex Hyde)

River Bradford that then survived droughted conditions. For a species that is protected under European and British legislation, this is a sorry state. Several of the White Peak SSSIs include rivers cited for healthy populations of White-clawed Crayfish as a qualifying feature, only for this to be lost within 20–30 years. The South West Peak HLF Landscape Partnership and other projects have established White-clawed Crayfish in new safe ark sites and these are breeding successfully. Hopefully, they can be introduced into sites after removal of the invaders in the future.

Cold springs emanating from the limestone slopes and caves are a unique habitat in the White Peak. Many demonstrate thermal stability which is distinctive biologically. A Wye Valley study found the river depends on at least 18 bankside springs as well as groundwater discharge from the river bed adding to its otherwise quite moderate flow from the uplands to the west (Smith, 2002). Many springs were found to harbour dense populations of Freshwater Shrimp, midge larvae (Chironomids) and *Crenobia alpina*, a flatworm which is rare nationally. Species that only occurred in the spring water were a dark brown diving beetle *Agabus guttatus* and the caddis *Beraea maurus*, although both are widespread generally. Their presence was related to flow regime rather than chemical parameters. The detailed fauna in each spring was quite distinctive though, partly as some were seasonal, but also determined by the moss cover, variation in temperature and presence of woody debris.

A fly assemblage is also associated with calcareous springs and wet tufa. The elegant little soldierfly, Hill Soldier (*Oxycera pardalina*) is a good example, known from Monk's Dale, Lathkill Dale, Dove Dale, several cliff seepages around Matlock and a tufa spring at Priestcliffe Lees. Larvae develop in wet moss with a specific water chemistry. The small rare cranefly *Dicranomyia aquosa* also breeds here and is a very localised northern species of springs and seepages.

Freshwater invertebrates are important in their own right, integral to the river's biodiversity, but they also support the high fishing value of the key White Peak rivers. Celebrated famously in Izaak Walton's *The Compleat Angler* (1653), it was Charles Cotton, born locally at Beresford Hall above the River Dove, and a contributor to Walton's book, who made fly fishing famous. His Fishing House, standing by the River Dove in Beresford Dale, is a small square building in the Artisan Mannerist style. It is a scheduled monument and listed building, with both authors' initials inscribed above the door. Charles Cotton was a poet and writer who loved the Peak District. Although fly fishing has changed since

FIG 215. Female Rainbow Trout from the River Wye showing the striking colouring. (Don Stazicker)

the 17th Century, some of Cotton's advice is still useful. He provides detailed descriptions of numerous river flies and lures to imitate them. Today, trout, especially, but also grayling fishing are renowned in White Peak Rivers. The Wye is said to be 'the finest insect-driven fishery in Britain' (Don Stazicker pers. com.) and can boast good populations not only of wild Brown Trout but is also renowned world-wide for the only major self-sustaining population of Rainbow Trout that grow to a good size. Being able to catch a wild-living Rainbow Trout is a unique experience (Fig. 215).

Generally the fishing is managed by Fishing Clubs. Compared with the past, most fish caught are returned to the river and there is little or no stocking carried out. In general, the better and more diverse the invertebrates, the better quality the fishing. Fly fishing focuses on attracting trout with lures reflecting the invertebrates on which they are feeding. A diverse fauna offers greater choice and therefore a greater challenge to find the preferred fish prey for that day to cast the right lure. Brown trout are visual predators and will lock onto a particular prey item if good numbers pass by and then focus on them. These could be adult insects stuck on the water meniscus, larvae hatching or other aquatic animals but are mostly mayflies, stoneflies, caddisflies and freshwater shrimps. Sometimes they might focus on ants or aphids falling into the water, depending on the adjacent habitat. Fishermen often sample the invertebrates present to select the right 'fly' for attracting the fish. This also enables them to sample the overall river diversity and the Angler's Riverfly Monitoring Initiative is collating these records to help understand water quality and provide a rapid response to pollution events. More detailed, robust data is now being collected in a new national 'SmartRivers' monitoring programme (Salmon and Trout Conservation) undertaken by Peak District river stakeholders to obtain species level data, which is already yielding better insights into river condition like the Dove (Everall *et al.,* 2019). It is the observant fisherman that has added much to our knowledge of the behaviour of target fish over time, started perhaps by Charles Cotton.

Trout can see colour and learn to avoid negative features, even for up to six months at a time. Their retina and eye pupil are not circular and they can see close-up and distance simultaneously. In addition, as juveniles, they can detect ultra-violet light. They sense movement and contrast between objects to find food, more in the water than above the surface, although animals depressing the meniscus can be readily detected. They are also good at detecting smells. Brown Trout are quite competitive and establish a hierarchy, expanding their gill covers to appear larger in a threatening gesture to secure the best feeding spots. They feed mostly singly and are often visible pointing upstream under a bridge or by groups of rocks, wood or overhanging vegetation. Trout lay their eggs in winter in the riffle sections as they require well-oxygenated, cold water with minimal sediment to hatch. These habitat requirements limit where they naturally live to the upper reaches of most rivers, as in the Peak District. Interestingly, Brown Trout can also change colour using chromatophores (pigment-containing and light-reflecting cells) in their skin and are often darker below a canopy than in the open. They can be distinguished by their brown background colour decorated with medium-sized black and red spots. Rainbow Trout are more colourful with a generally blue-green or yellow-green background and a pink streak along the sides (Fig. 215).

Grayling (*Thymallus thymallus*) is another game fish in the Salmonid family (Fig. 216). Clubs used to remove them believing they competed with Trout, but more recent research shows segregation of their chosen microhabitat and diet (more bottom feeding) and grayling are now seen as important additions to the aquatic ecosystem. Unlike Trout, they breed at the same time as coarse fish (March to June), so can be fished during the winter Trout close-season. Grayling numbers are greatest in the lower part of the Wye, from just above Ashford-in-the-Water downstream. There is also an excellent grayling population in the Derwent in its middle reaches, whilst coarse fish take over here on the edge of the Peak District below Matlock as temperatures rise, oxygen levels reduce, riffles are fewer and eutrophication often higher. Salmon (*Salmo salar*) have not appeared yet owing mostly to a large weir at Belper on the Derwent to the south, although they have reached the Dove south of the Peak District.

There are other fish not targeted by fishermen, about which far less is known. Bullheads, also known as Miller's Thumbs, are the only freshwater cottid in the UK and frequent the similar fast-flowing stony rivers as Brown Trout. Unlike

FIG 216. Female European Grayling from the River Wye. (Don Stazicker)

most fish, they have no swim bladder so are negatively buoyant; well adapted to their bottom-living habit. They are also remarkable for providing parental protection of the eggs by the male in a 'nest', in producing audible sounds when nest-making under a large stone (which might also be territorial behaviour) and having a unique shape with a large flattened head and tapering body 8–12 cm long. They are fairly sedentary and crepuscular, although depend on sight to catch Shrimps, Hoglice plus a wide range of insects in the summer (Tomlinson & Perrow, 2003). They are widely distributed in England and Wales but absent from Scotland, living largely in stream headwaters, but are included in the Habitats Directive Annex II, for which they are a qualifying feature in the Peak District Dales SAC (which includes all the main limestone dales).

The other equally protected and unique species is Brook Lamprey (*Lampetra planeri*), the smallest British Lampreys at only 10–17 cm long. It is a primitive, jawless fish which resembles an eel in having no fins or scales, but has a round sucker-like disc surrounding the mouth. It has no bones, replaced by strong, flexible cartilage. It obtains oxygen through gills that are arranged in seven gill pores behind each eye, but with no gill cover as in other fish. It shares Brown Trout's preference for clean gravel beds, but slower flowing ones for spawning when the water temperature reaches 10–11°C. On hatching, the ammocoetes (the young, semi-translucent, elongate larvae) swim or are washed downstream before burrowing into tunnels in marginal silt. Here they spend the next few years feeding on organic particles, microscopic organisms and algae, making them hard to detect. They metamorphose into adults synchronously over a dramatic few weeks, before migrating upstream to spawn, where they are both more visible and more vulnerable to predation. Brook lampreys do not feed as adults, but die after spawning. Like Bullhead, Brook Lamprey is a qualifying feature in the Peak District Dales SAC and is regarded as a vulnerable species across Britain and Europe where it has declined steadily. It occurs in several White Peak Rivers like the Dove, Hamps and Manifold as well as the Wye.

A range of factors determine river quality which affects the plants and animals and therefore the river's nature conservation value. Pollution is an ever-present threat here more than in any other habitat. Pollutants are primarily nutrients, especially nitrogen and phosphorus, sediments from soils, road runoff which can also contain de-icing materials, and chemicals derived from agrochemicals or manufacturing/processing works as well as effluent from sewage treatment works. Nutrients cause eutrophication; an elevation of the natural nutrient levels. This, in turn, can result in algal blooms, loss of pollution-sensitive invertebrates and negative changes in the overall biodiversity. In the worst cases, sewage fungus (actually a mass of filamentous bacteria and algae) coats surfaces over which nutrient-rich water flows, as from some waste water treatment works outfalls. This also has a characteristically nasty smell. Other sources of nutrients are from farmland where slurry (which is especially toxic) or fertiliser runoff passes into interconnected drains or ditches or directly to rivers. Point sources of pollution are generally managed and regulated to a lesser or greater extent (although many old road drains pass directly to rivers and not via sewage treatment works) by the Environment Agency. The non-point pollution sources can be very damaging collectively, but individually are not necessarily easy to pinpoint. These are essentially those from farmland, septic tanks and incorrectly connected drains.

Weirs also affect river habitat quality. If large enough, they prevent fish and invertebrate migration upstream, or fish become exhausted trying to leap them, with reduced breeding capability subsequently. Weirs can also fragment populations thus reducing genetic interchange. They trap sediment behind them, possibly becoming anoxic and silty, changing the natural stream functionality

FIG 217. Small weirs on the River Dove in Dovedale, some of which are being removed. The tree in the water has been felled to provide woody debris which significantly enhances the habitat.

and structure quite radically. Multiple weirs have an accumulative effect. The River Dove in its limestone section from Hartington to Ilam is one of the worst affected in this respect, with over 170 small weirs identified by the River Dove Catchment Partnership (Fig. 217).

These various threats to river quality are addressed in different ways. Significant improvements were engineered at Buxton sewage treatment works in 2004 which benefited the River Wye, although there are still impacts from Tideswell's treatment works. The efficacy of the improvements are reflected in monitoring results using freshwater invertebrates (Fig. 218). Each invertebrate family from standardised samples is scored according to their pollution tolerance and the scores added to give a total per sample called the BMWP score (see Glossary). The higher the score the better quality the water, largely related to the degree of eutrophication rather than other pollutants.

The surveys of the Rivers Dove and Manifold catchments showed significant depression in freshwater invertebrate abundance and diversity owing principally to non-point sources of pollution, particularly sediment and phosphates (Everall, 2010). Compared with the River Wye, these key White Peak rivers showed poorer quality, although this was patchy. One accountable difference is the Wye's catchment character, being buffered from agricultural inputs and in a gorge for

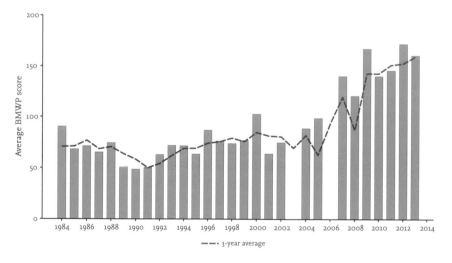

FIG 218. The average BMWP scores and three-year running mean from 1984 to 2013 showing the significant improvement of aquatic invertebrates sensitive to pollution in the River Wye in Miller's Dale after upgrading Buxton sewage works in 2004. (Data from Cressbrook and Litton Club)

most of its length from Buxton to Ashford-in-the-Water, whereas the Dove and Manifold directly drain more intensively managed farmland in their catchments upstream of the limestone. A programme of interventions led by the Dove Catchment Partnership (hosted by the Trent Rivers Trust) provides advice on catchment sensitive farming to farmers and land managers. Even simple actions like fencing livestock from river edges, providing water troughs instead of river drinking access, good soils management and, critically, avoiding slurry and fertiliser runoff into water, facilitates significant progress in addressing the issues.

The River Dove Catchment Partnership website shows that phosphate levels in the Upper Dove were reduced by 2015 through the programme, but the status for phosphate was still good (rather than high) for the Manifold and Hamps, whilst the Bletch and Bradbourne Brooks which join the Dove in Ashbourne, still required significant upgrading. The improvements also matches the 2016 Riverfly census results for White Peak Rivers collated by Salmon and Trout Conservation. Although issues remain on most surveyed rivers, the River Wye performed better compared with the Dove, which was still being impacted by phosphates and sediment. The exigencies of the European Water Framework Directive drive many of our river water improvements, but these are also beneficial in their own right to restore an important habitat loved by all.

Many measures enhancing river habitat quality also assist in flood risk reductions downstream and improve drinking water quality. The River Dove Catchment Partnership adopts similar measures to the South West Peak's Slowing the Flow project (Chapter 11); working with land managers to re-meander straightened rivers, removing weirs to improve habitat functionality, placing large woody debris in rivers to add habitat diversity, implementing targeted tree planting to reduce overland flows and filter runoff, and using leaky dams to store more water for longer in upper reaches as natural flood management techniques. Already, several weirs have been removed in Dovedale and selected water's edge trees felled and positioned in the water (Fig. 217), tree planting, leaky dams and river meandering are all designed or being undertaken. More is planned and the results are being monitored. Volunteers are welcome if you are interested.

An additional feature of many rivers is the number of non-native invasive species needing control. Himalayan Balsam is the main plant culprit and concerted efforts are being made to remove it; successfully on the River Wye thanks to the fishing club volunteers. Other invaders like American Mink that have been responsible for the demise of Water Voles (and nesting birds) and Signal Crayfish have already been mentioned. Mink, established after escapes for example from fur farms, is now widespread on all main rivers, especially the Dove, Goyt, lower Wye, Derwent and many of their tributaries, including small streams up to 500 m (Mallon *et al.* 2012). Again, the River Wye Fishing Clubs have successfully controlled this unwanted predator.

All these measures enhance habitat quality, which is then also reflected in the other river-side animals. The river bird life is rich and diverse, as might be expected with good fish and invertebrate populations to sustain them. The principle ones are Kingfisher, Dipper, Grey Heron (*Ardea cinerea*), Grey Wagtail (*Motacilla cinerea*) and different ducks (Frost & Shaw, 2013). Kingfishers are more frequently associated with the River Wye (plus rivers in the northwest), but suffer significantly in severe winters, taking up to six years to recover numbers. The lack of suitable steep nesting banks along some White Peak rivers could restrict their numbers. They were persecuted incessantly by fishing interests in the 19th Century, but no longer, and are legally protected under the Wildlife and Countryside Act 1981.

Dipper (Fig. 116) numbers are much higher on White Peak compared with other Peak District rivers, with breeding densities of 0.48 to 0.49 territories per kilometre on the Derwent, Dove and Wye Rivers for example. These are essentially upland stream birds and a welcome sight as they scour streams under water for caddis larvae on which they principally feed. They breed in sheltered spots under rocks or bridges, or over deep water. Young birds tend to disperse locally, no more than about 10 km, although severe winters can push birds downstream.

It would be rare not see a Grey Heron along a quiet White Peak River, although they do feed elsewhere (ponds, other rivers and even dry ground as their diet is wider than fish). They travel some distance in the breeding season as heronries are scarce in the Peak District. Grey Wagtails are a familiar sight, beautifully yellow and grey, their wagging tails improving flying agility while in pursuit of insects (Fig. 219). Although not rare in the lowlands, clean upland streams are more important habitats and the Wye (especially in Chee Dale), Lathkill, Bradford and Dove Rivers all support good numbers of this recent addition to the Red List. They are not confined to the White Peak though, and occur on the Derwent and around the reservoirs as well as in the northwest Peak District. Although they usually breed close to rivers and are largely resident, they do move downstream in severe winters and sometimes gather at sewage works like those in Hathersage, Stanton-in-Peak or Baslow.

Of the ducks, only Mallard is ubiquitous, breeding alongside rivers or in nearby woods. Winter numbers are high on the Wye in Bakewell; up to 715 in September 2004; but larger numbers are more familiar on lowland water bodies south of the Peak District. The other, much more exotic duck, is the Mandarin (see Chapter 11). Breeding was proven on the Dove in 1991 when five ducklings appeared and numbers increased to about 30 pairs by 2011, mostly between Crowdecote and

FIG 219. Grey Wagtail feeding juvenile in mid-river. (Kev Dunnington)

Thorpe in the early days, but now expanding along the Derwent, Wye, Bradford Dale, Monk's Dale and beyond. They nest in tree holes and even owl nest boxes. Wintering numbers are mostly higher outside the White Peak, but there were 76 on the River Wye at Rowsley in 2010, with fewer noted near Bakewell.

The last wetland bird of note is the Little Grebe or Dabchick (*Tachybaptus ruficollis*), which breeds in good number on White Peak rivers, but is scarce otherwise in the Peak District. *Tachybaptus* translates as 'fast' and 'to sink under' which fittingly describes its diving habit searching for small fish and aquatic invertebrates. It prefers well vegetated waters as in the White Peak rivers. Look out for it where disturbance is low, perhaps carrying its young on its back. High wintering numbers mostly occur just outside the Peak District at Carsington reservoir (Frost & Shaw, 2013).

As far as mammals are concerned, Water Voles are, or mostly, were an important species (see Chapter 9). Where Mink are controlled, as along the Wye, there is a thriving population, readily seen from the road to Litton (Fig. 220). However, none were found in the Dove catchment in the Peak District in

FIG 220. Water vole feeding amongst rushes. (Kev Dunnington)

Everall's surveys, illustrating their dire straits. Otters (see Chapter 11) have appeared since 1992 in the Dove and 1998 on the Wye-Derwent. Occasional sightings are known from the Manifold catchment too, and one splendid male was found as a road-traffic victim in Buxton in the mid-2000s. Their main strongholds are in the lower-lying river valleys south of the Peak District (Mallon *et al.*, 2012).

With the abundance of flies associated with White Peak Rivers, it is not surprising to find bat species specialising in river-side foraging (Mallon *et al.*, 2012). Daubenton's Bats have already been mentioned (Chapter 11), and they frequent the Rivers Wye (especially), Lathkill and Dove. Natterer's Bats and Soprano Pipistrelle also favour river valleys for foraging, although they utilise deciduous woods too. The Bradford River, the Via Gellia and below the cliffs at Water-cum-Jolly on the Wye are known feeding spots for Natterer's along with Noctule (*Nyctalus noctula*), while Soprano Pipistrelles and Leisler's Bat frequent the Wye Valley especially. Most of these bats roost in tree holes, buildings, or for Daubenton's, nooks and crannies under bridges, rock faces and tunnels.

CAVES AND WELLS

Little is known about dry caves as a habitat, although bat hibernation in some has been mentioned. Some moths and caddisflies are known to overwinter in caves and some use them to pass through a period of suspended development (diapause). In general terms, aquatic cave-life in the Peak District, and indeed, much of Britain, has been poorly studied. Without any primary production (plants) or organic carbon sources, most subterranean communities demonstrate low abundance and diversity. However, a two year study of freshwater animals in the Peak-Speedwell cave system around Castleton produced some interesting results. The caves form part of over 20 km of explored passages, Peak Cavern passages being fed by rainwater infiltrating through local soils and from springs, whilst Speedwell Cavern's water comes mostly from streams disappearing into swallets below Rushup Edge (Chapter 4). The water travels through passages, down major drops up to 20 m, and along flooded channels before emerging some 14 days later. This all produces a physically testing environment (Gunn *et al.*, 2000). Twenty-eight invertebrate taxa were found in the passages, only 10 common to both cave systems, which contrasts with a more diverse 50 taxa recorded from the swallets and springs prior to cave entry. Eight species were only found underground: one ostracod, six copepods (small crustaceans) and one beetle. This is a quite different from White Peak river invertebrates, where insects predominate. Diversity was lower in the Peak Cavern system, possibly related to the lack of stream inputs. One interesting species, though, is the flatworm *Crenobia alpina* only found in Speedwell Cavern in the study, but which was one of the special species in the cold springs along the Wye already described.

The most numerous group in the Peak-Speedwell system was the Chironomid *Rheocricotopus fuscipes* with 350 /m² in Speedwell Cavern in May 1997. Worms (Oligochaetes) were the only other group represented in each sampling month, although some were terrestrial species obviously washed into the caves. A stable population of the bivalve *Pisidium nitidum* was also found on one pool in Peak Cavern; notable as live bivalves are quite rare in caves. The diving beetle *Hydroporus ferrugineus*, a Nationally Scarce species, appeared almost 900 m into Peak Cavern, upstream of any stream water input and, with larvae and adults recorded, it appears to be breeding. It is probably the top predator in the assemblage. Merritt (2006) also considers it a subterranean species, with most local records coming from springs, pools in caves (including Lathkill Head cave), springs on the River Wye and in Milldale, but also unusually from a spring on the Eastern Moors. As a species that cannot fly, its powers of dispersal are limited, but it has been found in some interesting places like wells, with *Agabus guttatus* (abundant in cold spring water in the Wye).

Overall, the evidence points to a fauna that is largely derived from external sources, changing seasonally linked to outside patterns, but with some interesting features. These cave faunas can be scientifically fascinating as they exist in an energy-poor environment, their mode of colonisation and their survival and sustainability raising many interesting questions. Just like White Peak rivers, these underground streams and pools are susceptible to pollution, only in this case, ingress may be more obscure. Management of fields above, or effects of mineral extraction nearby can have unwanted but possibly unexpected effects, only disentangled by understanding the local hydrological links. Nutrients and sediment can have similar consequences as in the open water environment, overuse by cavers, lighting which can increase temperatures and algae or bacterial growth, can all change the ecological relationships. SSSI protection may not extend to cover all the hydrological inputs into the caves. There is much more that needs to be understood about both the ecological interest of the White Peak's many and lengthy cavern systems and the effects of land management above them.

Just as drying out of underground caverns, springs and White Peak rivers during droughts would have consequences for the animals they support, so continuous running of springs and wells in the limestone area were critical for people. The age-old fear and worship of watergods and spirits was presumably the foundation for well dressings. In earlier times, floral contributions, wreaths and garlands would have been caste into wells or springs to assuage the spirits and ensure continuing supplies of water during dry periods (Porteous, 1978). One version of the story (Merrill, 1988) suggests well dressings date back to the Black Death (1348–9) when nearly all the local priests died, except in Tissington. This was attributed to the pure, uncontaminated water in Tissington's five wells which were then dressed to offer thanks. However, what sounds like a well dressing was not described by travellers until the early 19th Century: a picture made of flowers stuck into clay. Well dressings are a Derbyshire custom now adopted by neighbouring areas even off the limestone, and most Peak District villages now have one in a rota from May to September.

The design is often based on a religious text, but may celebrate local or national events. A local artist often prepares the drawing. Soft clay is thoroughly pummelled and then spread out into large trays that have been adequately soaked. There are often several trays, some attached as a triptych, others displayed separately. Large trays can be more than 2 m high. The design is

FIG 221. A particularly fine well dressing in Peak Forest in 2018 showing the village church.

pinned onto the clay and the pin marks drawn in. The shapes are filled with a wealth of different natural materials from seeds and cones, to bark, crushed egg shells, peppercorns, flower petals, leaves and mosses. The experienced dressers press in all the materials and creating the design, whilst the novices ransack gardens to find suitable materials. It is a village community event, particularly amongst women.

The resulting well dressings are erected in a prominent place, blessed by the church and then maintained for a week for all to see (Fig. 221). Local schools may produce their own. In many villages, the well dressing is tied into fetes, or fayres (as in Elton and Peak Forest), competitions, sporting events (Longnor Races), carnivals (the Wells Queen in Hope or Carnival Queen in Great Longstone) or flower festivals (Earl Sterndale and Chelmorton). Look closely at the next well dressing you see to identify its materials. The RHS Chatsworth Flower Show (modelled on the Chelsea Flower Show) displays numerous well dressings made specifically for the Show by different villages.

LEAD RAKES AND SPOIL HEAPS

Moving onto dry land, lead rakes occur as lines or masses of hummocks and hollows striding across plateau fields, dale or around lead mines (see Chapter 6 and Fig. 58). The rake is covered mostly by limestone grassland on waste rock, plus unique communities where metal levels are higher at the surface supporting metal-tolerant plants adapted to high levels of, for example, lead, chromium, zinc or copper. Investigations have shown lead concentrations as high as 60,000–75,000 ppm, and zinc concentrations between 8,000 and 13,500 ppm, in surface soil layers – levels definitively toxic to most plants. In addition, lack of organic matter owing to the sparse vegetation (exacerbated during drought conditions) and poor soil structure, along with low nutrient levels, together provide a challenging habitat. Calaminarian grasslands is the technical term for these metalliferous communities, which tend to be sparse, allowing space for slow growth: conditions that have persisted for perhaps centuries in some places.

The exact nature of the vegetation is determined by the ratio of limestone waste to toxic metalliferous waste, the levels of heavy metals involved and where

FIG 222. Spring Sandwort or Leadwort growing on lead waste material in Tansley Dale beside the public footpath.

FIG 223. Lovely diverse floral display on lead waste mounds at Gang Mine, Derbyshire Wildlife Trust Reserve with Small Scabious, Burnet Saxifrage, Bird's-foot-trefoil, Red Clover, Harebells, Wild Thyme and many other plants.

they lie in the soil profile. This can vary on a small-scale and is visible in the plant communities' response. Sheep's-fescue and Common Bent form the community's core, and both can produce genetically adapted races tolerant of toxic conditions. Heavy metal-tolerant populations of Bird's-foot-trefoil and Common Sorrel are also known. The classic heavy metal tolerator is the pretty white-flowered Spring Sandwort (*Sabulina verna*) often called leadwort (Fig. 222). This species' strongholds are the Derbyshire and Yorkshire Pennine lead mining areas and base-rich volcanic rock in North Wales. It is quite frequent in the White Peak and a good indicator of high metal levels. Look for it in open vegetation on rakes, but one accessible spot is the raw waste heap in Tansley Dale that is also a rabbit warren beside the public footpath. The plant has been shown to be able to accumulate concentrations of lead and zinc respectively to 2,700 and 25,000 ppm without harm, which are exceptionally high. Not as widespread and often in smaller numbers is another metallophyte, Alpine Penny-cress (*Noccaea caerulescens*), which is also confusingly called Leadwort. This community also supports some rare lichens and a range of mosses as on Longstone Moor.

Where toxicity is lower and soils more stable, the vegetation thickens and increases in Sweet Vernal-grass, White Clover and Wild Thyme are visible (Fig. 223). Eyebrights, Kidney Vetch and Glaucous sedge also appear with more Fairy Flax in the gaps. Several orchids like Pyramidal, Fragrant and Frog Orchids provide colour and drama to these swards, which are often very pretty.

The wealth of flowers attract a range of invertebrates and provide seeds and forage for many other animals. The sparse vegetation patches are literally 'hotspots' for insects and associated mine shafts provide safety for many bats and hibernation sites for amphibians and other animals. The 'Lead Rake' Robberfly, *Leptarthrus brevirostris* is one very special animal, more widely known as the Slender-footed Robberfly and found more on chalk grassland in Southern England, but occurring also in Scotland and western England. This specialised habitat provides a warm, sandy, friable soil ideal for breeding. Often lead rakes have areas overgrown by scrub and open woodland with sunny glades and these habitats are also frequented by *Leptarthrus*. Although it does occur in some dales, its ideal habitat is on rakes with open sunny, short diverse vegetation and longer stems on which to perch ready to pounce on smaller flies. Its hump-backed shape and lazy flight make it distinctive (Whiteley, 2011a).

Calaminarian grasslands are internationally important and are listed in the Habitats Directive, with the best protected as SACs as well as SSSIs. They are very restricted in Europe and therefore considered to be of international importance. Each rake tends to be unique as well, reflecting its detailed history. As well as their unique flora, these grasslands are closely associated with our cultural and mining history (Chapter 6, Barnatt & Smith, 2004, Barnatt & Penny, 2004).

When the Peak District BAP for these grasslands was first produced in 2011, the area of lead rakes (not clearly defined, but assumed to be the areas with the richest vegetation) was estimated at only 41 ha, with losses to agricultural reclamation and new quarrying known to be threatening much of the archaeological and ecological resource. The Derbyshire ore-field surveys in 2011–12, revealed a miniscule 14 ha of Calaminarian grassland remaining, all but 1 ha in the National Park. The overall area of lead rakes is larger (about 203ha, nearly all in the White Peak), but without the Calaminarian community. The grassland was also in very variable condition, being particularly poor where not protected and where agricultural improvements have been attempted, and sites were often fragmented, both within a rake and between other calcareous grasslands. Of the 14 ha recorded, 6 ha lies within SSSIs and is better protected. Habitat restoration works have been undertaken on 12 sites, two outside the National Park and three in SSSIs. There is only about 100ha of Calaminarian grasslands nationally, showing how important the White Peak is in the wider context. Sites are concentrated in Bonsall, Castleton, Bradwell, Elton, Winster, Monyash, Middleton-by-Youlgreave and Brassington parishes. Public footpaths provide good access across Bonsall Mines and to Gang Mine, a Derbyshire Wildlife Trust reserve near Middleton-by-Wirksworth.

Calaminarian grasslands are a product of a complex history which cannot be emulated easily, and current extraction does not result in new metal-rich subsoils or waste on which the unique vegetation community can establish. It may be possible to undertake trials to rejuvenate some rakes and expose old metal-rich materials where these have been lost through natural succession or covered with other materials. In general, however, conserving those left is more urgent and important. Equally critical are the networks needed to connect tiny sites with other flower-rich grasslands through habitat creation, enhancement and management. The White Peak project (See Chapter 15) will hopefully address some of these issues over time.

QUARRIES

The legacy of the White Peak's quarrying and lime-burning history (Chapter 6) is dales and plateau pockmarked with small quarries and, in some places, relic lime-ash waste hillocks. The later industrial-scale extraction led to larger quarries, some of which are still being worked. This results in disused quarries of differing scales, depths, substrate and age. The oldest have revegetated, and the extent and composition of this is linked to proximity of seed sources and the ability of plants and animals to reach them. The overburden mounds and former soils usually contain more nutrients, so colonise readily with more vigorous plants like Rosebay Willowherb and typical local grasses. The barer limestone debris and rubble screes are more likely to support some of the special Hawkweed populations (as described for natural cliff ledges in Chapter 13). Early flowering Colt's-foot (*Tussilago farfara*) followed by summer lemon-yellow of Mouse-ear-hawkweed are regular companions in sparse vegetation.

Some of these hummocks and waste heaps in the longer-established quarries are rich environments for orchids, although they are not generally confined to this habitat. The populations of Bee and Fragrant Orchids have already been described in different contexts (Chapters 7 and 13), but others like Frog Orchid and, rarely, Fly Orchid also occur. Cauldon Quarry on the southwestern edge of the White Peak holds large numbers of Common Spotted Orchid. Additional seeding of flower-rich grassland by the Staffordshire Wildlife Trust, along with rescue of some of the richest areas around this large, active quarry as part of its planning conditions are ongoing and will over time create a larger area of flower-rich grassland and seek to connect the remaining fragments.

Cliff faces are a harsher environment and take longer to revegetate. Experimental research to ameliorate these in Tunstead limestone quarry trialled selective high-explosive blasting to engineer dale-side landforms with buttresses, headwalls and screes. Several different treatments aimed to increase vegetation establishment and reduce the eyesore of a newly worked-out large-scale quarry. The invertebrates and plants that colonised, compared with older quarries and more natural dale-sides showed that the engineered sites, even after 10 years, supported less plant cover, more bare ground and fewer animals than the older quarries and dale-sides. Time is the magic ingredient, of course, as the invertebrates especially have to colonise unaided, but some plants rarely grew on the blasted sites, despite seeding, like Salad Burnet, Cowslip, Wild Marjoram and Glaucous Sedge, all of which need more soil to persist. Only Red Fescue appeared throughout in any abundance. Of the invertebrates, greater numbers occurred on the dale-side grasslands than the disused or engineered quarry habitats. Molluscs (herbivores) and woodlice and millipedes (decomposers) were particularly sparse on the blasted sites. Spiders and mites showed the same pattern, but were more numerous (Wheater & Cullen, 1997).

Some of the invertebrates and plants colonising new sites like quarries will be early successional specialists, whilst these are overtaken through time by others preferring vegetated ground. It is valuable then to have examples of all age ranges and degree of nudity. Indeed, it has been cautiously postulated that the nature conservation gains from quarries over centuries in the White Peak have so far out weighed the losses (Anderson & Shimwell, 1981). Part of this equation are Quarries like Miller's Dale and Hopton (on the Via Gellia) that are now SSSIs, but others like Hoo and Peak Quarries at Longcliffe are newer quarries that are now Derbyshire Wildlife Trust reserves and Eldon Hill Quarry is quietly accumulating species (including Bee and marsh orchids) as the restored site develops. This conclusion does not mean that we need more, new quarries in the Peak District though.

LIMESTONE HEATH

Looking across the limestone plateau of neatly walled grassy fields, it is hard to imagine it covered with heathland on wastes and commons until the late 18th and early 19th centuries. Shimwell (1977) thought these heathlands developed earlier than many in the Dark Peak and to have been extensive from the Sub-Boreal period onwards, i.e. for some 2000 years, although the true extent of heath, acid grasslands and scrub might have fluctuated with climatic variations and human activities. Reconstruction of the historic landscape shows open wastes and common, incorporating heathland, predominating across the White Peak plateau in 1650 and 1750, but largely lost by 1850. Groups of medieval open fields surrounded villages that were established where there were reliable springs. These fossilised remains of medieval fields were later enclosed. Animals grazed the wastes and commons, which also supplied turves, peat and firewood; important resources for the commoners and lords alike (Barnatt, 2019).

Contemporary writers described the scene. Young (1770) recorded land between Tideswell and Castleton as 'recently reclaimed from black ling', while Farey (1815) writes of the heath that 'abounds on … most of the remaining commons in this county, in a less or greater degree', and that, 'within a few years past all the fine limestone hills between Ashbourne and Buxton were occupied by heath'. However, these same commentators could not wait to be rid of the heaths. Farey's view was that 'when the only remaining commons or unimproved lands of this calcareous district … shall have been enclosed, pared, burnt and limed etc, this noxious and useless plant [i.e. heather] will I hope disappear from the district.' He quotes several instances of farmers improving their land so that the heath 'on the limestone hills between Buxton and Ashbourne is now happily becoming rather scarce there'. Methods varied, but usually involved mixtures of: burning the heath, paring the land (turves removed with the peaty humus and roots), and breaking the land with ploughs and liming. A crop of turnips or oats was then grown or the land sown with White Clover, trefoil and hay seeds. Many of the resultant fields are large – larger where the common rights were fewer and the land higher, with straight, often rectangular enclosures across the landscape, contrasting with the smaller scale of the older strip fields' enclosures (Fig.56). Barnatt (2019) also offers the possibility of some walls being only 300 to 400 years-old, with the former landscape containing more hedges, as old hedge banks are evident occasionally where Estate maps show enclosure boundaries. Some of our famously austere, stone-walled, mostly open plateau landscape could therefore be relatively recent in its origin, although one that stretched across several waves of reclamation associated with landowner enclosure, followed by Parliamentary Enclosure Acts.

The outcome, though, has been the near destruction of our limestone heaths. Moss (1913) mapped what was left, showing several small sites south of Peak Forest, south and east of Chelmorton and larger areas on Bradwell and Hucklow Moors. The smaller areas have since gone, but those on Bradwell Moor and around Berrystall Lodge are still present, although reduced in extent. That on Longstone Moor is also smaller than it was in 1913. In 1998 the NPA updated a 1984 limestone heathland analysis I undertook and found only some 100 ha remained, a paltry 0.002 per cent of the White Peak, most on just one site, Longstone Moor, an SSSI for its heathland. Two sites were lost between these surveys in only 14 years (Buckingham & Chapman, 1999). Collating data for the Peak District's State of Nature report in 2016 found the total for heathland plus acid grassland and other incorporated habitats still to be a very small 216 ha.

There are two types of limestone heath. Primary heaths are remnants of the former wastes and commons such as on Longstone and Bradwell Moors. There are also a few former small plantations where trees have failed enough for heathland either to have survived or to have colonised. Secondary heaths are where heathland plants have colonised developing vegetation after disturbance as on some of the former railway embankments, quarry works, or roadside verges. The older primary heaths can be separated into those on siliceous soil materials such as the loess or chert layers (see Chapter 7) and those directly on the limestone. In the latter case, excessive leaching over time has resulted in the development of organic-rich soils, even peaty and wet in places,

FIG 224. Longstone Moor showing a mixture of heathland with abundant Heather intermixed with limestone grassland on old lead workings. Harebell and Lady's Bedstraw grow with Marsh Thistle on the mounds on the foreground next to Mat-grass tussocks.

which support acid-loving species like Heather and Bilberry. Relic populations of Bilberry in particular can survive in short turf when the other heathland plants have all but disappeared.

In all these though, diversity is enriched by rocky limestone outcrops of various shapes and size, the remains of small quarries or lead rakes, and metalliferous soils and waste heaps (Fig. 224). The result is a potpourri of soils and opportunities for different plants and animals in close juxtaposition. You may ask what is special about limestone heaths as we have already met most of the species on the moorlands and limestone grasslands. The answer lies in this habitat diversity and, for the older heaths, their long history. The admixture of species of different soils and habitats is unique on a small-scale. Thus there are swathes of Heather, sometimes with Bell Heather, often accompanied by Bilberry, occasionally with cottongrasses, Cross-leaved Heath or Purple Moor-grass and some of the moorland sedges where the organic-rich humus holds more water as on Bradwell and Longstone Moors. Next to these could be limestone soils where Wild Thyme, Bird's-foot-trefoil, Quaking Grass, Hoary Plantain, Fairy Flax or Wild Strawberry (*Fragaria vesca*) grow. Plant richness on each site, even when small, can be quite high, with annuals or ruderals appearing on recently colonised sites and the lead rake specialists adding to this. The 1984 survey found 42 vascular plants in just the 0.25 ha of Tideswell Moor Quarry for example, and 84 on the larger 9.5 ha of Earl Rake (a site lost by the 1998 survey). These are high numbers, and more than would occur on heathland in the Dark or South West Peaks (excluding flush species).

This high number is augmented on older sites by many more lichens and bryophytes than in a typically burnt heather moorland, owing to the long history of grazing and less intensive or no burning management. This is exemplified on Longstone Moor where at least 50 lichen species, including several rare ones, have been recorded, including Iceland Moss *Cetraria islandica subsp. islandica* (very rare in lowland Britain). Although many of these are associated with lead rakes, the heathland all round has sheltered the rakes from fertiliser drift or other agricultural improvements. There is also a number of *Cladonia* species, the Cup Lichens, including *C. unicialis*, which normally grows on strongly acidic soils and also has a restricted distribution in lowland England. Gilbert (1980) concluded that the richness of the lichens on this site represented a relict assemblage.

The 1984 survey also found that species like Cowberry and Crowberry, 17 of the 50 or so bryophytes found such as *Sphagnum fallax* and *Aulacomnium palustre* and 13 out of 19 lichens recorded seemed to be restricted to the relict heaths. In general there were more heath (twice as many on average) and calcareous grassland species on the older heaths too. Moss gives a list of species characteristic of the limestone heaths, of which the majority were also found in the 1984 survey, but some seem now to be far less common such as the diminutive fern Moonwort and Autumn Gentian (*Gentianella amarella*).

I always find unexplained features fascinating. Farey (1813) refers to one of these as 'mossy hillocks' found on limestone heaths at higher altitudes, which he thought seemed to be 'worn out or decayed ant-hills covered with moss or lichen', which caused problems for those trying to reclaim the land. Shimwell (1977) considered these mounds to be fossilised periglacial frost hummocks termed thurfur fields. Those still present on Parwich Moor (an SSSI for this feature and its remnant heathland, but with no access), demonstrate the typical 30–70 cm height and 2–3 m scale of the hummocks which are generally closely spaced and sometimes aligned. There are similar earth hillocks on Dartmoor. The general explanation is that they formed through a type of cryoturbation when local patchy freezing occurs in pore water in the active soil layer, with frost heaving repeatedly leading to the development of a small mound. This is an active process visible now in the subarctic in places like Iceland and Alaska but would have occurred in periglacial times in Britain.

However, there are many questions related to this theory for our hummocks. Most hummock fields listed by Fairy no longer survive, having been reclaimed for productive farmland, but I recently assessed mounds on the north slope behind Solomon's Temple, on High Edge and the neighbouring Upper Edge, all south of Buxton which led to some trial excavations with some colleagues. The superficial organic matter on the mound's centre is clearly more acidic with an undecomposed mor humus (one that breaks down slowly typical of heathland) supporting heathland plants like Bilberry, Mat-grass and sometimes Wavy Hair-grass (Fig. 225). This merges into a mull humus (thin layer of less acidic well decomposed material) between the hummocks supporting common neutral meadow not heathland plants. The accumulated undecomposed mor humus forms most of the hummock. There was no sign of any periglacial sorting in the fine, sandy loess beneath, which tended to be deeper below the hummock.

It is possible that heathland plants have played some part in building the mounds as it has already been described how Wavy Hair-grass growth helps develop a mor humus and podsol, which would help build the organic layer and possibly contribute to hummock formation. Mat-grass may also contribute and occurred on many of the newly found hillocks. Its abundance is more determined by past grazing management, so could have replaced Wavy Hair-grass and other heathland species in the last 50 to 100 years perhaps. It is difficult to establish a time line for mound development and their vegetation and this does not explain their regularity and size. Further investigations are planned.

FIG 225. A trench cut through a soil hummock on Upper Edge. The section shows a deep layer of undecomposed, raw organic matter in the upper horizon in the centre which supports acid grassland. The humus layer thins and becomes better decayed at the far end of the trench where the mound reduces near the shoe. There is a more leached A horizon in the centre. The whole lies over undifferentiated loess material.

The limestone heaths are precious and of high conservation value owing to their unique features. They greatly enhance the biodiversity of the limestone plateau. The additional mining and quarrying historical interest adds to the overall effect. Those that are SSSIs have some protection, although in their 2014 assessment, Natural England listed Parwich Moor as in unfavourable declining condition, which does not bode well for its longer term conservation. Several others are under conservation management or voluntary agreements or owned by the NPA. All are fragile, small, spread out, fragmented and very vulnerable to change. Some sites are still in need of protection. There is an urgent need to attempt limestone heathland restoration, although this is only likely to be successful on suitable soils. The White Peak Natural Character Profile includes an ambition to conserve, restore and where appropriate, create limestone heath.

DEW PONDS AND MERES

Natural ponds are rare in the White Peak, and it is only where drainage is impeded that meres formed. These were important for stock watering, often associated with villages, as in Monyash, on the hilltop above Taddington and at Heathcote. Others are now filled and forgotten. However, with the enclosure of the wastes and commons, fields were separated from springs and streams in the dales. Cows and pregnant ewes need to drink daily. The answer was the dewpond. Each is a circular hole, originally limed to discourage worms, then covered with puddled clay with straw to act as a binder. Stones were added to prevent damage from animal feet (Fig. 226). Each is fed essentially by rainwater and thus sometimes placed in a mini-catchment to augment this. Others may also have small springs. They often served more than one field, being carefully placed at wall junctures (Barnatt, 2019). Most are on the plateau, but several lie within the dales too as in Biggin Dale and Deep Dale Taddington.

Digitally counting all the open water marked on the OS maps for the White Peak gives a total of 5,368 ponds covering a total of 170.1 ha. This will not include

FIG 226. A larger than average dewpond with a cover of Common Pondweed and narrow patchy fringe of Soft Rush.

garden ponds which could add significantly to the total. The average size is only 23m², which is smaller than for ponds in the Dark and South West Peaks, reflecting the generally restricted size of dewponds (Anderson, 2016). This total may not all be ponds as ground-truthing is required to check them. The Peak District BAP suggests a total of around 1,500 BAP-quality ponds in the White Peak. Functional pond number will be somewhere between these totals. The OS mapping total will be seriously depleted by the number of former ponds that have been abandoned, infilled, damaged so they leak or contaminated by nutrient-rich runoff. Most loss is believed to have occurred in the 1970s and 1980s, with potentially as many as 50 per cent gone. This loss continues, possibly more slowly, but partly offset by pond restoration by landowners and conservation organisations.

Dewponds are an important part of our cultural history, integral with limestone heath loss. As the only still-water habitats (along with the few remaining meres and other ponds), they are important for nature conservation, but comprehensive wildlife data for them is lacking. Other ponds within the White Peak are often more recent, like those in the bottom of quarries. There are several small ones in Hoo and Peak Quarries near Longcliffe for example, and much larger ones lie in the bottom of some functional quarries, as at Dove Holes or in restored areas nearby.

The habitat quality in these ponds varies. Many dewponds hold no vascular plants, but may be green with algae in the warmer months. A blanket of floating Common Duckweed (*Lemna minor*) is often the only higher plant present, particularly where grazing animals have access. Better habitats reflect a good mix of submerged and emergent pond plants. A tangle of Common Pondweed (*Potamogeton natans*), our commonest pondweed, provides a good habitat for example (Fig. 226). Common Water-starwort might be present, and there are some related and much rarer Water-starworts that remain to be found. Similarly rare, Thread-leaved Water-crowfoot (*Ranunculus trichophyllus*) might occur as in the millpond at the bottom of Cressbrook Dale. Creeping bent and Sweet-grasses often form floating leaf mats spreading from pond edges. The richest floras occur in clay-lined rather than concrete-lined dewponds (Warren *et al.*, 1999), where emergent plants can also root in marginal mud. Soft Rush or Hard Rush (*Juncus inflexus* – distinguished by its glaucous stems and leaves) may find a footing round some, or tall herbs like Great Willowherb or even Common Nettle or Goosegrass might fringe ponds where field soils meet the water body.

Some of the ponds in the old quarries are much more diverse. New ponds can support beds of Stoneworts (*Chara* spp.), which are characteristic of calcium-rich clear water bodies, as in Hoo Quarry. These might disappear as the pond matures and accumulates leaf litter and sediment. Common Spike-rush (*Eleocharis palustris*) can spread in from the edges or Bulrush establish in greyish tall clumps which are difficult to contain, although this is more likely in larger ponds as in Peak Quarry. Great Horsetail (*Equisetum telmateia*), twice as tall and robust than the more common horsetails, forms extensive beds and fringes ponds in Peak Quarry for example. It is not common in the Peak District, being more often found in the Dark Peak's northwest fringe than in the White Peak.

White Peak dewponds are especially important for breeding Great Crested Newts (*Triturus cristatus*), which is legally protected under the European Habitats Directive and the Wildlife and Countryside Act 1981. Although not uncommon in parts of Britain, its rarity at a European scale has led to its high conservation value. At about twice the size of other native newts, its warty dark brown or black skin with bright orange blotchy underside is distinctive. Although no ponds are known to support large numbers (owing mostly to the small size of the dewponds), the total population across a large number of ponds is highly valued in the White Peak. Adults can move some 500 m, so closely located ponds can support more animals in a network, provided there are also suitable terrestrial habitat and hibernation sites. Great Crested Newts have so far been recorded in 131 ponds, of which 72 are dewponds. In a survey in 2008, 70 of 171 ponds supported Great Crested Newts, suggesting that there are likely to be more in as yet unsurveyed ponds (Anderson, 2016).

White Peak dewponds are also where to find Smooth Newts in clear water during the breeding season. Since this species prefers alkaline water with a pH above 7, it is more abundant in White Peak ponds. Interestingly, it also occurs at generally higher elevation than Palmate Newts, although their normal nationwide distribution is a reverse of this. The high altitude of the White Peak dewponds accounts for this, whilst Palmate Newt's more acidic ponds are usually, in the Peak District, at lower elevations.

Common Toads usually breed in deeper water (as in reservoirs, Chapter 11), which in the White Peak is in the larger meres. Unlike Common Frogs that gradually appear at their breeding ponds, Toads exhibit an 'explosive breeding' behaviour where all sexually mature animals migrate from their hibernation sites to breeding grounds within a few days, often crossing roads where many are squashed. To help them (and other amphibians), many villages operate Toad patrols. That in Monyash has the distinction of safeguarding the highest number recorded in Derbyshire: 3,378 rescued in 2017 (plus Frogs, Great Crested Newts and Smooth Newts), and nearly the same in 2018 (DARG, 2018, 2019). The population is estimated to be a very high 10,000, which is hugely positive against a background of population declines, particularly in lowland England. Toads feed on a wide variety of invertebrates – more particularly those that are in good populations and sharing the same habitat. They wait motionless until the prey passes within tongue-distance. Analysis of droppings of a Toad trapped temporarily in a local greenhouse revealed woodlice, beetles and spiders as the main diet, including quite large beetles like Violet Ground Beetle (*Carabus violaceus*), which are 20 – 30 mm long, and a large number of smaller rove beetles (Whiteley, 1997).

The White Peak ponds are also important for their invertebrate assemblages. No comprehensive data are available, but a 1992 survey of 40 dewponds comparing concrete and clay-lined ponds between Bakewell and Peak Forest, revealed 70 different invertebrate taxa, with water beetles (more than 24 species), water bugs (12 species) and fly larvae (at least 9 taxa) numerically the most abundant. Only four species of water snails, four of leeches and a range of micro-crustaceans such as Ostracods and water fleas were found, plus representatives of few other aquatic invertebrate groups. In general, as for plants, the clay-lined ponds were richer than the concrete-lined ponds, although some, like water bugs, were better represented in the latter. Although 70 different taxa is modest, most ponds did not support many species (Warren *et al.*, 1999). Other larger and more mature, well vegetated White Peak ponds support more aquatic invertebrates, as Merritt (2006) notes, with up to 48 (an exceptional number) of water beetles and 18 water bugs in some of the larger and better vegetated ones.

The key for most pond animals is mobility. Some are strong flyers and can find new ponds and move between ponds relatively easily. This applies to dragonflies and damselflies (Odonata) and some water beetles and bugs. Of the Odonata, the commonest are as in the Dark Peak: Large Red Damselfly, plus the smaller Azure and Blue-tailed Damselflies (*Ischnura elegans*), all preferring well-vegetated ponds. The Broad Bodied Chaser, in contrast, seeks out new ponds with bare margins where it whizzes round defending its territory. Look out too for the Emerald Damselfly, smaller, green-coloured and with a weak fluttery flight that prefers ponds with luxuriant emergent marginal vegetation.

There are species that are early colonisers of new ponds, which appear in quarries, garden ponds or even just wheel ruts. The water beetles and bugs provide good examples, but unfortunately there are no recognised English names. *Agabus nebulosus* and *A. bipustulatus* were characteristic dewpond species in Warren's survey, along with three *Helophorus* species, *H. brevipalpis*, *H. aequalis* and *H. grandis*. Merritt (2006) also found *Agabus nebulosus*, a medium-sized widespread beetle, favours small ponds. Rarer beetles, *Hygrotus confluens* and *Stictonectes lepidus*, a Nationally Local species, favour bare mineral substrates in new or temporary ponds as in quarries. Species of well vegetated ponds include *Hydroporus umbrosus*, which seeks out grassy dewponds or other marshy ponds and is largely a White Peak species in the East Midlands whilst *Hygrotus inaequalis* is much more widespread, but prefers well-vegetated water, not just in ponds. In contrast, *Coelostoma orbiculare*, a small black beetle, lives in mosses at the shallow margins of base-rich water bodies and is more common in the White Peak, but found less frequently elsewhere.

Water bugs show similar patterns of colonisation to the water beetles. The pond skater, *Gerris thoracicus* is a Local species that seeks out muddy or silty shallow water, including pools in quarries, whereas *Gerris lacustris* is the commonest pond skater found in a wide range of water bodies throughout the Peak District. Similarly, there are back swimmers and water boatmen with equivalent preferences. *Notonecta glauca* is the commonest backswimmer found in a wide range of water bodies, including White Peak ponds, whilst *Notonecta olbiqua*, a Local species, is more a Peak District specialist utilising ponds that are mildly acidic to mildly alkaline. The water boatmen, *Sigara fossarum* (a local species) and *Sigara lateralis* (sometimes occurring in abundance) are early colonising species in new ponds, including dewponds. In contrast, *Sigara limitata*, a Nationally Scarce species, prefers well vegetated, more mature dewponds and quarry ponds (Merritt, 2006).

There are many other animals associated with White Peak ponds like Pond snails *Lymnaea* species and Ram's-horn Snails *Planorbis* species, which will be widespread and are important grazers on algae in White Peak ponds. A range of ponds of different types and ages are needed to support the full suite of species. How snails reach new ponds is not really clear; they could be transported as eggs on birds or amphibians moving between ponds and ponds close together often share more species than those farther apart. Leeches and flatworms also favour ponds. The distinctive greenish coloured Horse Leech (*Haemopis sanguisuga*) has been found in a few dewponds and may be more widespread. It is misnamed as it cannot pierce mammalian skin and feeds on smaller animals like midge larvae and snails, sometimes reaching 15 cm in length. There are several craneflies that breed in the pond surrounds in soft mud– although not around ponds where this is lacking. So far, observations suggest seven species associated with damp muddy margins or more vegetated or silted dewponds. Two of these are useful habitat indicators but are scarce regionally: *Trimicra pilipes* and *Ptychoptera contaminata*, the latter more usually in the lowlands.

WHAT NEXT?

Although home to a wonderful assortment of plants and animals, whether they are charismatic, renowned, common, rare, colourful, enigmatic or just beautiful, the White Peak's habitats are generally rather fragmented, limited in extent and small. We need urgently both to help reverse the biodiversity crisis and to temper effects of climate change. These topics are explored in the next chapter.

CHAPTER 15

What of the Future?

Hopefully, the splendour and drama of the Peak District's landscape emerges from the preceding chapters in all its glory. Landscape here is not just what you see, but the natural beauty moulded by the rocks, physical processes, wildlife and people over millennia. I have tried to portray its distinctive nature, integrate it with its cultural heritage, reveal its spirit of place and show off some its wonderful wildlife. I have also identified its benefits for those beyond its borders through water supplies, flood alleviation, carbon storage and food; to name but a few of these wider environmental values or ecosystem services. The Peak District, though, remains a lived-in landscape; and one that is treasured by the millions of visitors who benefit from its special qualities.

However, although we have celebrated the Peak District's wildlife and have much more than many of our neighbouring areas, we are not isolated from the increasing pressures that are affecting us all and there are two that have to be addressed: the biodiversity crisis and the climate emergency, the solutions to both being intimately linked. These, along with diseases like Ash-dieback, are the biggest threats to the Peak District's wildlife into the future.

THE BIODIVERSITY CRISIS

Taking the biodiversity crisis first, I have alluded to losses, to issues of pollution, pressures on some birds of prey, degraded environments through fire, inappropriate management, invasive species, the dearth of some habitats and species and the loss of wildflower-rich grasslands especially to more intensive agriculture, all despite the National Park status of much of the Peak District. There will be other losses too that we have not yet quantified, particularly related to invertebrates. Research in Germany (Hallmann *et al.*, 2017) demonstrates a shocking 76 per cent loss of invertebrate biomass seasonally in protected areas, regardless of habitat type, over a 27-year study. Similar losses are happening in Britain too, including in the Peak District. As a biomass loss indicator, how often do you clean squashed insects off the car windscreen now compared with the past? Since insects are at the heart of every food web, responsible for pollinating most crops, keeping soils healthy, breaking down plant material to recycle nutrients and controlling pests, we cannot afford to lose them. Buglife has started a 'No Insectinction' campaign and action is needed in the Peak District too.

We have also seen the variable level of habitat protection in the region from a high 46 per cent in the Dark Peak to only 6 per cent in the White Peak. The White Peak's habitats are more fragmented, narrower, with a higher edge to bulk ratio than in some other upland regions or National Parks. There is an urgent need to expand and repair what we have, to add new opportunities for wildlife, rescue disappearing species and habitats, restore some missing species and basically promote the mantra of Lawton et al. (2010) of 'bigger, better, more and joined up' to reverse biodiversity losses. The benefits to society would be enormous.

The next step then is to engage with the crystal ball and gaze into the future for this impressive and beautiful region. What lies ahead is fairly predictable short-term by looking at current projects, strategies, policies and trends, assessing their impact and efficacy as we go. But the longer Peak District journey is critical too: the pressures and opportunities, what might be done, what needs to be done and the likelihood of any of it happening within say 50 or 100 years.

THE CLIMATE EMERGENCY

Considering the climate emergency at the same time as the biodiversity crisis is essential since many of the issues and solutions are integral. I have already highlighted some climate changes to date in Chapter 7 and mentioned species that are responding to warmer conditions, particularly insects and some birds. The Meteorological Office's predictions for the end of the 21st century are for all areas of the UK to have warmed, more so in summer than winter compared with a 1981–2010 baseline in the most recent modelling results (UKCP18). Different scenarios relate to whether greenhouse gases are vigorously reduced or not. At the reduced emissions level, predictions for our region have a 50 per cent probability for average summer temperatures by the 2080s to be around 1–2 °C warming, but 4–5 °C at the high emissions scenario. Winter warming based on the same scenarios could be 1–2 °C or a higher 2–3 °C. Snow lie and the amount of snow are also likely to reduce significantly. The chances of having hot, dry summers like in 2018 are expected to become more regular; as much as a 50 per cent probability compared with only 10 per cent in the past. Under the high emissions scenario, hot summer day temperatures could increase between 3.7 and 6.8 °C plus increases in the frequency of hot sunny spells.

Rainfall totals are predicted to decease in summer and increase in winter, but with more intense storms. Again, the modelled predictions for our region suggest a possible 10–20 per cent reduction in summer rainfall at lowered emission levels or around 20–30 per cent at the higher emissions level. Winter (December to January) levels could be 0–10 percent or 10–20 per cent more in each emissions scenario respectively.

The predictions are based on models and the likelihood of effects provided as probability of occurrence. These are averages over time, not for weather in any particular year, so there will still be some cold winters and dull summers as climate varies from year to year. The changes detected so far (Chapter 7) generally reflect these predictions, but this ignores the diverse topography and local variations in rainfall, aspect, slope, shadow and exposure which could mean that at the micro-scale, future change might be patchy, with refugia from intolerable conditions available in climatic nooks and crannies for different species. The next question is what effect climate change is having or might have on the Peak District's wildlife?

The nature of change being detected is varied, but have implications for ecosystem functioning and species diversity. First, species distributions are predicted to alter. Nationally, some species could become extinct as suitable climate space is lost, while others could expand their range. A wide variety of species are moving polewards and up in altitude, but there is limited evidence so far of southern range edges contracting. As part of the Peak District's distinctiveness is the mix of southern and northern species, the anticipated changes have grave implications for our local rich biodiversity. Secondly, changes in the timing of biological events (phenology) and mismatch between interdependent species are also evident with longer growing seasons and reduced snow cover. Finally, in combination, these changes could affect many ecological relationships such as food webs, nutrient cycling and soil dynamics, although there is little information on these aspects to date. We will examine the evidence for change so far.

Starting with plants and soils, there are several possible outcomes. One of the more dramatic is the link with heavy storms that can result in landslides, which are already visible in many cloughs. Having witnessed the effects of the 2002 dramatic summer storm in Longdendale, these can be damaging, bringing tonnes of soil, stone and debris down hillsides, leaving scars that then erode,

FIG 227. A land slippage in the upper reaches of William Clough below Kinder Scout resulted from storm flows. The crumbly shales are easily eroded. Stabilisation and re-colonisation are difficult.

contributing sediment to streams and rivers, and causing localised flooding as in Glossop. These changes affect the landscape and destroy habitats, although they also provide new colonisation opportunities. It is difficult to see how to restore such steep, fragile, eroding ground (Fig. 227).

On a larger scale, Evans (2009) summarises evidence for the Mam Tor landslide increasing its instability with climate change, suggesting that the slope system is highly sensitive to small changes in climate based on future increased winter rainfall. He also suggests that landslips are conditioned not only by their initial slippage, but also by the ongoing interaction of climate with slope processes over time. It is possible that it can take 8,000 years or more for stabilisation of deep-seated landslips in the Peak District. Thus those that are younger than this (the Cown Edge Rocks massive landslide in the Ashop Valley for example), could still move, although the longer the gap since the initial slippage, the more hydrological forcing is required to lubricate parts of a landslide again.

The effect of repeated summer droughts like that of 2018 on peatlands could be particularly detrimental. Peat is the largest terrestrial store of carbon in the

FIG 228. More hill tops could look like this with peat eroding and lost.

country and accounts for just over half of all soil carbon. If dried and aerated, peat decays and carbon is lost, thus adding significantly to climate change. The rate of peat decay increases with elevated temperatures and drought conditions. Peat pipes could increase due to cracking and storms could result in more active erosion (Holden, 2009). All these would result in further peat loss and therefore attrition of our peatland habitats, particularly blanket bog and its associated plants and animals. The overall loss of our peat over time cannot be discounted (Fig. 228).

If dry, peat is then very vulnerable to wildfires as witnessed in 2018 and results in highly damaged and eroding ground (Chapter 8, Figs. 89–92). Modelling of fire incidence and climate change suggests that summer seasons will become more hazardous owing to longer, hotter dry spells (Albertson *et al.*, 2010). If greenhouse gases are reduced significantly, the model suggests little change in wildfire risks before 2070, but with a marked deterioration subsequently. Under the high emissions scenario, changes would be seen in the next 20–30 years. Beyond 2070 the models predict an average five fires each week in summer, which could be catastrophic if not contained. However, the models may not materialise fully as they take no account of changes in vegetation when peatland dries or in human numbers and behaviour (currently most fires start through arson or carelessness). Fires are complex, being dependent on a fuel supply (like tall, woody Heather occupying large areas), a source of ignition and appropriate weather conditions, and these all vary, leading to different wildfire patterns. The predicted changes demonstrate the

FIG 229. Birch colonising unmanaged heather moor on Woodhead, Longdendale.

importance of peatland hydrological restoration to improve resilience for as long as possible (Chapter 8) and to public engagement in the battle against wildfires and the destruction that can ensue.

If the hydrological restoration is inadequate in the face of climate change predictions then the blanket bogs of the Peak District will change, perhaps intermittently as the driest areas and those at lowest altitude are affected first. Dry ground species like Heather, Bilberry and Crowberry will replace cottongrasses, wet ground mosses and *Sphagnum* species that require high water tables. If ungrazed, trees could invade such as willows and birches as is already evident on Dove Stones and south of Longdendale (Fig. 229), where stock were removed to restore eroding peat. Such scrub could reduce the incidence of wildfires but would change the habitat totally, depending on stock levels and Mountain Hare or Red Deer populations.

It is difficult to envisage the effect of climate change predictions on dwarf-shrub heath and acid grasslands. Wet heath and Purple Moor-grass swards might suffer, depending on the level of summer drying. Greater susceptibility to disease or insects such as the Heather Beetle might be manifest. There could be a re-balancing of species abundance and distribution which would then have knock-on effects on dependent invertebrates like Bilberry Bumblebee (Fig. 125).

Of some concern is the future of the fungi-rich acid grasslands. There was a notable dearth of fruiting bodies after the 2018 drought but there is no relevant research on the long term effects on these fungi.

Possible climate change effects on limestone grassland is the subject of long term studies in Buxton established by Sheffield University. The effects of applying single or in combination treatments of summer drought (no rain in July and August), summer rainfall supplements (20 per cent above average) and heat treatments (3 °C above ambient in winter) have been explored since 1993 on a steep, west facing limestone slope. So far these have not shown large scale changes in plant abundances (unlike a similar site near Oxford with identical experimental treatments), but this masks more subtle alterations, which over time could have significant effects.

The greatest response is seen in the summer-droughted plots where small-scale variation in soil depth and character allow species affected on the shallowest soils to persist in local pockets of deeper soils more resilient to drought; Bird's-foot-trefoil for example. Some shrubs like Common Thyme increased, whilst forbs such as Tormentil declined, although Sheep's-fescue remained resistant. The diversity and productivity of summer droughted plots also declined significantly from 17 to 15 species/100 cm². Drought-sensitive species included shallow rooted sedges like Carnation and Flea Sedges (both usually frequent damper soils), and Common Dog-violet. Quaking Grass and Common Rock-rose both declined with winter-heating treatments. These responses occurred within the first year of the experimental treatments and then mostly persisted (Grime et al., 2008).

There were also other subtle changes. Ribwort Plantain, for example, diverted more resources into reproduction than into biomass to avoid stress, particularly when growing on shallower soils under drought treatment. This shows the plant is sufficiently plastic (in for example leaf length and numbers, rosette numbers and reproductive activity) to respond to drought conditions through changes in resource allocation, although it is suggested that some of the variation might be genetic (Ravenscroft et al., 2015). Detailed experimentation using Sheep's-fescue seeds from the Buxton treatments did reveal genetic changes. Those receiving the drought treatment flowered earlier, had reduced pollen production and the resulting seeds germinated later, possibly thus avoiding drought as otherwise they usually establish soon after falling. Sheep's-fescue therefore showed rapid evolutionary changes in reproductive activity and its abundance expanded under the droughted treatment (Trinder et al., 2020).

Hence, although there was some long-term resistance to large-scale change from these experimental effects, subtle modifications occurred. This is partially explained by species being buffered at the population level against site extinction by local genetic and phenotypic diversity showing adaptive responses; a trait well known in ancient calcareous grasslands. Perennials may also be able to persist through bad years owing to accumulation of reserves in good growth years (Buckland *et al.*, 2001). The micro-scale variation in soil depth and character is another possible stabilising mechanism coupled with a low influx of plant seeds or other propagules and limited opportunities for establishment in the sward. Limestone grasslands do not in general contain persistent seed banks of typical grassland plants (Akinola *et al.*, 1998), although Thale Cress (*Arabidopsis thaliana*), a common, small white-flowered annual, increased as drought produced colonisation gaps and hence increased its seed bank after drought conditions (Buckland *et al.*, 2001).

Further research from the same site shows that some mosses and liverworts are also potentially affected by climate change (Bates, *et al.*, 2005). Although effects after seven years of continuous treatments were generally modest, drought reduced cover of some of the commonest grassland mosses, *Calliergonella cuspidata* and *Rhytidiadelphus squarrosus*, while species richness declined too. Winter warming increased other species, whilst the common mosses responded more positively to supplementary summer rainfall. Dewfall was considered to be a possible factor allowing most mosses to persist under conditions of altered climates along with the natural ability of many to withstand repeated desiccation.

Possibly more concerning is the finding of marked shifts in fungal and microbial taxa in the summer droughted plots after 17 years of treatment on the same Buxton site (Sayer *et al.*, 2017). Although the dominant microbial taxa were largely unaffected (suggesting they are the widespread generalists with a broad range of tolerances), losses were of subordinate fungi (32 per cent were missing from the summer drought treatments), and to a lesser extent, bacteria, believed to be largely decomposer and pathogenic species. As low fertility, long-established calcareous grassland soils are a fungi-dominated habitat, this could have significant consequences. Previous studies on the site showed reductions in the density of some mycorrhizal hyphae in the soils too. The changes in plant responses to different climate change treatments could account for changes in fungal taxa, particularly with more drought-resistant plants growing better in the droughted plots compared with the generally faster turnover of the more productive species in the warmed and rain-enhanced plots (Fridley *et al.*, 2011). These types of changes alter the soil carbon : nitrogen ratios (and therefore bacterial activity) as well as leaf dry matter content (which would affect fungal growth and activity).

Applying these detailed findings to our limestone grasslands' future is not easy. The experiments are all with one aspect, slope, and altitude and with treatments imposed uniformly. This does not entirely reflect likely climate change patterns, which will be more variable. Nevertheless, the research indicates the scope of changes and, combined with observations in 2018's severe drought, provides a glimpse of what we might expect. Summer drought had the greater effect of all the Buxton treatments, with rain excluded from plots in July and August, but no additional heating. Several of our droughted summers recently have been much drier earlier (2018's drought brought little rain from May to August, whilst 2020's was shorter in April and May). We do not know whether droughts at different times of the growing season will have different effects, but as most growth tends to be in spring and early summer, it could. In addition, warmer temperatures (as in 2018), as well as lack of rain would have a greater effect than drought alone as, unless well adapted, plants would be more vulnerable to wilting or even death.

The experiments suggest that drought-stressed plants are likely to withdraw from the most affected places – that is the south-facing slopes on the thinnest soils. In wetter growing seasons, some recovery might be expected. Summer drought incidence intervals would be critical in determining the extent and persistence of effects. This scenario was played out in 2018 when south-facing vegetation on shallow soils around and on rock outcrops in the whole of the White Peak suffered badly. There was large-scale die-back and plant death, including of Sheep's-fescue, which the research suggests should be more tolerant. 2018's drought obviously exceeded its water stress threshold on a significant scale. The slopes became grey and brittle with dead material persisting through autumn 2018 and into spring 2019, some of which were still partly bare in 2020. Moreover, it was noticeable that decay slowed or stopped everywhere at the drought's height. Spread manure and dead leaves did not start decaying until it started raining. This ties in with the observations on changes in the soil fungal and bacterial assemblage.

The bared areas were then colonised by a variety of annuals and biennials such as sowthistles (*Sonchus* spp.), Hairy Bittercress and Musk Thistle (*Carduus nutans*), which is not generally very common but appeared en masse in places (Fig. 230), sometimes to the consternation of some farmers who tried to herbicide them out. Other plants taking advantage of the opened sward were Shining Crane's-bill, Herb Robert, some uncommon species like Beaked Hawk's-beard (*Crepis vesicaria*) and other more typical rock-edge species. Mosses that had appeared completely shrivelled after the drought had mostly recovered and Sheep's-fescue was beginning to re-establish, but the gaps had not closed by the end of the year. Common Thyme and Common Rock-rose, which both looked rather stressed with wilting leaves in 2018, appeared healthy and recovered in 2019. An equally severe drought two year's running could have been a different story.

FIG 230. Masses of Musk Thistle have colonised the drought-stricken limestone grassland after the 2018 drought in mid-Dovedale.

Also vulnerable would be our more northerly species like Globe Flower (Fig. 202), Melancholy Thistle and Blue Moor-grass (*Sesleria caerulea*). My Globe-flower long-term counts did not reveal any major decline in 2018, but Melancholy Thistle was another story (Fig. 75). North-facing populations were sheltered from the peak effects, but where plants had spread into other areas in wetter years, they suffered badly. One patch had spread in 20 years to about 24 m^2 then retrenched to a sparse 3 m^2 spot in partial shade. Blue Moor-grass is a northern species found only in Monk's Dale in the Peak District. I mapped six patches in 2016, the largest 4 x 5 m and packed with a dense cover of plants. The 2018 drought saw this large patch thin to about 35 per cent cover by summer 2019, even though it was on a west rather than south-facing slope and therefore supposedly less susceptible.

There were other causalities in the 2018 drought that might be a taste of the future. There were populations of plants in different places that withered and appeared to die on a large scale. Fragrant and Common Spotted Orchids wilted and died back, with lower numbers in affected patches in 2019. Hay meadow monitoring across a range of soil types showed reductions of Red

FIG 231. A small meadow created in 1998, showing diverse cover in 2017 (above left), with abundant Meadow Buttercup, Yellow Rattle, Bird's-foot-trefoil, Oxeye Daisy and Red Clover, together with Southern Marsh Orchid hybrids. Compare this with patchy bare ground and vegetation affected by damaging droughts in 2018 and spring 2020 (below left). Both photographs taken in June. The meadow is managed annually by cutting and removing the arisings in autumn plus an early April cut before cowslips flower to remove winter growth.

Clover, Creeping Buttercup, Sweet Vernal-grass, Yorkshire Fog and Ribwort Plantain that appear to be drought-related (Fig. 177). These changes were extreme in a little meadow I created on skeletal soils, formerly a horse ménage. Strong drought effects occurred in 2006, 7, 10 and 13, with 2018 the most severe, marked by depression in plant cover. 2019 showed variable recovery which was reduced in 2020's spring drought (Fig. 231). However, moss cover expanded after droughted periods as it temporarily replaced vascular plants. Common Spotted Orchid numbers declined dramatically in 2018 by 85 per cent and again in 2020, anecdotally repeated elsewhere in the White Peak.

These data show the effects are complex and depend on drought timing, its severity and any combination with higher temperatures. Thus different plant populations expand and contract in response to the prevailing climate, more so where they are less buffered, for example by aspect or slope. The prognostications for the future based on these observations suggest more northerly and westerly species declining with similar droughts and higher temperatures in the future, but common and widespread species may also be reduced, temporarily at least. At the same time, there may be more typically southern species that could spread. Stemless Thistle (Fig. 74) is one perhaps (see Chapter 7), along with Clustered Bellflower (*Campanula glomerata*), Field Madder (*Sherardia arvensis*), a pink-flowered little annual) or Basil Thyme (*Clinopodium acinos*). All these are rare or uncommon in our limestone grasslands currently. If winters are warmer, then Bee Orchids (Fig. 200) could well expand their populations too (see Chapter 7).

The general expectation is for southerly species to spread northwards. As there are no calcareous grasslands near enough for most species to reach here naturally, and as seed production is limited in stress-tolerating species that form the bulk of these communities, the chances of us gaining many more plants without human assistance seem slim. The prospect is, rather, for our supremely rich limestone grasslands to diminish in diversity over time with climate change. The more drought-resistant species will spread and those less well adapted will disappear gradually once the small-scale refugia that might escape the worst effects are also lost. We do not yet know what effects there might be on our specialist grassland fungi. Similarly, there has been little research on potential woodland changes for our region. General predictions are for changes in species composition of trees, shrubs, ground flora, epiphytic mosses and lichens, and potential loss of wet woodlands, either through climate change or indirectly owing to different diseases becoming more prevalent.

The evidence for change to our fauna differs from that of the plants. There are no obvious signs from Peak District mammals yet, possibly relating to their ability to control their body temperature, seek shelter or shade and the fact that many are nocturnal. Currently, habitat losses or changes are the main factors affecting them. There is the possibility of more southerly bat species like Greater Horseshoe Bat (*Rhinolophus ferrumequinum*) and Barbastelle (*Barbastella barbastellus*) expanding their range, possibly as far as the Peak District (Thomas et al., 2011). At the local scale, seasonal weather affects the diet of some mammals. Badgers, for example, focus on earthworms at particularly times of year, which are hard to find in droughts. Digging for Pignut tubers was evident everywhere, possibly as an alternative food, during 2018's drought.

Some birdlife is responding or is predicted to respond to climate change in the future. These can be species retracting or moving northwards as bioclimatic-envelopes move or mismatched timing of habitat needs occur. Willow Warblers vacating southern Britain has already been mentioned and we are on the boundary of their now more northerly distribution (Chapter 10). Increases in Stonechats mirrored by the decline in Whinchats are also believed to be climate-related. Other southern species like Dartford Warbler have appeared (until a recent cold period) and could return. Populations of other species like Wren, Kingfisher and Long-Tailed Tit (*Aegithalos caudatus*) could expand as winters become milder and their numbers are not reduced by severe cold. The British Trust for Ornithology's nest record scheme shows that 51 UK birds show trends over 25 years of earlier egg laying. This adaptation is successful if food is adequate for earlier chicks. There is evidence though for an increasing mismatch between breeding Pied Flycatchers (Fig. 138), for example, and their peak food supply; moth caterpillars that are developing earlier than the birds are breeding. This can lead to reduced breeding success and survival and significant declines have been noted in some areas.

Research suggests that some moorland breeding birds could decline or eventually disappear from the Peak District as climate warms because they are essentially northern species. Studies model species' distributions as a function of climate (called climate envelope modelling) and then use future climate projections to model potential change predicated on the species closely following these to highlight the potential magnitude of future conservation issues. Climate change effects do not operate in isolation though and alterations in prey availability, predation risk and disease, plus shifts in agricultural and forestry management could all be indirect factors with complex consequences (Pearce-Higgins, 2010, 2011; Pearce-Higgins *et al.*, 2010).

For the upland birds, beetles and flies, especially craneflies and their relatives, are the most important food sources. Cranefly abundance could reduce significantly in moorland and pastures with summer drought and increasing temperatures (see Chapter 8). Midges (that are also important dietary components for some upland birds) might also be reduced by climate change. Moreover, many of the upland birds feed on pastures when not at the nest and depend on earthworms, the abundance of which has been shown to be positively correlated with soil moisture and which could decline or move to inaccessible depths in dry periods. Warming could increase the survival of some beetles, moths and caterpillars but reduce others, dependent on whether they can adapt, so these groups as food for upland birds may be less critical. Mismatch in timing could be important though if caterpillar abundance does not coincide with bird breeding.

The conclusions from these investigations are that upland birds like Dunlin, which depend on craneflies, midges and earthworms most, may be more sensitive to climate change than Golden Plover and Snipe, but these are more likely to decline than some other species. Red Grouse and Meadow Pipits might be sensitive as the young chicks depend heavily on cranefly emergence. A negative correlation between Ring Ouzel numbers and summer temperature suggests potential further population decline too. As one of our iconic species, Golden Plover productivity is correlated with the abundance of adult craneflies that are cold adapted, which in turn is linked to August temperatures in the previous year, when they are eggs, or small larvae, and more vulnerable to desiccation in hot, dry summers. Modelled scenarios with future increases in August temperatures predict a significant risk of Golden Plover loss from the southern edge of its range in the next 100 years. However, the likelihood and severity of this both depends on the magnitude of warming and the extent of adaptive management to reduce the impact. These findings demonstrate how critical moorland re-wetting is in buying ourselves time (Chapter 8). Monitoring has shown that this has resulted in enhanced cranefly biomass and increases in population and breeding success of some of the critical moorland birds like Golden Plover and Dunlin, which will help buffer climate change impacts for a while.

FIG 232. Ringlet on Wild Thyme: one of the butterflies that has expanded its range and numbers in recent years with climate warming.

The responses of other invertebrates to climate change is likely to be very variable, with winners and losers. An assessment of a wide variety of British plants and animals across a range of taxonomic groups suggests that 35 per cent might be classified as being at high or medium risk of being negatively affected by climate change, whilst 42 per cent are more likely to have a medium or high opportunity for expansion. The groups with the highest number of potential beneficiaries are ants, bees, centipedes, some beetle groups, hoverflies and wasps. Those with the least are millipedes, soldier beetles and their allies, dragonflies and damselflies, as well as Bryophytes and vascular plants. Upland habitats are the only one where the majority of species assessed are at risk of decline (Pearce-Higgins et al., 2017).

More specifically, nearly half of British butterflies are expected to expand or continue expanding their range with climate change, including Silver-spotted Skipper which has not quite fully reached the Peak District yet. Other butterflies that have spread into or become commoner in the Peak District are described in Chapter 13, including the Essex Skipper, Brimstone, Small Skipper (*Thymelicus sylvestris*), Purple Hairstreak, Speckled Wood and (Fig. 232), all of which have colonised or spread mostly during the 1980s and/or 1990s. Future colonists might include the Silver Washed Fritillary and White Admiral (*Limenitis camilla*), which has reached southern Derbyshire (Machin, 2018).

Those species with food plants or specific habitat requirements not found outside their current range are unlikely to extend further without human assistance, such as Chalk-hill Blue (*Polyommatus coridon*) whose larvae feed on Horse-shoe Vetch (a White Peak rarity). It is possible there may be larger numbers of some butterflies at times in the future, provided there is suitable habitat, as some are now breeding earlier and can fit in additional broods in warm years. There may also be a mismatch between butterflies and their parasites and predators, which has implications for food web interrelationships and population numbers. Additionally, severe droughts that reduce the quality and quantity of food plants could affect populations negatively.

Another group that might largely benefit from a warming climate is dung beetles. Thirteen Peak District moorland dung beetles have been found, representing around 20 per cent of British dung beetles, with species more likely to be found at lower than higher altitudes except *Agoliinus lapponum*. Most species are likely to expand into higher altitudes as it warms, apart from *A. lapponum*, which is more likely to contract its range (Menedez & Birkett, 2011). Bumblebees, in contrast, are struggling with weather extremes. The less predictable seasons, summer droughts, prolonged flooding, and asynchrony with the flowering plants on which they depend are all causing population stresses. Bumblebees are adapted to cold climates with their round shape and thick fuzzy coats. They are showing substantial declines (from habitat loss as well as climate warming) at the southern range edges nationally, but little expansion northwards, suggesting widespread declines across Europe and parts of Britain and this pattern is expected to continue. As pollinators of many crops and wildflowers a widespread decline or loss could be critical.

There are other gains though. The Slender Ground-hopper, a small grasshopper-like species, was first found in the Peak District in 2002 after spreading into the surrounding lowlands to the south and east. It is now in the White Peak in some of the dales, in Abney Clough and near Bradfield where it favours ditches and muddy pool edges (Whiteley, 2004). A similar pattern of spread is being witnessed for the Lesser Marsh Grasshopper (*Chorthippus alobomarginatus*). There are other grasshoppers and crickets that are spreading northwards. Roesel's Bush Cricket (*Metrioptera roeselii*), for example, was largely confined until the early 1980s to the Thames and Solent estuaries. Warm summers stimulate more long-winged forms that disperse further and it is now widespread throughout England from the Humber to the Severn Estuary. Being now almost on our doorstep, it could appear soon in the Peak District, more possibly at lower altitudes in rough grassland.

One last habitat to consider is freshwater. It has already been shown (Chapter 14) how cool water springs in White Peak rivers are likely to counteract warming water and thus protect some of the more sensitive freshwater animals (Fig. 233). Outside the White Peak this safety valve is missing, so the upland stream fauna is more vulnerable, particularly cold-water and generally northerly species. Higher temperatures reduce oxygen concentrations in the water, which could affect some species directly. For others, their thermal tolerances could be exceeded, possibly at different phases of their life cycles. Some sensitive species have already been mentioned (Chapters 11 and 14). Similarly, in the White Peak rivers, Brown Trout as a cold-water fish could survive in the cold spring-fed waters for some time, but would be more exposed to warmer waters in the Dark and South West Peak rivers, particularly at lower altitudes. Sudden changes during extreme events are a higher risk than gradual changes to which some species might evolve adaptations.

In conclusion, it is clear that climate change will and is already affecting a wide range of plants and animals in the Peak District, as it is more widely. The extent to which they can cope or adapt will depend on numerous factors likely to be species-specific. These could relate to phenological plasticity and ability to adapt, short-term evolution of traits that might be beneficial, the availability of climate refugia within our topographically and geologically varied area, and

FIG 233. Wormhill Springs on the River Wye in Chee Dale.

habitat suitability. However, wildlife does not occupy a habitat in a vacuum, but is part of the great interdependence of nature. Thus ecosystem functioning could change in both subtle and more dramatic ways as food webs re-organise themselves, different diseases and parasites alter their niches, and breeding patterns change as species move north or uphill, expand their ranges or succumb to temperatures exceeding their thermal limits.

This analysis gives impetus to the imperative to slow climate change as much as we can and to increase habitats' and species' resilience so that effects are minimised or delayed as far as possible. This is also the essence of the Climate Change Act and its recent modifications, which set a target to reduce greenhouse gas emissions by at least 100 per cent of 1990 levels (net zero) by 2050. Others argue that this is too late and should be 2030. Whatever the date, action is urgent and everyone can contribute to and, indeed, it is vital that everyone supports, these goals. Measures to reduce and adapt to climate changes essentially revolve around reducing carbon dioxide and other more potent greenhouse gas outputs (nitrous oxide from agriculture and sewage works, and methane from landfill, livestock and biomass burning) through personal, community and political action. There is ample advice elsewhere on how.

However, reducing release of carbon dioxide into the atmosphere will not be enough nor fast enough to keep climate change within manageable limits. Measures to increase carbon capture in vegetation and soils, thus sucking more carbon dioxide out of the atmosphere, are essential too. These will not only create a more resilient environment to future change, but also help arrest and reverse the biodiversity losses. There are numerous ways this can be achieved in the Peak District. First, soils store most carbon, even to considerable depths. Indeed, globally soils contain more than three times as much carbon as vegetation and twice as much as the carbon levels in the atmosphere, showing how powerful and critical they are. Carbon is also stored at depth: 60 per cent or more in grassland soils occurs below 30 cm, for example (Ward et al., 2016). Peat is the best store consisting essentially of carbon-rich undecomposed plant matter, but only when it is waterlogged (Chapter 8), so re-wetting dry peat is the best approach to maintaining and enhancing this ancient carbon store. Podsols (see Chapter 7); those soils under heathland, Bracken or acid grassland, and clay soils that are wet and organic-rich store the most carbon. This gives added impetus to the numerous local examples of heathland restoration from overgrazed grasslands. Woodland soils are next in line, with grasslands on other soils next. Soils under flower-rich grassland, especially those with abundant red clover, store more carbon than improved pastures. Arable soils fall at the bottom of the list and indeed, if regularly ploughed, they lose more carbon annually than they store, thus contributing to carbon dioxide levels in the atmosphere, particularly on organic-rich soils.

Carbon storage is dependent on above ground vegetation trapping carbon and it then moving into the soils through litter accumulation and incorporation. Vegetation on the whole does not hold much carbon except in trees and shrubs that accumulate woody material. What is more important is the rate of carbon sequestration; that is the process by which carbon sinks remove carbon dioxide from the atmosphere. The numbers in Table 9 are indicative and variable related to the condition and age of the vegetation. Vegetation may not continue to be a net trap of carbon once established, but could be an important store. The total amount of carbon that can be sequestered will depend on the area of habitats that can be created or restored as well as their efficiency at trapping carbon. The subject is complex and there is much still to discover.

Thus the evidence indicates that restoring habitats to maintain their carbon stores and promote carbon capture and creating new ones that are most efficient at sequestrating carbon, all at a large scale, are essential. Including other nature-

TABLE 9 Some carbon sequestration rates in vegetation and soils in different habitats

Habitat	Carbon capture (tonnes/ha/year)	Comment
Peatland	0.105 to 1.01	Depending on wetness and Sphagnum abundance, if not restored would be losing carbon (which can be very high, 10–30 tC/ha/yr depending on level of damage)
Heather moorland	0.6 to 3.45	Varies with Heather age
Woodlands – young plants or scrub	0.6	Early stages of establishment
Woodlands	2.4 to 5.6	Varies with age and species, greater when younger rather than old and depends on whether trees harvested or not and their subsequent uses
Species-rich grassland	0.15 to 3.17	Much higher levels for flower-rich grasslands, especially with red clover, compared with monocultures.
Acid grassland	1.61	Example on podsols – heathland higher on similar soils
Ponds	0.8 to 2.5	Few measures available, small ponds, depends on silt accumulation rates

based solutions, for example increased natural flood measures protecting those living downstream, will also add to the habitat diversity and carbon capture. We will deal with all of these together as there are many inter-related implications.

TRANSFORMATION FOR BIODIVERSITY LOSS AND CLIMATE CHANGE

Second guessing the future direction is difficult as there is much uncertainty at the time of writing related to Government policies and actions on biodiversity losses; spread of diseases (not just Coronavirus, but ash die-back for example, and what might be the next challenge); the effect of leaving the European Union; new agri-environment schemes not yet in place, or the future for agriculture and grouse-moor economics in our uplands, to name but a few. Many of these are interrelated as well. There may be new solutions and strategies that we have not even thought of yet, while others may be adapted to deliver new outcomes. We will delve into some ideas to try to paint a picture of what might happen.

First, policies and recommendations at a national level generally support actions to counter climate change and the biodiversity crisis, although the scope needs greater ambition. The Government's 25 year Environment Plan sets out policy and aspirations for enhancement and restoration of precariously placed habitats and species and commits the government to the principle of public money for public goods coupled with environmental protection. Its locally relevant targets include: improving water quality in specially protected water bodies; restoring soil health; restoring and protecting peatlands; creating or restoring 500,000 ha of wildlife-rich habitat (which sounds high but is actually only 3.8 per cent of England's land area); increasing woodland cover; taking action to avoid extinction of important species; and applying natural flood management solutions. Nature Recovery Networks are the key mechanism for achieving these goals and initial work has already begun on what such Networks might look like locally, based on opportunity mapping linked to access for people. A new Nature Strategy is due in 2021, which will also add to the requirements. The 25 year Environment Plan's targets have already been increased. The Government's Statutory Committee on Climate Change (2018) recommends restoring up to 70 per cent of the UK's peat bogs, expanding woodland cover from 13 to a much larger 19 per cent plus deep reductions in agricultural emissions.

Added momentum comes from the Glover Report (2019), which reviewed protected landscapes (National Parks and Areas of Outstanding Natural Beauty) in England. Among many recommendations, Glover wants to see these National Landscapes as special places for nature – much more than they are currently. He sees them leading the response to the nature crisis, which would also start reversing the UK's status as amongst the most 'nature-depleted nations in the world'. A renewed mission is essential for large scale nature recovery and enhancement to form the backbone of the Nature Recovery Networks and Glover sees them as ensuring a 'gold standard' in providing coherent and resilient ecological networks. He calls for creating wilder areas to produce a better wilderness experience, whilst ensuring cultural traditions are safeguarded. This is not necessarily about letting nature take its course everywhere, but more about choosing diverse approaches with some more and others less 'wild'. Glover is looking for places where people can capture a closeness to nature, freedom, solitude and even a sense of danger and challenge. He wants National Landscapes to take a leading role in the climate change response through re-vamped Management Plans. The priorities are peatland restoration and much more woodland creation at a faster pace than currently: simultaneously providing biodiversity gains and benefits for downstream flooding and water quality.

The Peak District is well placed to play its role in all these deliverables. At the local level, the Peak District National Park covers most of the wider Peak District.

It also seeks to work beyond the National Park boundaries so its policies and plans are pertinent, bearing in mind that they will be soon out of date and others will take their place reflecting the pressures, urgencies and responses of the day. We must assume no future change to National Parks' two primary purposes to conserve wildlife, landscape and cultural interest and to promote opportunities for the understanding and enjoyment of the special qualities. It is germane, therefore, to gauge National Park documents for their ambition and scope for future change. The Management Plan (NPMP – a statutory requirement, reviewed on a five yearly cycle) is a partnership commitment to enhance and develop the National Park's key qualities, which have been characterised to define what is distinctive and significant about the Park. These special qualities are:

- Beautiful views created by contrasting landscapes and dramatic geology;
- Internationally important and locally distinctive wildlife and habitats;
- Undeveloped places of tranquillity and dark night skies within reach of millions;
- Landscapes that tell a story of thousands of years of people, farming and industry;
- Characteristic settlements with strong communities and traditions;
- An inspiring space for escape adventure, discovery and quiet reflection; and
- Vital benefits for millions of people that flow beyond the landscape boundary.

This formalising of what is celebrated throughout this volume holds no surprises and it is clear that they are often intimately interlinked and interdependent. However, as already shown, their condition is not always as good as it could be. The heart of the NPMP is therefore to conserve and enhance them. The 2018–23 NPMP sets out six areas of action, some focused on a special quality, others cutting across them, which, for the natural environment, seek to reduce the effects of climate change; ensure a future for farming and land management; and manage landscape conservation on a large scale.

The NPA's Corporate Strategy 2019–24 sets out targets which enable its contribution to the NPMP, some with ambitions extending to 2040. Key ones for wildlife are listed in Table 10. Some of the initiatives have already been mentioned and there are others instigated by partner organisations. Moors for the Future (MfF) was established in 2003 to implement conservation works on many moors across the Peak District and focuses on blanket bog restoration and moorland enhancement working with its partners such as the water companies and the National Trust at a landscape scale. So far £30 million has been spent largely on revegetating bare peat and restoring its hydrology, also benefitting water quality and downstream flood amelioration (see Chapter 8). When the

TABLE 10 Summary of NPA's Corporate Strategy 2019–24 related to wildlife

Target outcome: A sustainable landscape that is conserved and enhanced	
2024 target	**2040 target**
At least 10 per cent more of PDNP in environmental land management schemes	95 per cent of PDNP land in environmental land management schemes
Net enhancement as identified by landscape monitoring	Increased net enhancement as identified by landscape monitoring
3,650 tonnes net decrease in carbon emissions from moorland	Focus on wider range of habitats to reduce further net carbon emissions and increase carbon storage
Restoration activities on 1,500 ha of degraded blanket bog	8,233 ha (30 per cent) of active blanket bog
Sustain at least 5,000 ha of non-protected, species-rich grassland	Sustain 10,000 ha through retention, enhancement and creation
Create at least 400 ha new native woodland	Create 2,000 ha new native woodland
Restore birds of prey breeding pairs in moorlands to at least pre 1990s levels	Enhance a number of different priority species in key areas

project began, the scale of the restoration works was the largest of its kind in Europe – reflecting the poor condition of our moorlands. There are now similar large scale upland projects across Britain, many learning from this earlier work. The programme is set to continue; indeed it has to continue as peatland restoration is a process, with functionality improving over time as hydrology is restored. MfF estimates that there is at least another 30 years of restoration works needed in the Peak District and South Pennines. Similarly, the market for providing high quality evidence for understanding the peatland landscape, for the interpretation of science and raising awareness to support responsible tourism and reduce wildfire incidence is expanding. With its moorland edge geographical location, the Peak District is recognised as an important place to study the future impacts of change.

Other landscape-scale projects already in action are the South West Peak and Sheffield Lakeland; both multimillion pound HLF Landscape Partnership programmes. The South West Peak project (hosted by the NPA) runs to the end of 2021 and embraces the whole of the South West Peak Character Area, although with hotspots for different projects. Its dedicated website provides more information. From the wildlife perspective, key projects are:

- habitat enhancement for declining curlew numbers based on suitable management and restoration and research into critical features like soil condition and invertebrate populations in rushy pastures;
- flower-rich and fungi-rich grassland surveying, restoration and enhancement;
- establishing thriving White-clawed Crayfish ark sites; and
- slowing runoff through natural flood management measures.

The Sheffield Lakeland Landscape Partnership, managed by the Sheffield and Rotherham Wildlife Trust as a partnership, started its five year implementation phase in 2018. It embraces a large part (145 km²) of the mid-eastern side of the National Park set within Sheffield City District, where many of the reservoirs and tumbling feeder streams lie in steep, wooded valleys leading to the industrial heartland of Sheffield. The project seeks to achieve a more natural and resilient landscape through co-ordinated, long-term landscape-scale management plans that would effect a bigger, better and more joined-up natural environment for people and wildlife. Better recorded and valued cultural heritage celebrated, understood and cared for by people is also central to the project. Wildlife projects evolving out of this vision include improving and creating new woodland habitats, thus enhancing connectivity and complementing flood protection measures; improving management in existing nature reserves and creating new ones; identifying and enhancing important small interconnecting sites, and creating opportunities for habitat management for important species like Goshawk, Osprey and Nightjars.

These programmes are short-term projects, so we need to ensure they continue and new approaches emerge. There are other ideas and partnerships in the making. The embryonic White Peak Partnership, will be applicable to the whole of the White Peak Character Area. An NPA-led partnership vision has been prepared, seeking to revitalise the White Peak's natural and cultural heritage, working in harmony with the landscape and joining up and enhancing habitat corridors into better networks which will tie in with Nature Recovery Networks. At the same time, it will ensure that the palimpsest of landscape history remains readable and conserved and promote sustainable land management to provide high quality food and environmental benefits. This vision recognises the high degree of fragmentation the White Peak's habitats suffer despite their rich natural and cultural heritage. For example, only 5 per cent of limestone grasslands are rich in flowers. Moreover, as already highlighted, ash-dieback (see Chapter 13) is likely to have significant negative effects on both the wider landscape and many of the woodlands here. There are no detailed strategies or plans yet to implement this vision, although some approaches are being trialled, so its success is for the future depending on funding and landowner participation.

Natural England has undertaken opportunity mapping in the White Peak to identify suitable soils and locations where new habitats would be best placed to reduce the current fragmentation and linearity. This could include more woodland, wood pasture, shrubs and grasslands of different sorts reflecting the aspects, slopes and soils. Habitat creation to link Cressbrook Dale's habitats with those in neighbouring Hay Dale are in progress: providing links across the plateau top which are hard to achieve on good quality farmland soils. The opportunity mapping has also resulted in the National Trust purchasing some fine plateau and dale-side farmland where the future of groups of flower-rich meadows on acid soils, limestone pavement and calcareous grassland have all been secured, providing important links with neighbouring dales.

The County Wildlife Trusts are all seizing various initiatives. The Living Landscapes programme, for example, has been adopted for some years across all the Trusts, working with partners and others to restore wildlife habitats, join up damaged and fragmented areas and generally reconnect and enrich the countryside and towns. In the Peak District, The Derbyshire Wildlife Trust is applying the Living Landscapes principles to country centred on the Wye Valley and its tributaries, southeast of Buxton and to more north of Hathersage focused around existing Trust Reserves. Through its Wilder Derbyshire programme, the Trust has goals to restore natural processes and the habitats they support, such as flower-rich grasslands, rivers and wetlands, peat bogs and woodland; with aspirations for at least 25 per cent of the National Park to become wilder with minimal intervention where people can enjoy wild places. The Dark and South West Peaks are in line for developing this theme. These goals support Glover's recommendations described earlier. Such a programme could include re-introduction of missing or lost species such as Red Squirrel, Pine Marten, Beaver, Salmon, Osprey, Black Grouse and Golden Eagle. Pine Marten may be able to colonise unaided as one was found dead locally in 2018, after travelling all the way from a re-introduction programme in North Wales – the first record here in over 17 years. Beavers are being introduced into Willington in south Derbyshire and Delamere Forest in Cheshire by the respective Wildlife Trusts. Reintroduced Beavers elsewhere show how they are natural ecosystem engineers; creating pools, expanding wetlands, helping hold up and cleanse water and improving the habitats significantly. They would be a welcome addition in suitable locations. Cheshire Wildlife Trust in the Peak District is also focusing on the Upper Dane Valley, providing support and advice to help improve water quality, reduce downstream flooding and enhance wildlife. This Trust is also a major player in the South West Peak HLF Landscape Partnership.

There are other pressures that will need to be resolved. One is grouse moor management. If Red Grouse decline as expected with climate change, a new management model will be needed for grouse moors. There may be pressure in the first instance for increasingly intensive management (which would not benefit peat or wildlife) to counteract population declines; grouse moor management might be abandoned altogether; or new models might emerge offering a broader moorland experience that includes wildlife watching and some walk-up shooting or stalking. Consideration will also need to be given to current nature conservation targets, features and objectives. Currently, special features for each protected area are listed and targets for their condition set within the British and European context. If species are lost, move, or new ones arrive, this wildlife protection approach will need to be more fluid and flexible, with ecosystem health perhaps more critical than individual features.

Between them, all the initiatives and projects outlined could start to rebuild biodiversity, capture more carbon and restore habitats and species on the scale required. Glover recommended greater funding for National Landscapes, and most of the aspirations outlined need resources and support. It is worth considering where these might be found. MfF's funding depends on outside sources and although this is never assured, the partner organisations have committed support into the future. Due to be completed in 2021, MfF secured the largest ever LIFE fund for a UK-based conservation project of 12 million euros, co-financed by the local Water companies, bringing the total budget to 16 million euros (£12 million). LIFE programme is the EU's funding instrument for environment and climate action used mainly for restoring European protected habitats and species, but will not be available to GB in the future, which leaves a major gap in funding opportunities, at least in the short-term. The IUCN UK Peatland Programme has recently calculated in a cost-benefit analysis that the societal benefits outweigh the costs of peatland restoration, and that the latter compare favourably with other mitigation options such as afforestation or biogas (Moxey & Morling, 2018). A benefit:cost ratio of 4:1 is regarded as typical but this could be as high as 12:1 for a 300 year period, depending on the location and condition of the peat. This strongly supports the rationale for funding such restoration activities.

The future of MfF will therefore be dependent on securing large scale funding from other sources against a background that should be conducive to such inputs. As peatland restoration is THE most important action to combat climate change through conserving ancient carbon and sequestrating new carbon, as well as increasing the peat's resilience to drying, some alternative funding sources might include carbon offsetting, implementation of the carbon

FIG 234. Field-edge planting with native trees and shrubs to form a future wooded belt with woodland ground flora added. Much more of this type of habitat creation along with new woodland and grassland enhancement are needed.

code and government funding to secure restoration for all its ecosystem services. This might be in the form of special projects or through agri-environment schemes. Many carbon offsetting schemes currently invest in projects overseas, but as trapping and sequestrating carbon are becoming more urgent, the time is ripe for a rapid expansion of investment in peat bog restoration in Britain. However, the public nature of the many ecosystem services enhanced and the practical difficulties of restricting these to individuals or businesses paying for them, lends further support to future funding from general taxation rather than private sources (Moxey & Morling, 2018). Some has already been provided by Defra from which the Dark Peak has benefited. It remains to be seen whether more, and of the scale required, is forthcoming, as the Committee on Climate Change recommends.

An alternative pathway for restoration funding would be agri-environment schemes. These have helped to deliver significant restoration works, especially in the North Peak's (1988 to 2005) and South West Peak's (1992 to 2005) Environmentally Sensitive Area schemes and the subsequent Higher Level Stewardship schemes. With our withdrawal from the EU, a new Environmental Land Management Scheme (ELMS) is being developed and tested ready for official launch in 2024. Based on the principle of public money for public goods, the consultation policy suggests three tiers, one for local action such as wildflower or tree planting margins in fields (Fig. 234), the second for locally targeted outcomes based on opportunity planning and the third geared to landscape scale action across a number of landholdings. This would provide the scope for large scale woodland creation and peatland restoration. However adequate funding commitments are essential and are not allocated at the time of writing. Severe budget constraints not only from Brexit but also now the Coronavirus pandemic and other competing demands are likely to influence the overall funding available for some years.

These thoughts do not take us 50 let alone 100 years into the future, but at least suggest the route we are expecting to take is appropriate. I can only represent a snapshot in time on a moving escalator of climate change and habitat loss. You, the reader, will be in the best place to make that judgement, especially if you are reading this decades from now. To what extent have the predictions set out here come to pass? How altered is our wildlife? Is there significant recovery from losses? Is it still highly distinctive and special? I can only hope so.

Gazetteer

Location	Grid Reference	Notes
Abbey Brook	SK18 92	Tributary of Upper Derwent Dale
Abney & Moor	SK19 79	The moor is west of the village
Agden Reservoir & bog	SK25 92	Agden Bog on north side of the reservoir, Sheffield and Rotherham Wildlife Trust Reserve
Alport Castles, Alport Dale	SK14 91	Large landslide above Alport Dale, a tributary of Woodlands Valley
Alport Dale, White Peak	SK22 64	The River Bradford joins the River Lathkill in Alport
Alport Moor	SK11 93	Moor west of the Dark Peak Alport Dale
Apes Tor	SK09 58	In the Manifold Valley beside the lane from Hulme End to Ecton Mine
Arbor Low	SK16 63	Stone circle off Long Rake
Arnfield Moor and Reservoir	SK01 97 SK02 98	Northwest end of Longdendale
Ashford-in-the-Water	SK19 69	Off the A6 west of Bakewell
Ashop Valley	SK09 90	Tributary of Lady Clough at upper end of Woodlands Valley (A57 valley)
Ashover	SK33 63	At the southern end of Matlock Moor
Audenshaw Clough	SE11 00	East end of Longdendale
Axe Edge	SO3 70	South of Buxton alongside the A53 road to Leek
Back Dale	SK09 70	Joins Horseshoe Dale to form Deep Dale, Topley Pike
Back Forest	SJ98 65	Part of the Roaches Estate
Back Tor	SK 19 90	Derwent Edge
Bakewell	SK21 68	Main National Park town
Ballidon Quarry	SK20 55	Limestone quarry north of Ballidon village
Bamford	SK20 83	Bamford Moor and Edge to the north of the village
Barmoor Clough	SK08 80	Astride the A623 to Sparrowpit
Baslow	SK25 72	Baslow Edge north of the village
Beeley	SK26 67	On the River Derwent, Beeley Moor to the east
Big Moor	SK27 76	Part of the Eastern Moors
Biggin Dale	SK14 58	Tributary of Dovedale in the limestone
Birchover	SK24 62	Village on gritstones near Stanton Moor

Location	Grid Reference	Notes
Black Brook	SK 02 64	Staffs Wildlife Trust reserve, with Brund Hill & Gib Tor
Black Moss	SK03 08	Part of National Trust Marsden Estate, S of A62 cross-Pennine road
Black Rocks, Brassington	SK21 54	Between Brassington and Ballidon, S. edge of Peak District
Blacka Moor	SK 28 80	Owned by Sheffield City Council, managed by Sheffield and Rotherham Wildlife Trust
Blackden Brook	SK12 88	Clough on northeast side of Kinder Scout
Blake Brook	SK06 61	Part of Leek Moors SSSI , important fossils in edges
Blake Mere	SK25 58	West of Bonsall on Bonsall Moor
Bleaklow	SK10 97	Large moorland block north of Kinder Scout
Blue John Cavern	SK15 83	Show Cavern below Mam Tor
Bobus	SK03 09	Hill near Marsden
Bonsall	SK28 58	Village east side of Peak District in limestone
Bosley Minn	SJ94 66	Hill in Cheshire section of Peak District
Bradbourne	SK20 52	Just outside the National Park to south
Bradwell	SK17 81	Large village on the limestone, Bradwell Dale passes south
Brassington	SK23 54	Village in the White Peak
Bretton Clough	SK20 78	North of Eyam Moor and east of Bretton
Broadbottom	SJ 99 94	West of Glossop
Broomhead Reservoir	SK26 96	East side of Peak District south of Stocksbridge
Brough	SK18 82	North of Bradwell, in Hope Valley
Buckstones	SE01 13	Just north of the A640 in the South Pennines Moors
Bull Clough Head	SK18 96	Howden Moors
Bunster Hill	SK14 51	West of the bottom of Dovedale
Burbage Moor/Valley/ Rocks	SK26 82	Burbage Moor to the east of Burbage Brook and Burbage Rocks
Burr Tor	SK18 79	Hilltop Enclosure Scheduled Ancient Monument, above Great Hucklow, no public access
Butterton	SK07 56	Village South West Peak
Buxton	SK05 73	Main Peak District town
Buxworth	SK02 81	West of Chinley
Cales Dale	SK14 82	South of Castleton
Calver	SK23 74	On the edge of the mid Derwent Valley
Carl Wark	SK26 81	On Hathersage Moor
Carsington	SK25 53	Carsington Reservoir to the south
Castle Naze	SK05 78	Northern end of Combs Moss above Chapel-en-le-Frith
Castleton	SK15 82	Large village in Hope Valley
Cat's Tor	SJ99 76	Above and west of the Goyt Valley
Cauldon Low	SK08 49	Southern edge of the White Peak
Cavendish Mill	SK20 75	On Middleton Moor
Chapel-en-le-Frith	SK06 80	Small town north of Buxton
Charlesworth	SK00 92	West of Glossop
Cheddleton	SK96 52	South of Peak District and Leek
Chelmorton	SK11 70	White Peak linear village
Chew Valley/ Reservoir	SE03 02	Chew Reservoir in high moorland above Dove Stones Res.

Location	Grid Reference	Notes
Chinley	SK04 82	Dark Peak village, Chinley churn quarry to the north
Chisworth	SJ99 92	Small village fringing western Peak District
Chrome Hill	SK07 67	Limestone peak in Upper Dove Valley
Chunal Moor	SK04 90	Moorland south of Glossop
Churnet Valley	SK02 61 to SJ98 50	Churnet River starts east of the Roaches and the A53 and continues south of Leek
Cistern Clough	SK03 69	On Axe Edge next to the A53
Combs Moss	SK05 76	Between Buxton and Chapel-en-le-Frith
Combs Reservoir	SK03 79	Just west of Chapel-en-le-Frith
Coombs Dale	SK22 74	Near Stony Middleton in White Peak
Cown Edge	SK01 91	Above Charlesworth
Cowns Rocks	SK12 90	North side of Woodlands Valley
Cracken Edge	SK03 83	Above Chinley
Cressbrook Dale	SK17 73	Limestone dale in the NNR
Cromford Canal	SK30 56	Starts in Cromford in the Lower Derwent Valley
Crowden Brook	SK10 87	Edale
Crowden Great & Little Brooks	SE07 00	Join just north of Longdendale
Crowden Tower	SK09 87	South side of Kinder Scout
Curbar Edge	SK25 75	Above Curbar, between Froggatt and Baslow Edges
Dale Dike Reservoir	SK24 91	Below Bradfield Moors on the east side of the Dark Peak
Dam Dale	SK11 77	Upper end of Peter and Monk's Dale near Peak Forest
Damflask Reservoir	SK27 90	Below Bradfield east side of Peak District
Dane River	SK01 71 to SJ93 63	Starts at top of Goyt Valley, flows pass Wincle
Darley Bridge	SK26 61	In lower Derwent Valley north of Matlock
Deep Dale, Monsal	SK16 69	Plantlife Reserve, tributary of Wye
Deep Dale, Topley Pike	SK10 71	Limestone dale, Derbyshire Wildlife Trust reserve
Derwent Edge/Moors	SK19 89	East side of Upper Derwent Dale
Derwent Valley	SK17 92 top SK24 78 mid SK29 60	Upper Derwent Valley with the reservoir string, Mid Derwent Valley between Bamford and Darley Dale, lower Derwent Valley south of this (unofficial, approximate nomenclature for this book)
Disley	SJ97 84	Large village on A6 to Stockport
Dove Valley & Dovedale	SK03 68 to SK16 50	Upper Dove Valley is Axe Edge to about Crowdecote, Mid Dove Valley extends down to south of Hartington, Dovedale starts at the junction of Wolfcotes and Biggin Dales through the limestone south to Thorpe
Dove Holes	SK07 78	North of Buxton on A6
Dove Stones	SE04 03	The whole moor above Chew and Dove Stone Reservoir managed for United Utilities by RSPB
Doxey Pool	SK00 62	On the Roaches
Earl Sterndale	SK09 67	Small village in White Peak
Ecton and Copper mine	SK0958	Manifold Valley in the White Peak
Edale Cross & Edale Head	SK07 86	South end of Kinder Scout
Edale & Valley	SK12 85	Valley south of Kinder Scout
Edensor	SK25 70	In the mid Derwent Valley near Chatsworth
Eldon Hill	SK11 81	Worked out quarry south of Rushup Edge

Location	Grid Reference	Notes
Elton	SK22 60	Small village in White Peak
Errwood Hall	SK00 75	Derelict hall in Goyt Valley
Ewden Valley	SK23 96	Starts Upper Commons, flows into Broomhead Reservoir
Eyam	SK21 76	The plague village on the gritstone
Fairholmes Car Park	SK17 89	Below Derwent Reservoir Dam
Featherbed Moss	SK08 92	South of Snake Pass on A57
Fin Cop	SK17 70	Above Monsal Dale and A6
Flash	SK02 67	Small village in South West Peak near A53
Friden and brickworks	SK17 60	On the High Peak Trail
Froggatt Edge	SK25 76	Northern end of Curbar/Baslow Edges
Furness Clough	SK00 83	Furness Vale
Gang Mine	SK28 55	Near Wirksworth, Derbyshire Wildlife Trust Reserve
Glossop	SK 03 94	Town on western edge of moorlands
Goldsitch Moss	SK01 64	Small moss north of Roaches
Goyt Moss/Valley	SK01 72 to SK01 77	Goyt's Moss blanket bog at upper end of valley, two reservoirs in centre, plantations and woodland
Gradbach, Gradbach Hill	SJ99 65	Hamlet in South West Peak, the Hill to the east north of Back Forest
Gratton Dale	SK20 60	Limestone Dale
Greave's Piece	SK29 76	Small moorland east of Big Moor
Green Clough	SK15 92	Tributary of Upper Derwent Dale
Greenfield Valley	SE02 04	Greenfield west of Dove Stones and valley to east
Grindleford	SK24 77	Gritstone village in mid Derwent Valley
Grindsbrook	SK11 87	One of the Pennine Way routes onto Kinder Scout
Grindslow Knoll	SK1 87	Above Grindsbrook
Gun Moor	SJ97 63	Southern edge of the NPA in South West Peak
Haddon Hall	SK23 66	Above River Wye southeast of Bakewell
Hangingstone	SJ97 65	West of Back Forest
Harewood Moor	SK30 67	East of Beeley
Harthill Moor	SK21 62	Southeast of Youlgreave
Hartington	SK12 60	Larger limestone village, old station to the east
Hassop	SK22 72	North of Bakewell
Hathersage	SK23 81	Large village, Hathersage Moor to the east
Hayfield	SK 03 87	Village west of Kinder Scout
Hen Cloud	SK01 61	East end of the Roaches
Heyden Brook	SE09 03	Tributary of River Etherow by A6024 road to Holme Moss
High Edge	SK06 68	Limestone hill south of Buxton
High Tor, Matlock	SK29 58	Just east of Matlock
High Wheeldon	SK10 66	Limestone hill with cave, National Trust
Hinkley Wood	SK12 50	Near Ilam, National Trust
Hogshaw Brook, Lightwood	SK05 74	North side of Buxton below Combs Moss
Holling Dale Plantation	SK22 91	Edge of Bradfield Moors
Hollingworth	SK01 96	West end of Longdendale
Hollinsclough	SK06 66	Small hamlet west of Longnor

Location	Grid Reference	Notes
Holme Moss	SE09 04	Blanket bog with BBC aerial landmark
Holme Valley	SE11 06	East of Holme Moss
Holmfirth	SE14 08	East of Home
Hope/Valley	SK17 83	In the Hope Valley
Houndkirk Moor	SK28 81	Southeast end of Burbage Moor
Howden Moors	SK18 93	East of Upper Derwent Valley
Hulme End	SK10 59	Upper end of Manifold Trail
Hurdlow	SK12 66	On the High Peak Trail
Ilam	SK13 50	On Manifold River just before it joins River Dove
Ipstones	SK0352	Just south of the Peak District
Isle of Skye	SE05 06	A public House now gone, but name given to the A365
Jagger's Clough	SK14 87	Southeast end of Kinder Scout
Jenkin's Chapel	SJ98 76	Small chapel near Saltersford
Kerridge	SJ94 76	Small ridge near Rainow
Kettleshulme	SJ98 97	Small village
Kinder Scout	SK08/09 88	Large upland plateau of blanket bog
Ladybower Reservoir	SK19 86	The lowest of the Derwent Reservoirs
Langsett and Moor	SE21 00	Large moor to the west of the village
Lantern Pike	SK02 88	Small National Trust moor west of A624 near Hayfield
Lathkill Dale	SK17 65	Long west-east limestone dale
Leash Fen	SK29 73	Part of Eastern Moors Estate
Leek	SJ98 56	Town to the southwest of the Peak District
Lineacre Reservoir	SK33 72	Eastern edge of the Peak District
Little Don	SE19 00	Starts on Langsett Moors and flows to Sheffield
Little Hayfield	SK03 88	Just north of Hayfield
Little Hucklow	SK16 78	West of Great Hucklow
Litton	SK16 75	Small limestone village
Litton Mill	SK15 72	In Miller's Dale by River Wye
Long Causeway	SK25 84	Extends from Sheffield to Stanage Pole
Long Clough	SK03 92	Near Glossop, Derbyshire Wildlife Trust Reserve
Long Dale – two different dales	SK13 61 & SK19 60	One slopes down to just above Hartington and another is north of Pikehall
Longdendale	SK06 98	Large valley of the River Etherow with chain of reservoirs
Longnor	SK08 65	Gritstone village in South West Peak
Longstone, Great	SK20 71	Village near Bakewell, below Longstone Edge limestone
Losehill Ridge	SK15 85	Ridge between Hope and Edale valleys
Loxley Valley	SK29 89	Flows to Sheffield
Lud's Church	SJ98 65	Landslip in Back Forest
Lyme Park	SJ96 82	Disley, National Trust Estate
Magpie Mine	SK17 68	Old lead mine near Sheldon
Mam Tor	SK12 83	At end of Rushup Edge
Manifold Valley	SK03 68 to 14 50	Starts on Axe Edge, flows past Longnor & Hulme End to Ilam, South West Peak
March Haigh Reservoir /Hill	SE01 12	On Close Moss, South Pennines, just north of Peak District
Marple	SJ95 98	Small town to west of Peak District

Location	Grid Reference	Notes
Marsden	SE04 10	Northern edge of Peak District
Masson Hill	SK28 58	Above Matlock
Matlock	SK30 60	On River Derwent, southeast edge of Peak District
Matlock Bath	SK29 58	Downstream from Matlock
Meltham	SE10 10	Northern edge of Peak District
Merryton Low	SK04 61	Northern end of Morridge, South West Peak
Middle Seal Clough	SK10 89	East side of Kinder Scout
Middleton Moor	SK20 74	West of Stoney Middleton
Midhope Reservoir	SK22 99	East side of Peak District
Milldale	SK14 54	In Dovedale
Miller's Dale	SK14 73	Section of the Wye Valley
Millstone Edge	SK24 80	Above Hathersage
Minninglow	SK20 57	Hill top tumuli, White Peak
Mixton	SK04 57	Outlying area of limestone, South West Peak
Monk's Dale	SK13 74	Tributary of Wye Valley, NNR
Monks Road	SK03 90	Starts at Hollinworth Head leading to Charlesworth
Monsal Dale	SK17 71	Lower section of Wye Valley west of Bakewell
Monyash	SK15 66	Village in White Peak
Morridge	SK02 58	Long ridge forming National Park boundary in southwest
Moscar	SK23 88	East of Derwent reservoirs
Mosley	SD98 02	Town on west side of Peak District
Narrowdale Hill	SK12 57	White Peak near Alstonefield
New Mills	SK00 85	Town on west side of Peak District
Noe River	SK13 85	Flows through Edale
Noe Stool	SK08 86	Southern edge Kinder Scout
Offerton Moor	SK20 80	South side of Hope Valley
Ogden Clough	SK02 99	Northwest side of Arnfield Moor
Onecote	SK04 55	Village South West Peak
Orchard Farm	SK02 69	Axe Edge
Over Haddon	SK20 66	Small village above Lathkill Dale
Owler Bar	SK29 77	Edge of Eastern Moors
Oyster Clough	SK11 90	Tributary of Woodlands Valley
Padley Wood & gorge	SK25 79	East side of Peak District
Parkhouse Hill	SK08 67	Limestone hill next to Chrome Hill, upper Dove Valley
Parkin Clough	SK19 85	Above Ladybower Dam
Peak Forest	SK11 79	Village in White Peak
Peak Forest Canal, Buxworth	SK01 82	Manchester to Buxworth Basin
Perry Dale	SK10 80	Dry dale to Peak Forest
Perryfoot, Rushup Edge	SK10 81	In valley below Rushup Edge
Peveril Castle	SK14 82	Above Castleton
Pott Shrigley	SJ93 79	West side of Peak District
Pule Hill	SE03 10	Northern edge of Peak District
Quarnford	SK01 66	Parish, includes Flash
Rainow	SJ95 76	Village west side of Peak District

Location	Grid Reference	Notes
Rainster Rocks	SK22 54	Brassington
Ramshaw Rocks	SK01 62	Part of Roaches Estate
Ramsley Moor	SK28 75	Part of Eastern Moors Estate
Redmires Reservoir	SK26 85	East side of Peak District
Revidge	SK07 59	Small heather moor
Ricklow Quarry	SK16 66	Lathkill Dale
Ringinglow	SK28 83	Sheffield to Burbage Moor and Stanage
Roaches	SK00 63	Estate on southwest edge of Peak District
Robin Hood	SK28 72	Near Baslow, former coal mine
Robin Hood's Stride	SK22 62	Harthill Moor
Robinson's Moss	SK04 99	Above Longdendale
Rose End Meadows	SK29 56	Derbyshire Wildlife Trust reserve near Cromford
Rowlee Bridge	SK14 89	On River Ashop in Woodlands Valley
Rushup Edge	SK11 82	East of Chapel-en-le-Frith to Mam Tor
Saddleworth & Moors	SD99 06	Moors to the east of the town
Salt Cellar	SK19 89	Derwent Edge
Saltersford	SJ98 76	West of Goyt Valley
Seal Stones	SK11 88	Edge of Kinder Scout
Shacklow Wood (Great)	SK17 69	Above Wye west of Ashford-in-the-Water
Sheen	SK11 61	Small village
Shelf Moor/Moss	SK08 94	Shelf Moss north end of Shelf Moor
Shire Hill	SK05 94	Glossop
Shutlingsloe	SJ97 69	West of Wildboarclough
Small Dale	SK09 77	North of Buxton
Snels Low	SK11 81	North of Eldon Hill
Sparklow	SK12 65	On High Peak Trail
Sparrowpit	SK98 80	On A623 out of Barmoor Clough
Speedwell Cavern	SK13 82	Bottom of Winnats Gorge
Sponds Colliery	SJ96 79	Between Kettleshulme and Pott Shrigley
Stalybridge	SJ96 98	West side of Peak District
Stanage Edge/Pole	SK24 84	Edge continues north & south just west of Pole
Standedge	SE01 10	Just north of Peak District
Stanton Moor	SK24 63	Southeast Peak District
Stoke Brook	SK23 75	East of Stoney Middleton
Stoney Middleton	SK23 75	Village on edge of White Peak
Strines	SK22 90	Strines and Dale Dike Reservoirs, east side of Peak District
Swellands Reservoir	SE03 09	On Wessenden Moors in north
Swineholes Wood	SK05 50	Staffs Wildlife Trust reserve on edge of Peak District
Swythamley Grange	SJ97 64	South West Peak
Tegg's Nose Country Park	SJ94 72	Near Macclesfield
Thorpe Cloud	SK15 51	Above lower Dovedale
Thors Cave	SK09 55	Manifold Valley near Wetton
Three Shires Head	SK01 69	Three counties meet on River Dane
Tideswell / Dale	SK15 75	One of larger villages in White Peak, Dale passes to south

Location	Grid Reference	Notes
Tintwistle High Moor	SK03 99	North of Longdendale
Tissington	SK17 52	White Peak village
Tittesworth Reservoir	SJ99 59	Edge of South West Peak
Toddbrook Reservoir	SK00 80	Whaley Bridge
Torside Reservoir	SK06 98	Longdendale
Totley Moss	SK27 79	North of Big Moor
Treak Cliff	SK13 83	Cliff and show cavern below Mam Tor
Trentabank Reservoir	SJ96 71	In Macclesfield Forest
Turn Edge	SK01 67	South West Peak
Underbank Reservoir	SK24 99	On River Don near Stocksbridge
Upper Elkstone	SK05 59	South West Peak on Warslow Brook
Upper Hulme	SK01 60	Just off A53 below Roaches
Via Gellia	SK25 56	A5012 running through Griffe Grange Valley to Cromford
Wain Stones	SK09 96	West end of Bleaklow
Warcock & Cupwith Hills	SE03 14	North of Marsden
Warslow	SK08 59	Village in South West Peak
Water-cum-Jolly	SK16 73	By River Wye
Watford Lodge	SK00 86	New Mills, Derbyshire Wildlife Trust Reserve
Wensley Dale	SK26 60	East of Winster
Wessenden Reservoir/ Moor	SE06 08	Large moor west of Reservoirs
Wetton Hill /Mill	SK10 56	Above or in Manifold Valley in White Peak
Whaley Bridge	SK01 81	Small town north of Buxton
Wheel Stones	SK20 88	Derwent Edge
Whetstone Ridge	SK00 70	Upper end of Goyt Valley
White Hill	SD99 12	East of Denshaw
Wigber Low	SK20 51	South of Bradbourne
Wildboarclough	SJ98 68	North of A54 Congleton road
Win Hill	SK18 85	West of Ladybower Reservoir
Wincle	SJ96 66	South of A54 Congleton road
Windgather Rocks	SJ99 78	South of Kettleshulme
Windy Knoll	SK12 82	Below Rushup Edge
Winnats Pass	SK13 82	Steep pass in limestone
Winster	SK24 60	White Peak village
Wirksworth	SK28 54	On southern edge of White Peak
Woo Dale	SK09 72	Tributary of Wye Valley
Woodhead Tunnel	SE12 00	East end of Longdendale valley
Woodlands Valley	SK14 89	Main valley with River Ashop and A57 from Snake Pass
Wormhill	SK12 74	Small White Peak village
Wye River/valley	SK06 73 to 25 65	Main valley is east of Buxton to Rowsley
Wyming Brook	SK27 86	Flows from Redmires to Rivelin Reservoirs
Youlgreave	SK20 64	White Peak village above River Bradford

Nature Conservation Explained

SITE VALUE AT THE NATIONAL AND INTERNATIONAL LEVEL

Sites of Special Scientific Interest, SSSIs, are designated for their biological or earth science (geological, landform or soils) interest by the Statutory Nature Conservation Agency; Natural England in our area. These are sites that are considered to be of special interest at a national scale. Within each area, a representative series of the best examples of each significant natural habitat may be notified, or for rarer habitats all examples may be included. Sites may be for specific taxonomic groups like birds, butterflies etc. Some sites are designated for both their earth science and biological features, as in many Peak District examples. Figure 3 shows all the Peak District SSSIs. Designation does not necessarily mean that each is in ideal condition, but Natural England works with landowners to improve this. There are controls on what activities can be undertaken on SSSIs in order to protect their features and their protection is supported by planning policies.

National Nature Reserves (NNRs) are usually SSSIs for much of their area, but may include additional land used for management purposes. They were established in particular to protect some of our most important habitats, species and geology but also to provide outdoor laboratories for research and public experience of nature conservation. The Peak District has three NNRs: Kinder Scout and Dovedale managed by the National Trust and the Derbyshire Dales managed by Natural England. There are also a number of Nature Reserves in the Peak District managed by charities such as Plantlife and the County Wildlife Trusts. RSPB owns Combes Valley in the Churnet Valley just to the south and also works with others to manage large areas of our moorland. Many of these sites are also SSSIs. The National Trust, Water Companies, NPA and Sheffield City Council own and manage large areas of the Peak District SSSIs as well as some other land holdings with nature conservation part of their objectives.

At the European Scale, there are two more designations: Special Area of Conservation (SAC) is defined in the European Habitats' Directive (the Directive on the Conservation of Natural Habitats and of Wild Fauna and Flora) to protect species and habitats of Community Importance. The value of the proposed site is assessed in relation to the whole of the national resource of each habitat type and species. The Habitat's Directive lists the habitats of European interest with blanket bog and dwarf-shrub heath as well as the lime woods in the Derbyshire Dales being the key ones in the Peak District. Figure 3 shows the extent of the South Pennine Moors SAC, which covers much of our moorlands, and the limestone dale woodlands and flower-rich grasslands, which lie in the Peak District Dales SAC. Some Habitat's Directive key species protected by SACs in the Peak District are Brook Lampreys, Bullhead and White-clawed Crayfish.

Special Protection Areas (SPAs) complement SACs but are specifically for birds and are designated under the 1979 European Union Directive on the Conservation of Wild Birds for both migratory and breeding birds. They represent areas used regularly by 1 per cent or more of the population of an Annex I species or 1 per cent or more of the biogeographic population of a regularly occurring migratory species. In the Peak District, the South Pennine Moors SPA covers the main moorland areas with important numbers of breeding waders and some birds of prey in particular. These, together with SACs, form the European NATURA 2000 network. SACs and SPAs are already SSSIs, but there are additional levels of protection for European sites in relation to site activities and planning policies.

OTHER NATURE CONSERVATION ACTION

At a level below SSSIs, the County Wildlife Trusts identify sites of county importance within their areas. There are usually planning policies to protect these, but they are not as strong as those for nationally or internationally important sites. The Trusts work with landowners as far as possible to protect and enhance these sites, but this may be limited. There are also Regionally Important Geological and Geomorphological Sites (RIGGS) that local Geological groups have identified. These all provide a second tier of nature conservation sites. In addition, however, there are many habitats and species occurring outside these in the wider countryside which are dependent on sympathetic and knowledgeable land owners to protect or manage them.

This wider nature conservation need has been partly addressed by the UK Biodiversity Action Plan (BAP). The first one published in 1994 was a response to the Convention on Biological Diversity (CBD) which was agreed in Rio de Janeiro in 1992. Action plans for the most threatened species and habitats were produced. With devolution, the UK countries then developed their own BAPs and these were updated and the approach developed after new CBD agreements in Aichi (Japan) in 2010. Local BAPs were also prepared, including by the Peak District NPA, which collated existing information and produced targets and action plans, some of which are quoted in this book. UK BAP priority species and habitats were those identified as being the most threatened and needing conservation action. These have been updated and are now part of the UK Post-2010 Biodiversity Framework published in 2012. The priority lists are enshrined under Section 41 of the Natural Environmental and Rural Communities Act 2006 in England. BAP habitats and species are referenced in this book where relevant. Action for their future conservation often focuses on sites outside the SSSI network. This is undertaken by a range of charities, water companies, NPA and others, but also, especially in the Peak District as a rural area, as part of agri-environment schemes.

As part of the reporting of progress towards our Biodiversity and CBD targets, over 70 conservation, government and research organisations combine to analyse monitoring data on habitats and species. The results are presented in regular State of Nature Reports (e.g. Hayhow et al. 2019) that show how national wildlife is faring. I prepared a local version for the Peak District in 2016.

RARITY

There is a strong link with measures of rarity and the BAP and CBD process as set out above. Many species are declining rapidly on a national scale and updated information is needed to highlight their conservation needs. However, there are different systems for deciding rarity or relative rarity for different plant and animal groups which can lead to some confusion. IUCN (International Union for Conservation of Nature) criteria are being followed now and conversion to them is in place. The following explains phrases and words used in this book.

The rarity status is complicated by the species that are protected by law. The Wildlife and Countryside Act 1981, Schedules 5 and 8 (updated by quinquennial reviews) lists animals and plants respectively that are protected from various levels of damage, disturbance or possession under Section 9. Amongst these, a higher level of protection applies to the following Peak District species: White-clawed Crayfish, Adder, Grass Snake (*Natrix natrix*), Slow-worm, Great Crested Newt, Common Lizard, Dormouse, Water Vole, Otter and all bats. There is also a list of birds in Schedule 1 that are protected from disturbance in the breeding season, although nearly all breeding birds and their nests are also protected from loss or damage.

Plants

A species that is common, widespread, abundant or similar are generally taken from their national or regional distributions. The comments I give might be derived from the County Floras, the National Atlas of plant distributions or other identification books. There are no clear criteria on which such abundance or frequencies are based, but experience helps. Descriptors are added when looking at distribution maps where there are distinctive patterns.

For plants with more restricted distributions, the IUCN criteria for rarity are applied (Stroh *et al.*, 2014). Nationally, these are:

IUCN categories and criteria	Derbyshire Flora categories and criteria
Internationally Rare	
Nationally Threatened in 4 categories: Critically Endangered, Endangered, Vulnerable or Near Threatened.	
Nationally Rare – in 15 or less GB hectads	
Nationally Scarce – in 16 to 100 GB hectads	
	Locally Rare – in 3 or less monads
	Locally Scarce – in in 4 to 10 monads

The threat categories relate to population reductions, geographic range, small population sizes and restricted locations. The plant species covered by the criteria in the IUCN column constitute the Red List for England. These categories replace Red Data Book species that used to be classified as those in less than 10 UK hectads. A hectad is 10 × 10 km square. A monad is a 1 × 1 km square. The older county floras use a previous system as they were published prior to the incorporation of the IUCN criteria. There are a number of Nationally Threatened and Nationally Rare or Scarce species in the Peak District which you might find, some of which are mentioned in the text, but the Derbyshire Flora lists them all. Any other descriptors using words associated with rarity without capitals are based on experience or other books.

Birds

The rarity of birds is based on the same kind of threat criteria: historical declines, trends in population and range, population size, localisation and international importance of each species as well as their global and European threat status. The outcome is the Birds of Conservation Concern produced by a partnership of bird and nature conservation organisations (Eaton *et al.*, 2015). There are equivalent lists for Europe. The lists are regularly updated and categorise birds as on the Red, Amber or Green lists of Conservation Concern and the most recent assessment shows increases in those on the Red and Amber lists compared with previous ones, including species characteristic of the Peak District like Curlew and Pied Flycatcher. Where birds mentioned in the book feature on the Red list, this is mentioned.

Invertebrates

The status of invertebrates is generally more difficult to determine unless they are well-known groups like butterflies or dragonflies. They are not always recorded on the same grid scale as other species groups and there is often far less known about populations, trends and habitats. In general the same IUCN threat levels are used with Critically Endangered, Endangered, Vulnerable and Near Threatened all regarded as Red List species. In addition, Nationally Rare species are those found in 1 to 15 hectads and Nationally Scarce are found in 16 to 100 hectads within a reasonable date class. The latter is subdivided in older publications into Nationally Notable A and B species defined as occurring in 16 to 30 and 31 to 100 hectads respectively. Local status is given to those in 101–300 hectads, and this may be qualified by a geographical distribution. The IUCN criteria are based on tetrads (4 km squares) not hectads which complicates the process, and are regionally based (Great Britain is a region in this context). Nationally Rare or Scarce species are not part of the IUCN system which is for a whole region. Of the estimated 37,000 macro-invertebrates in Great Britain, some 27 per cent have been assessed of the better known groups, although this number is expected to reach 50 per cent (Webb and Brown, 2016).

Glossary

Acrocarpous Of mosses, mostly unbranched where stems are erect and capsules are at the tips of stems or branches.

Acrotelm Upper rooting layer of peat overlying the catotelm, which has a lower bulk density and higher hydraulic conductivity.

Aeolian deposits Windblown deposits.

Agaric A fungus with a fruiting body like that of a mushroom with a cap and gills on the underside.

Ammocoetes Larval stage of the lamprey, a blind, worm-like animal that burrows in silt.

Anaerobic Without oxygen.

Apomixis Asexual reproduction whereby there is no cross fertilisation but the plant develops seeds as a maternal clone which are identical to the parent and assures reproduction in the absence of pollinators.

BAP Biodiversity Action Plan, see Appendix 2.

BCE/CE Before Current Era and Current Era.

Benthic Bottom-living, as of freshwater invertebrates.

Biomass The total biological material of a plant or animal or of populations or communities.

BMWP Biological Monitoring Working Party is an index of water quality based on families of freshwater invertebrates that are used as pollution indicators. Higher scores are given to those least tolerant of pollution. .

Calcicole Plant restricted to lime-rich soils with a high pH.

Calcifuge A plant that will not normally grow in alkaline soils.

Calaminarian grassland Occurs on soils with high levels of heavy metals such as lead, zinc, chromium and copper that are toxic to most plant species.

Capitulum of *Sphagnum* The top of the plant which has compact clusters of young branches forming the head.

Carapace The hard upper shell of a crustacean.

Catotelm The lower horizon in well-structured peat which is darker, anaerobic, more compacted and wet, but with low hydraulic conductivity.

Chromatophores Pigment-containing cells or groups of cells which allow shifts in body colouration and pattern.

Cottid A fish of the family Cottidae typically having a large head and tapering body with spiny fins, such as the Bullhead.

Crepuscular Active mostly at dusk.

Crustose lichen One that forms a crust on a surface such as rock.

Cuticle In a plant this is the protective filmy surface covering the epidermis of leaves, young shoots and other organs.

Defra Department of Environment, Food and Rural Affairs.

DNA (Deoxyribonucleic acid) The foundation of life in cells that consists of two polynucleotide chains that coil to form a double helix that carries the genetic instructions in organisms.

Ecosystem A community of living organisms with their non-living environmental components like soils, interacting together as a functioning system, linked by nutrient cycling and energy flows.

Ecosystem services The environmental benefits healthy ecosystems can provide for humans such as clean air, flood control, mental and physical well-being, pollinators and clean water.

Elytra The modified, rigid forewings of a beetle which cover the hindwings used for flying.

Epicormic A shoot growing from a previously dormant bud on a tree trunk or limb.

Epidermis Outermost layer composed of a single layer of cells covering plant parts (leaves, flowers, stems etc).

Epiphytic A plant growing on another for support. They get their moisture and nutrients from the air or water washing down the host.

Evapotranspiration The combination of water evaporating from surfaces such as soil and transpiration losses from plants.

Gizzard An organ in the digestive tracts of some animals without teeth such as birds where food is ground up, often with grit or small stones that the animal ingests.

Glacial erratic Rocks originating elsewhere dropped by ice sheets when they melt.

Glaucous A dull grey-green or blue colour.

Gleying The process of waterlogging in soils that causes anaerobic conditions, giving a sticky bluish-grey subsurface layer or spotting as iron changes from ferric to the ferrous form. This may appear seasonally.

Greenhouse gases The main gases are carbon dioxide, methane, nitrous oxide and fluorinated gases. These block the earth's heat escaping into space, thus contributing to the greenhouse effect.

Grips Usually herringbone patterns of drains excavated in peat in the last century to drain the ground.

Groughs Often shallow peat gullies that frequently anastomose around haggs of up-standing peat on flatter plateau tops where there is significant peat damage.

Hollow-way An old, often sunken track or path, often braided as new routes replaced muddy ones.

Inbye land Land taken in from rough grazing/moorland and used as more intensive pasture, sometimes in valleys near farms.

Ironpan A hard layer cemented with iron oxides which forms an impermeable layer in a podsol.

Karst Geological formations resulting from the dissolution of carbonate rocks like limestone creating sinkholes, sinking streams, springs, caves, fissures and other characteristic features.

Lanceolate Long and narrow shape e.g. of leaves or petals.

Leaching The movement of water-soluble nutrients, salts and minerals down the soil profile. These may be precipitated out in a lower layer.

Lepidopterous Belonging to or pertaining to the Lepidoptera The insect group that includes moths and butterflies.

Loess Wind-borne sediment predominantly of silt-sized particles forming a soil.

Mitochondrial DNA analysis Analysis of DNA located in mitochondria, which are cellular organelles that convert chemical energy into a form that cells can use.

Molluscs The second-largest Phylum of invertebrates, most of which are marine, but which include snails and slugs.

Mor humus A raw, acidic humus, usually dark-coloured where there are few decomposers like earthworms and low biological activity. It can form thick mats of largely undecomposed dead material that is not mixed into the soil. Found under conifers and heathlands.

Mull humus Well-decomposed organic matter that is mixed much more deeply in the underlying soil and does not accumulated thick layers on the surface. Usually contains many more earthworms and other decomposers. It is more prevalent under broadleaved woodland or neutral grasslands.

Mycorrhiza A symbiotic association between plant roots and beneficial fungi which enables the plants (mostly trees and shrubs) to gain water and minerals from the soil and the fungi to obtain sugars from green plants. The fungi produce a hyphal mat which can penetrate the roots (endomycorrhizal) or wrap round the outside (ectomycorrhizal) which increases the surface absorbing area for roots significantly.

NPA National Park Authority, PDNPA Peak District NPA.

Natural capital The stock of natural assets, such as rock, soil, air, water and living things, on which humans depend for a wide variety of environmental benefits called ecosystem services (see above).

Ombrotrophic All nutrients and water come from rainwater.

Orogeny A mountain-building event.

Ovule The structure in seed-bearing plants that develops into a seed after fertilisation.

Peat pans The name I have given to shallow depressions in bare peat trapped within vegetation a few metres across that hold water when wet, but dry out later. They are thought to be produced by wildfire. They are common prior to restoration on flatter blanket bog peat.

Periglacial Conditions and processes at the edge of ice sheets.

pH The measure of acidity on a logarithmic scale from 0–14. A low pH is acidic, a high pH is alkaline, 7 is neutral. Most natural soils and water bodies lie in the range 3 to 8.

Phenology The timing of periodic events in biological life cycles such as leaf emergence, flowering, egg-laying or migration. Some of these events are changing with climate warming.

Phenotypic The observable characters of an organism resulting from the interaction of its genetics and the environment, such as its physical structure and shape. Phenotypic plasticity refers to the flexibility an organism can show to different environmental stresses such as drought.

Photosynthesis The chemical reaction using sunlight in the pigment chlorophyll in green plants to make sugars using carbon dioxide and water.

Phreatic A geological term relating to the water table. A phreatic cave is full of water which can dissolve the limestone all around it. A phreatic eruption is a steam-blast when magma heats the ground or surface water.

Pleurocarpous Applied to mosses which tend to be flat and straggly on the ground rather than upright, with the capsules produced on the short, lateral branches rather than at the tips as in acrocarpous mosses.

Poaching Applied to soils where feet, stock or machinery have produced a bare surface.

Podsol An infertile, acidic soil which has a dark organic-rich surface layer over a white or greyish sublayer formed by minerals being leached into a lower darker stained layer. It especially develops under heathland.

Pollarding The practice of cutting trees at above the height of grazing animal's reach to obtain a crop of timber for particular purposes and to ensure regrowth of the tree from the cut faces.

Pollen analysis Pollen grains have a hard outer coat that resists decay, particularly in acidic and water-logged deposits. As the deposit builds up, layers of plant and animal remains and the pollen rain are conserved. Pollen grains are quite distinctive so an analysis provides an interpretation of the vegetation at the time of deposition. Interpretation takes into consideration the differing amounts of pollen species produced, rates of breakdown and distances that pollen might have travelled. Aging the layers using archaeological artefacts or radio-carbon dating gives a timeline.

Porrect Relatively long, forward pointing antennae; an entomological term.

Propagules Any structure that can form new plants, through sexual or vegetative reproduction e.g. suckers, seeds, spores or runners as in Creeping Buttercup or Strawberry plants.

Radio-carbon dating A method for determining the age of organic material by measuring the radioactive isotope of carbon.

Rendzina A lime-rich, shallow soil lying over calcareous rocks, often on a steep slope. Organic matter is concentrated at the surface and the soils are free-draining and often stony.

Rhizomes Underground modified stems, usually growing horizontally that are used mostly as food storage organs. They produce new buds and roots. Multiplication of rhizomes is a form of reproduction and some plants like Bracken can spread far through their rhizomes.

Rotational landslide Where the material slides down and outwards on top of a failed concave surface. See translational landslide.

SAC Special Area of Conservation, European designation for specific habitats. See Appendix 2.

Saprophytic Organisms that obtain their food by absorbing the products of dissolved organic material. Examples are some fungi and some plants with no chlorophyll.

Serrated Saw-like teeth along an edge as of a leaf. They can be small, large or variable in scale.

Sessile Without a stalk or stem.

Solifluction The gradual mass wasting slope process related to freeze-thaw activity.

Sough Local term for a drain, either for draining water in fields or from mines.

SPA Special Protection Area, European designation for birds. See Appendix 2.

Spadix A type of inflorescence with small flowers on a fleshy stem, usually surrounded by a large leaf-like curved bract called the spathe, as in Wild Arum.

Spathe See spadix.

SSSI Site of Special Scientific Interest, see Appendix 2.

Taxa (singular Taxon) is a biological term for a unit in classification. It is used as an overarching word to include genera, species, sub-species, races, hybrids and those organisms that cannot be fully identified in a sample.

Thallus of lichen This is the vegetative part of the lichen usually made of filaments of the fungal hyphae in which algal cells or cyanobacteria sit. They produce a diverse range of structures – tufts, flat, leaf-like, thin crusts or layers of powdery granules.

Translational landslides Where the mass of material slides downwards and outwards on top of an inclined planar surface resulting in accumulating material at the foot of the landslide.

Turbary A right to cut turf or peat for fuel, usually from common land.

Umbellifer A plant family that is characterised by having umbrella-shaped flowering heads called umbels consisting of many small flowers on short stalks.

Vadose The unsaturated zone as in a cave with water flowing on the bottom.

Vascular plants Those that produce xylem, a lignified water-conducting tissue that also provides support, and phloem, a food-conducting tissue. These are the vascular tissues. All flowering plants, conifers and ferns are vascular plants. Mosses and liverworts are not.

References

Abrahams, D. (2019). Observations from two decades of hay meadow restoration. *Conservation Land Management* 17 (3), 10–18.

Aitkenhead, N, Barclay, W. J., Brandon, A., Chadwick, R. A., Chisholm, J. I., Cooper, A. H. & Johnson, E. W. (2002). 2 *British Regional Geology: The Pennines and Adjacent areas (fourth edition)*. British Geological Survey, Nottingham.

Akinola, M. O., Thompson, K & Buckland, S. M. (1998). Soil seed bank of an upland calcareous grassland and management manipulations. *Journal of Applied Ecology* 35 (4), 544–552.

Albertson, K., Aylen, J., Cavan, G. and McMorrow, J. (2010). Climate change and the future of occurrence of moorland wildfires in the Peak District of the UK. *Climate Research* 45, 105–118. doi:10. 3354/cr00926.

Allan, S. A. (2004). *A macroinvertebrate focused ecological survey of the moorland stream network in the Peak District National Park.* MSc. Thesis, University of Manchester.

Alonso, I, Weston, K., Gregg, R. & Morecroft, M. (2012). *Carbon storage by habitat: Review of the evidence of the impacts of management decisions and condition of carbon stores and sources.* Natural England Research Report NERR043.

Anderson, P. (1986). *Accidental Moorland Fires in the Peak District: A study of their incidence and ecological implications.* Peak Park Joint Planning Board, Bakewell.

Anderson, P. (1997). Fire damage on blanket mires. In: *Blanket Mire Degradation, Causes, Consequences and Challenges* (Eds. Tallis, J. H., Meade, R. & Hulme, P. D.), 16–28. Conference Proceedings, University of Manchester, British Ecological Society Mires Research Group.

Anderson, P. (2016). *State of Nature in the Peak District: What we know about the key habitats and species of the Peak District.* On behalf of Nature Peak District. Peak District National Park Authority.

Anderson, P. & Shimwell, D. (1981). *Wild Flowers and other Plants of the Peak District.* Moorland Publishing, Ashbourne.

Anderson, P. and Yalden, D. W. (1981). Increased sheep numbers and the loss of heather moorland in the Peak District, England. *Biological Conservation* 20, 195–213.

Anderson, P., Tallis, J. H. & Yalden, D. W. (1997) *Restoring Moorland, Peak District Moorland Management Project, Phase III report.* Peak Park Joint Planning Board, Bakewell.

Andrews, J. E., Pedley, H. M. & Dennis, P. F. (1994). Stable isotope record of palaeoclimatic change in a British Holocene tufa. *The Holocene* 4, 349–355.

Anon. (2017). Look what we've found! *Parklife*, Autumn/Winter 23, 6, Peak District National Park Authority.

Archaeological Research Services Ltd. (2011) Fin Cop Hillfort. archaeologicalresearchservices.com/projects/site/index.html

Ardron, P. A. (1999). Peat Cutting in Upland Britain, with special reference to the Peak District – its impact on Landscape, Archaeology and Ecology. PhD thesis, University of Sheffield.

Armitage, S. (2007). *Sir Gawain and the Green Knight.* Faber and Faber Ltd, London.

Atherton, I, Bosanquet, S. & Lawley, M. (2010). *Mosses and Liverworts of Britain and Ireland: a field guide.* British Bryological Society.

Balme, O. E. (1953). Edaphic and Vegetational Zoning on the Carboniferous limestone of the Derbyshire Dales. *Journal of Ecology* 41 (2), 331–344.

Balmer, D. E., Gillings, S., Swann, R. L., Downie, I. S. & Fuller, R. J. (2013). *Bird Atlas 2007–11: the breeding and wintering birds of Britain and Ireland.* BTO Books, Thetford.

Banks, V. J., Jones, P. F., Lowe, D. J., Lee, J. R., Rushton, J. & Ellis, M. A. (2012). Review of tufa deposition and palaeohydrological conditions in the White Peak, Derbyshire, UK: implications for Quaternary landscape evolution. *Proceedings of the Geologist' Association* 123 (1), 117–129.

Barden, N. (2007). *Helianthemum* Grassland of the Peak District and their possible mycorrhizal associations. *Field Mycology* 8 (4), 119–126.

Barker, A. & Beck, J. S. (2010). *Caves of the Peak District.* Derbyshire Caving Association.

Barley, L. (2018). The road not taken: a case study in decision-making in nature conservation. *Conservation Land Management* 16 (4), 3–10.

Barnatt, J. & Smith, K. (2004). *The Peak District landscapes through time.* English Heritage, London.

Barnatt, J. & Penny, R. (2004). *The lead legacy: The prospects for the Peak District's lead mining heritage.* Peak District National Park Authority, Bakewell.

Barnatt, J. (2013). *Delving ever deeper: the Ecton Mines through Time.* Peak District National Park Authority, Bakewell.

Barnatt, J. (2019). *Reading the Peak District Landscape. Snapshots in Time.* Historic England, Swindon.

Bates, J. W., Thompson, K. & Grime, J. P. (2005). Effects of simulated long-term climatic change on the bryophytes of a limestone grassland community. *Global Change Biology* 11 (5), 757–769.

Benton, T. (2006). *Bumblebees.* The New Naturalist, Collins, London.

Bevan, B. (2007) *Sheffield's golden frame: the Moorland Heritage of Burbage, Houndkirk and Longshaw.* Sigma Press, Ammanford.

Bevan, B. (2005). *Conservation Heritage Assessment Edale Valley.* Report No. 3, Moors for the Future Partnership, Edale.

Blockeel, T. L, Vanderpoorten, A., Sotiaux, A. & Goffinet, B. (2005). The Status of the mid-western European endemic moss *Brachythecium appleyardiae. Journal of Bryology* 27 (2), 137–141.

Border, J. A., Massimino, D., Newson, S. E., Boersch-Supan, P., Hunt, M., Bosanquet, S., Ainsworth, M., Cooch, S., Genney, D. & Wilkins, T. (2018). *Guidelines for the Selection of Biological SSSIs. Part 2 Detailed Guidelines for Habitats and Species Groups, Chapter 14 Non-lichenised fungi.* JNCC, Peterborough.

Brancaleoni, G., Banks, V. J., Leoncini, C., Kirkham, M., Thorpe, J. & Castellaro, S. (2016). Peter's Stone, Cressbrook Dale, Derbyshire: landslide or paraglacial feature? *Mercian Geologist* 19 (1), 51–54.

Brian, M. V. (1977). *Ants*. The New Naturalist, Collins, London.

Bridgland, D. R., Howard, A. J., White, M. J., White, T. S. & Westaway, R. (2015). New insight into the Quaternary evolution of the River Trent, UK. *Proceedings of the Geologist' Association* 126, 466–479.

British Association for Shooting and Conservation. (Undated). *Grouse shooting and management in the United Kingdom: its value and role in the provision of ecosystem services*. BASC White Paper.

Bromley, J., McCarthy, B. & Shellswell, C. (2019). *Managing grassland road verges: A best practice guide*. Plantlife, https://www. plantlife. org. uk/uk/our-work/publications/ road-verge-management-guide

Brown, L. E. Ramchunder, S. J., Beadle, J. M. & Holden, J. (2016). Macroinvertebrate community assembly in pools created during peatland restoration. *Science of the Total Environmental* 569–570, 361–372.

Buckingham, H. & Chapman, J. (1997). *Meadows beyond the Millennium*. Peak District National Park Authority, Bakewell.

Buckingham, H. & Chapman, J. (1999). *Hidden Heaths. A portrait of Limestone Heaths in the Peak District National Park*. Peak District National Park Authority, Bakewell.

Buckland, S., Thompson, K., Hodgson, J. & Grime, J. P. (2001). Grassland invasions: effects of manipulations of climate and management. *Journal of Applied Ecology* 38 (2), 301–309.

Bull, J. (2012). *The Peak District: A Cultural History*. Signal Books, Oxford.

Bunting, J. (2006). Bygone Industries of the Peak. *The Peak District Journal of Natural History and Archaeology* 3.

Butterfly Conservation. (2013). *The State of Britain's Larger Moths 2013*. Butterfly Conservation.

Caporn, S. J. M. & Emmett, B. A. (2009). Threats from air pollution and climate change to upland systems: past, present and future. In: *Drivers of Environmental Change in Uplands*. (Eds. Bonn, A., Allott, T., Hubacek, K. & Stewart, J.), 34–58. Routledge Studies in Ecological Economics, London.

Caporn, S. J. M., Carroll, J. A., Studholme, C. & Lee, J. A. (2006). *Recovery of ombrotrophic Sphagnum mosses in relation to air pollution in the Southern Pennines*. Report to Moors for the Future Partnership, Edale.

Capper, M. (2001). Labrador Tea *Ledum groenlandicum* in the Upper Derwent Valley: A brief update. *Sorby Record* 37, 34–36.

Carr, G. & O'Hara, D. (2015). Breeding Golden Plovers in the Peak District National Park. *British Birds* 108, 273–278.

Carr, G. (2004). *Breeding Bird Survey of the Peak District Moors*. Moors for the Future Partnership, Edale.

Carroll, M. J. (2012). *The ecology of British upland peatlands: climate change, drainage, keystone insects and breeding birds*. PhD, University of York.

Carroll, M. J., Dennis, P., Pearce-Higgins, J. W. & Thomas, C. D. (2011). Maintaining northern peatland ecosystems in a changing climate: effects of soil moisture, drainage and drain blocking on craneflies. *Global Change Biology* 17 (9), 2991–3001.

Chapman. P. (1993). *Caves and cave life*. The New Naturalist, HarperCollins, London.

Cheshire Mammal Group. (2008). *Mammals of Cheshire*. Liverpool University Press, Liverpool.

Clapham, A. R. Ed. (1969). *Flora of Derbyshire*. County Borough of Derby Museum and Art Gallery.

Cliffe, S. (2010). *Derbyshire Cavemen*. Amberley Publishing, Stroud, Gloucestershire.

Cobbett, William. (1821). *Cottage Economy*. Reprinted at various times.

Committee on Climate Change. (2018). *Land use: Reducing emissions and preparing for climate change*. https://www. theccc. org. uk/ publication/land-use-reducing-emissions- and-preparing-for-climate-change/

Cooper, R. G. (2007). *Mass Movements in Great Britain*, Geological Conservation Review Series, No. 33, Joint Nature Conservation Committee, Peterborough.

Cranfield University. (2018). *Soils survey of England and Wales*. http://www. landis. org. uk/services/soilsguide/index. cfm

Crofts, S. (2011) Caddisfly Recorder's Report. *Sorby Record* 47, 72.

Culpeper, N. (1826). *Culpeper's Complete Herbal and English Physician*. Facsimile of original Cleave, J & Sons, Deansgate, Manchester by Gareth Powell Ltd.

Dalton, R., Fox, H. & Jones, P. (1999). *Classic landforms of the Dark Peak*. (Eds. Castleden, R & Green, C.). The Geographical Association.

Dalton, R., Fox, H. & Jones, P. (1999a). *Classic landforms of the White Peak*. (Eds. Castleden, R & Green, C.). The Geographical Association.

Davis, B. N. K., Walker, N., Ball, D. F. & Fitter A. H. (1992). *The Soil*. The New Naturalist. HarperCollins.

Defoe, D. (1726) *A Tour thro' the whole Island of Great Britain*. Volume 3. Reprinted Dent, 1962.

Defra (2018) *A Green Future: Our 25 year plan to improve the Environment.*

Derbyshire County Council, Derby City Council, & the Peak District National Park. (2013). *Local Aggregates Assessment Draft*. Peak District. gov. uk website.

Derbyshire Heritage. (undated). Gibbets rock, Peter Stone. https://derbyshireheritage. co. uk/misc/peters-stone-gibbet-rock-wardlow-mires/

Derbyshire Reptile and Amphibian Group. (2018). February Newsletter 26, https:// groups. arguk. org/DARG/

Derbyshire Reptile and Amphibian Group. (2019). January Newsletter 28, https:// groups. arguk. org/DARG/

Dilks, T. J. K. & Proctor, M. C. F. (1974). The pattern of recovery of bryophytes after desiccation. *Journal of Bryology* 8, 98–116.

Dranfield, K. (2008) *Goyt Valley Miner: Errwood Hall & Castedge Pit*. Published by Kevin Dranfield.

Duncan, I., Seal, P., Tilt, J., Wasley, R. & Williams, M. (Eds.) (2016). *Butterflies of the West Midlands*. Pisces Publications, Newbury.

Eaton, M., Aebischer, N., Brown, A, Hearn, R., Lock, L., Musgrove, A., Noble, D., Stroud, D. & Gregory, R. (2015) Birds of Conservation Concern 4: the population status of birds in the UK, Channel Islands and Isle of Man. *British Birds* 108, 708–746.

Edwards, K. C. (1962). *The Peak District*. Collins New Naturalist.

Ellis, S. & Robinson E. J. H. (2015). The Role of Non-foraging Nests in Polydomous Wood Ant Colonies. *PLoS ONE* 10 (11). https://doi. org/10. 1371/journal. pone. 0143901. g001

Evans, M. (2009). Natural Changes in upland landscapes. In: *Drivers of Environmental Change in Uplands*. (Eds. Bonn, A., Allott, T., Hubacek, K & Stewart, J.), 13–33. Routledge Studies in Ecological Economics, London.

Evans, S. (2004). *Waxcap-grasslands – an assessment of English sites*. English Nature Research Report no 555.

Everall, N. C. (2010). *The Aquatic Ecological Status of the Rivers of the Upper Dove Catchment in 2009*. Natural England Commissioned Report no NERC046. Natural England, Peterborough.

Everall, N. C., Johnson, M. F. Wilby, R. L. & Bennett, C. J. (2014). Detecting phenology change in the mayfly *Ephemera danica* responses to spatial and temporal water temperature variations. *Ecological Entomology* 40 (2), 95–105. Doi. org/10. 1111/ een. 12164

Everall, N. C., Johnson, M. F., Clarke, A. and Gray, J. (2019). The visual state of riverbeds and their associated invertebrate community biosignatures. *FBA News* 77, 11–15.

Farey, J. (1811–1817). *A General View of the Agriculture and Minerals of Derbyshire.* 3 volumes, B. McMillan.

Food and Environment Research Agency. (2011). *Evaluation of the Potential Consequences for Wildlife of a Badger Control Policy in England.* https://webarchive. nationalarchives. gov. uk/20130402173332/http://archive. defra. gov. uk

Ford, T. D. & Gunn, J. (2010). *Exploring the Limestone Landscapes: a walking and cycling guide.* BCRA Cave Studies Series 19. British Cave Research Association, Buxton.

Ford, T. D. (2002). *Rocks and Scenery of the Peak District.* Landmark Publishing, Ashbourne, Derbyshire.

Foster, R. & Leach, J. (2019). Rare Scottish hoverfly discovered at Longshaw. *Sorby Newsletter,* October LVI (8), 15–16, Sorby Natural History Society.

Franks, J. W. & Johnson, R. H. (1964). Pollen Analytical Dating of a Derbyshire Landslip: the Cown Edge Landslides, Charlesworth. *New Phytologist* 63 (2), 209–216.

Fridley, J. D., Grime, P. J., Askew, A., Moser, B. & Stephens, C. J. (2011). Soil heterogeneity buffers community response to climate change in species-rich grassland. *Global Change Biology* 17, 2002–2011.

Frost, R. & Shaw, S. (2013). *The Birds of Derbyshire.* Liverpool University Press, Liverpool.

Game & Wildlife Conservation Trust. (2021). Information on strongylosis in Red Grouse. www.gwct.org.uk/ search?keywords=strongylosis

Gannon, P. (2010). *Rock trails: Peak District; A hillwalker's guide to the Geology and Scenery.* Pesda Press Ltd. Caernarfon, Gwynedd.

Garton, D. (2017). Prior to Peat: Assessing the Hiatus between Mesolithic Activity and Peat Inception on the Southern Pennine Moors. *Archaeological Journal* 174 (2), 281–334.

Gartside, K. (2017). A rare, nationally scarce hoverfly, *Callicera rufa,* at Dovestone. *Sorby Newsletter,* Sorby Natural History Society.

GeoConservation Stafforshire. (2013). *The Hamps and Manifold Geotrail.* http:// srigs. staffs-ecology.org.uk/Geotrails/ HampsManifold/index.html

Gilbert, O. (1980). Effect of land-use on terricolous lichens. *Lichenologist* 12 (1), 117–124.

Gilbert, O. L. & Giavarini, V. J. (1997). The lichen vegetation of acid watercourses in England. *Lichenologist* 29 (4), 347–367.

Glover, J. (2019). *Landscapes Review.* Defra.

Glover, S. (1829). *The History of the County of Derby, Volume 1.* H. Mozley & Son.

Gosney, D. (2018a). A Survey of Nightjars and Long-eared Owls in SK29 (Stocksbridge) in 2018 with Observations of Increased Numbers of Barn Owls. *Sorby Record* 54, 18–22. Sorby Natural History Society.

Gosney, D. (2018b). Breeding Birds in SK27 in 2017 – changes in Numbers since 1988–90. *Sorby Record* 54, 29–50. Sorby Natural History Society.

Gosney, D. (2018c). Breeding Birds in SK29 in 2016 – Changes in Numbers since 1988–90. *Sorby Record* 54, 23–28. Sorby Natural History Society.

Grayson, B. (2017). Grazer selectivity: benefits for livestock, habitats and people. *In Practice* 96. Bulletin of the Chartered Institute of Ecology and Environmental Management.

Gregory, C, & Hines, S. (2019). *The Land that Made Us, The Peak District Farmer's Story.* Farming Life Centre & Peak District National Park Authority, Bakewell.

Grieve, M. (1976). *A Modern Herbal.* Penguin Books, London.

Griffiths, T., Rotherham, I. D. & Handley, C. (2013). *Sphagnum:* the healing harvest. In: War and Peat, (Eds. Rotherham, I. D. & Handley. C.). *Landscape Archaeology and Ecology* 10, 201–219.

Grime, J. P., Fridley, J. D., Askew, A. P., Thompson, K., Hodgson, J. & Bennett, C. R. (2008). Long-term resistance to simulated climate change in an infertile grassland.

Proceedings of the National Academy of Science 105 (29), 10028–10032.

Grindon, L. (1866). *Summer Rambles.* Palmer & Howe, Manchester.

Grosdidier, M., Scordia, T., Loos, R. & Marçais, B. (2020). Landscape epidemiology of ash dieback. *Journal of Ecology* 108 (5), 1789–1799, https://doi. org/10. 1111/1365-2745. 13383

Gunn, J. & Ford, T. (1990). *Caves and Karst of the Peak District.* Cave Studies series No. 3. British Cave Research Association.

Gunn, J. (1985) Pennine Karst Areas and their Quaternary history. In: *The geomorphology of North–west England.* (Ed Johnson, R. H.), 263–281. Manchester University Press.

Gunn, J. (1992). Hydrological contrasts between British Carboniferous Limestone aquifers. In: *Hydrogeology of selected karst regions,* (Eds Back, W. Herman J. S. & Paloc, H.). International Association of Hydrogeologists, International contributions to hydrogeology, 13, 25–42.

Gunn, J., Hardwick, P. & Wood, P. J. (2000). The invertebrate community of the Peak-Speedwell cave system, Derbyshire, England – pressures ad considerations for conservation management. *Aquatic Conservation: Marine and Freshwater Ecosystem* 10, 353–369.

Gunn, J., Lowe, D. J., & Waltham, A. C. W. (1998). The Karst Geomorphology and Hydrogeology of Great Britain. In: *Global Karst Correlation,* (Eds. Daoxian, Y. & Liu Zaihua, L,) VSP, The Netherlands, 109–135.

Hallmann, C. A., Sorg, M., Jongejans, E., Siepel, H., Hofland, N., Schwan, H., Stenmans, W., Müller, A., Sumser, H., Hörren, T., Goulson, D. and de Kroon, H. (2017). More than 75 percent decline over 27 years in total flying insect biomass in protected areas. *PLoS ONE* 12 (10), https://doi. org/10. 1371/journal. pone. 0185809

Hawksford, J. E. & Hopkins, I. J. (2011). *The Flora of Staffordshire.* Staffordshire Wildlife Trust.

Hayhow, D. B., Eaton, M. A., Stanbury, A. J., Burns, F., Kirby, W. B., Bailey, N., Beckmann, B., Bedford, J., Boersch-Supan, P. H., Coomber, F., Dennis, E. B., Dolman, S. J., Dunn, E., Hall, J., Harrower, C., Hatfield, J. H., Hawley, J., Haysom, K., Hughes, J., Johns, D. G., Mathews, F., McQuatters-Gollop, A., Noble, D. G., Outhwaite, C. L., Pearce-Higgins, J. W., Pescott, O. L., Powney, G. D. and Symes, N. (2019). *The State of Nature 2019.* The State of Nature Partnership. https://nbn.org.uk/stateofnature2019/reports/

Heath, A. (2003). *Prehistoric Settlement and Agriculture on the Eastern Moors of the Peak District.* PhD thesis, University of Sheffield.

Hey, D. (2014), *The History of the Peak District Moors,* Pen and Sword, Barnsley.

Hicks, S. (1971). Pollen-analytical evidence for the effect of prehistoric agriculture on the vegetation of North Derbyshire. *New Phytologist* 70 (4), 647–667.

Higginbottom, T. (2010). An Introduction to Plant Galls in the Sorby Area. *Sorby Record* 46, 22–27.

Holden, J. (2009). Upland hydrology. In: *Drivers of Environmental Change in Uplands.* (Eds. Bonn, A., Allott, T., Hubacek, K & Stewart, J.), 113–134. Routledge Studies in Ecological Economics, London.

Holland, P. K. & Yalden, D. W. (2002). Population dynamics of Common Sandpipers *Actitis hypoleucos* in the Peak District – A different decade: A report of the failure of a population to recover from a catastrophic snow storm. *Bird Study* 49 (2), 131–138.

Holland, P. K., Robson, J. E. & Yalden, D. W. (1982). The breeding biology of the Common Sandpiper *Actitis hypoleucos* in the Peak District. *Bird Study* 29 (2), 99–110.

Jacobi, R. M., Tallis, J. H. & Mellars, P. A. (1976). The Southern Pennines Mesolithic and the Ecological Record. *Journal of Archaeological Science* 3, 307–320.

Jarvis, M. S. (1960). *The influence of climatic factors on the distribution of some Derbyshire Plants.* PhD University of Sheffield.

Johnson, R. H. & Walthall, S. (1979). The Longdendale landslides. *Geological Journal* 14 (2), 135–158.

Johnson, R. H. (1985). The imprint of glaciation on the West Pennine Uplands. In: *The geomorphology of North-west England.* (Ed. Johnson, R. H.), 237–262. Manchester University Press

Johnson, R. H., Tallis, J. H. & Wilson, P. (1990). The Seal Edges Coombes, North Derbyshire – a study of their erosional and depositional history. *Journal of Quaternary Science* 5 (1), 83–94.

Jump, A. S. & Woodward F. I. (2003). Seed production and population density decline approaching the range-edge of *Cirsium* species. *New Phytologist* 160, 349–358.

Jump A. S., Woodward F. I. & Burke T. (2003). *Cirsium* species show disparity in patterns of genetic variation at their range-edge despite similar patterns of reproduction and isolation. *New Phytologist* 160, 359–370.

Kitcher, S. J. (2014). *Reconstructing Palaeoenvironments of the White Peak Region of Derbyshire, Northern England.* PhD thesis, University of Hull.

Knight, L. (undated). *Cave Life in Britain.* Freshwater Biological Association.

Lake, J. & Edwards, B. (2017). *Peak District National Park Farmsteads Assessment Framework.* Historic England, Locus Consulting, Peak District National Park, Bakewell.

Lawton, J. H., Brotherton, P. N. M., Brown, V. K., Elphick, C., Fitter, A. H., Forshaw, J., Haddow, R. W., Hilborne, S., Leafe, R. N., Mace, G. M., Southgate, M. P., Sutherland, W. J., Tew, T. E., Varley, J. & Wynne, G. R. (2010). *Making Space for Nature: A review of England's Wildlife Sites and Ecological Networks.* Report to Defra. https://webarchive.nationalarchives.

gov. uk/20130402170324/http://archive. defra. gov. uk/environment/biodiversity/ documents/201009space-for-nature.pdf

Leather, S. (1996). Biological flora of the British Isles. *Prunus padus* L. *Journal of Ecology* 84, 125–132.

Lee, J. (2015). *Yorkshire Dales.* The New Naturalist Library, William Collins.

Lees, F. A. (1888). *The Flora of West Yorkshire,* Facsimile edition 1978, E. P. Publishing.

Lindsay, R. (2010). *Peatbogs and Carbon: A critical synthesis.* RSPB Scotland.

Linton, W. R. (1903). *Flora of Derbyshire.* Bemrose & Sons, London.

Lloyd, J & King, E. (1771). An account of Elden Hole in Derbyshire; By J. Lloyd, Esq; with some observations upon it, by Edward King, Esq; F. R. S. ; in a letter to Matthew Maty, M. D. Sec. R. S. *Phil. Trans. R. Soc.* 61, 250–265.

Lloyd, P. S., Grime, J. P. & Rorison, I. H. (1971). The grassland of the Sheffield Region: I General Features. *Journal of Ecology* 59 (3), 863–886.

Long, D. J., Chambers, F. M. & Barnatt, J. (1998). The Palaeoenvironment and the Vegetation History of a Later Prehistoric Field System at Stoke Flat on the Gritstone Uplands of the Peak District. *Journal of Archaeological Science* 25, 505–519.

Lousley, J. E. (1969). *Wild Flowers of Chalk and Limestone.* New Naturalists, Collins, London.

Loveluck, C. P., More, A. F., Spaulding, N. E., Clifford, H., Handley, M. J., Hartman, L., Korotkikh, E. V., Kurbatov, A. V., Mayeswski, P. A., Sneed, S. B. & McCormick, M. (2020). Alpine ice and the annual political economy of the Angevin Empire, from the death of Thomas Becket to Magna Carta, c. AD 1170–1216. *Antiquity,* 94 (374), 473–490. https//doi. org/10. 15184/ aqy. 2019. 202.

Mabey, R. (1972). *Food for Free. A guide to the edible wild plants of Britain.* Collins, London.

Mabey. R. (1996). *Flora Britannica.* Sinclair Stevenson, London.

Machin, B. (2018). *The Butterflies of the Peak District*. Seven Stones Publishing, Leek.

Maitland, P. S. (2003). *Ecology of the River, Brook and Sea Lamprey*. Conserving Natura 2000 Rivers Ecology Series No. 5 English Nature, Peterborough.

Mallon, D., Alston, D. & Whiteley, D. (2012). *The Mammals of Derbyshire*. Derbyshire Mammal Group & Sorby Natural History Society.

Marren, P. & Mabey, R. (2010). *Bugs Britannica*. Chatto and Windus, London.

Matthew, F., Coomber, F, Wright, J. & Kendall, T. (2018). *Britain's Mammals 2018: The Mammal Society's Guide to their Population and Conservation Status*. Mammal Society.

McAlpine, J. A. I. (2014). *An Assessment of the Extent, Distribution and Change of Bracken* (Pteridium aquilinum) *in the Peak District National Park*. MSc. thesis, Dept. Geography, University of Leicester.

McFarlane, D. A., Lundberg, J., Rentergem, G. V., Howlett, E. & Stimpson, C. (2016). A new radiometric date and assessment of the Last Glacial megafauna of Dream Cave, Derbyshire, UK. *Caves and Karst Science* 43 (3), 109–116.

McGuire, S. (2016). Pole to Pole. *Archaeology and Conservation in Derbyshire* 13, 18–19.

McLean, A. S., Richards, J. P. & Whiteley, D. (2020) *Dragonflies of the Sheffield Area*. Sorby Record Special Series No. 17.

Mellanby, K. (1971). *The Mole*. William Collins Sons & Co. Ltd., Glasgow.

Melling, T., Thomas, M., Price, m & Roos, S. (2018). Raptor persecution in the Peak District National Park. *British Birds* 111, 275–290.

Menendez, R. & Birkett, A. J. (2011). *Effects of climate and land-use changes on dung beetle communities: predicting the consequences for insect biodiversity and function in British moorlands*. Yorkshire Peat Partnership and Moors for the Future Partnership.

Merrill, J. N. (1988). *Customs of the Peak District and Derbyshire*. J. N. M. Publication.

Merritt, R. (2006). *Atlas of water beetles (Coleoptera) and water bugs (Hemiptera) of Derbyshire, Nottingham and South Yorkshire*. Sorby Record Special Series no 14.

Merton, L. F. H. (1970). The History and Status of the Woodlands of the Derbyshire Limestone. *Journal of Ecology* 58 (3), 723–744.

Middleton, J & Middleton, V. (2017). Some observations on the 'draw-down mud flora' of Dale Dike & Damflask reservoirs during 2017. *Sorby Record* 53, 8–12.

Moss, C. E. (1900). Changes in the Halifax flora 1775–1900. *Naturalist Hull* 165–172.

Moss, C. E. (1913). *The Vegetation of the Peak District*. Cambridge University Press, Cambridge.

Moxey, A. & Morling, P. (2018). *Funding for peatland restoration and management*. IUCN Peatland Programme's Commission of Inquiry on Peatlands.

Musk, L. F. (1985). Glacial and Post-glacial climatic conditions in North-west England. In: *The geomorphology of North –west England* (Ed. Johnson, R. H.), 59–79. Manchester University Press, Manchester.

Natural England. (2012). *National Character Area Profile, 51 Dark Peak*. http://publications.naturalengland.org.uk/publication/

Natural England. (2013). *National Character Area Profile, 53, South West Peak*. http://publications.naturalengland.org.uk/publication

Natural England. (2014). *National Character Area Profile, 52. White Peak*. http://publications.naturalengland.org.uk/publication

Norman, D. (2008). *Birds in Cheshire and Wirral. A Breeding and wintering Atlas*, on behalf of the Cheshire and Wirral Ornithological Society. http://www.cheshireandwirralbirdatlas.org

O'Hara, D. & Carr, G. (2017). Recovery of a breeding Dunlin population in the Peak District in response to blanket bog restoration. *British Birds* 110, 109–121.

O'Regan, H. (2018). Caves were a bear necessity.

Archaeology and Conservation in Derbyshire 15, 6–7.

Onrust, J. (2017). *Earth, worms and birds.* Thesis, University of Groningen, The Netherlands, https://www. rug. nl/research/portal/ files/51447395/Complete_thesis. pdf

Parker, A. G., Goudie, A. S., Anderson, D. E, Robinson, M. A & Bonsall, C. (2002). A review of the mid-Holocene elm decline in the British Isles. *Progress in Physical Geography: Earth and the Environment* 26 (1), 1–45.

Peak District National Park Authority. (2018). *Peak District Bird of Prey Initiative – 2018 report.* https://www. peakdistrict. gov. uk/__data/assets/pdf_file/0008/1429766/ PDNP-Bird-of-Prey-Initiative-2018- Report-18–12–13. pdf

Pearce-Higgins, J. W., Dennis, P., Whittingham, M. J. & Yalden, D. W. (2010). Impacts of climate on prey abundance account for fluctuations in a population of a northern wader at the southern edge of its range. *Global Change Biology* 16 (1), 12–23, doi: 10. 1111/j. 1365–2486. 2009. 01883. x

Pearce-Higgins, J. W. & Yalden, D. W. (2004). Habitat selection, diet, arthropod availability and growth of a moorland wader: the ecology of European Golden Plover *Pluvialis apricaria* chicks. *Ibis* 146, 335–346.

Pearce-Higgins, J. W. (2019). *Trends in breeding bird populations of the Peak District Moorlands from 1990 and 2004/5 to 2018.* Report for Moors for the Future Partnership, Edale.

Pearce-Higgins, J. W. (2010). Using diet to assess the sensitivity of northern and upland birds to climate change. *Climate Research* 45, 119–130, doi:10. 3354/cr00920.

Pearce-Higgins, J. W. (2011). Modelling conservation management options for a southern range-margin population of Golden Plover, *Ibis* 153, 345–356.

Pearce-Higgins, J. W., Beale, C. M., Oliver, T. H., August, T. A., Carroll. M., Massiminoa,

D., Ockendon, N., Savage, J., Wheatley, C. J., Ausden, M. A., Bradbury, R. B., Duffield, S. J., Macgregor, N. A., McClean, C., Morecroft, M. D., Thomas, C. D., Watts, O., Beckmann, B. C., Fox, R., Roy, H. E., Sutton, P. G., Walker, K. J. & Crick, H. Q. P. (2017). A national-scale assessment of climate change impacts on species: Assessing the balance of risks and opportunities for multiple taxa. *Biological Conservation* 213, 124–134.

Penny Anderson Associates, (2015). *Sustainable Catchment Management Programme, Final Report.* United Utilities.

Phillips, B. B, Gaston, K. J., Bullock, J. M. & Osborne, J. L. (2019). Road verges support pollinators in agricultural landscapes, but are diminished by heavy traffic and summer cutting. *Journal of Applied Ecology* 56, 2316–2327.

Phillips, J., Yalden, D. & Tallis, J. (1981). *Peak District Moorland Erosion Study Phase 1 Report.* Peak District National Park Authority, Bakewell.

Piggott, C. D. (1969). The Status of *Tilia cordata* and *T. platyphyllos* on the Derbyshire Limestone. *Journal of Ecology* 57 (2), 491–504.

Piggott, C. D. (1968). Biological Flora of the British Isles, *Cirsium acaulon* (L) Scop. *Journal of Ecology* 56, 597–612.

Piggott, C. D. & Huntley, J. P. (1981). Factors controlling the distribution of *Tilia cordata* at the northern limits of its geographical range, III Nature and causes of seed sterility. *New Phytology* 87, 817–839.

Piggott, C. D. (1962). Soil formation and development on the Carboniferous limestone of Derbyshire I Parent Materials. *Journal of Ecology* 50 (1), 145–156.

Piggott, C. D. (1970). Soil formation and development on the Carboniferous limestone of Derbyshire II. The Relation of Soil Development to Vegetation on the Plateau near Coombs Dale. *Journal of Ecology* 58 (2), 529–541.

Piggott, C. D. (1983). Regeneration of Oak-Birch woodland following exclusion of sheep. *Journal of Ecology* 71, 629–646.

Pilkington, J. (1789). *A view of the present state of Derbyshire Vol 1*. J. Drewry, Derby.

Pilkington, M., Walker, J., Maskill, R., Allott, T. & Evans, M. (2015). *Restoration of Blanket bogs; flood risk reduction and other ecosystem benefits*. Final Report for the Making Space for Water Project: Moors for the Future Partnership, Edale.

Pollard, E. (1988). Temperature, rainfall and butterfly numbers. *Journal of Applied Ecology* 25, 819–828.

Pope, L. J. (2009). *Fate and Effects of Parasiticides in the Pasture Environment*. PhD, University of York.

Porteous, C. (1978). *The Well-dressing Guide*. Derbyshire Countryside Ltd, Derby.

Porter, L. (1984). *The Peak District: Pictures from the Past*. Moorland Publishing, Ashbourne.

Radley, J. (1965). Significance of Major Moorland Fires. *Nature* March 27, 1254–1259.

Ramchunder, S. J., Brown, L. E. & Holden, J. (2012). Catchment-scale peatland restoration benefits stream ecosystem biodiversity. *Journal of Applied Ecology* 49 (1), 182–191.

Ravenscroft, C, Whitlock, R & Fridley, J. D. (2015). Rapid divergence in response to 15 years of simulated climate change. *Global Change Biology* 21 (11), 4165–4176.

Redfern, M. (2011). *Plant Galls*. New Naturalist Library, A Survey of British Natural History. Collins, London.

Rewilding Britain. (2019). *Rewilding and Climate Breakdown: How restoring nature can help decarbonise the UK*. Rewilding Britain.

Richards, P. & Thomas, R. (1998). Woodlice and centipedes – new to the Region. *Sorby Record* 34, 78.

Richards, P. (1995). Millipedes, Centipedes and Woodlice of the Sheffield Area. *Sorby Record Special Series* no 10.

Roberts, A. F. & Leach, J. T. (1985). *The Coal Mines of Buxton*. Scarthin Books, Cromford.

Rodwell, J. S. (1991). *British Plant Communities Volume 2, Mires and Heaths*. Cambridge University Press.

Rodwell, J. S. (1991a). *British Plant Communities Volume 1, Woodlands and Scrub*. Cambridge University Press.

Rodwell, J. S. (1992). *British Plant Communities Volume 3, Grasslands and Montane Communities*. Cambridge University Press.

Rotherham, I. D. (1986). The introduction, spread and current distribution of *Rhododendron ponticum* in the Peak District and Sheffield Area. *Naturalist* 111, 61–68.

Rotherham, I. D. (2017). *Shadow Woods: A search for Lost Landscapes*. Wildtrack Publishing, Sheffield.

Rothwell, J. & Evans, M. (2004). *Flux of heavy metal pollution from eroding southern Pennine peatlands*. Moors for the Future and Manchester University Research Report.

Salmon & Trout Organisation. (2021) *Smart rivers monitoring programme*. https://www.salmon-trout.org/smart-rivers

Sayer, E. J., Oliver, A. E., Fridley, J. D., Askew, A. P., Mills, R. T. ER. & Grimes, J. P. (2017). Links between soil microbial communities and plant traits in a species-rich grassland under long-term climate change. *Ecology and Evolution* 7, 855–862.

Shimwell, D. W. (1977). *Studies in the History of the Peak District Landscape: I Pollen Analysis of some Podzolic Soils on the Limestone Plateau*. University of Manchester School of Geography Research Papers no 3.

Shimwell, D. W. (1981). Images of the Peak District 1150–1950. *Manchester Geographer*. Manchester Geographical Society.

Simmons, I. G. (2003). *The Moorlands of England and Wales: An Environmental History 8000 BC–AD 2000*. Edinburgh University Press, Edinburgh.

Smith, E. J. (2000). Coleoptera Recorder's Report for 2000. *Sorby Record*, 37.

Smith, H. (2002). The Hydro-ecology of Limestone Springs in the Wye Valley, Derbyshire. *Journal of CIWEM* 16, 253–259.

Smith, R. (2012). *A Peak District Anthology. A literary companion to Britain's first National Park.* Frances Lincoln Ltd, London.

Spray, M. (1981). Holly as Fodder in England. *The Agricultural History Review* 29 (2), 97–110.

Spray, M. & Smith, D. J. (1977). The rise and fall of holly in the Sheffield Region, *Transactions of the Hunter Archaeological Society* X, 239–51.

Stace, C. A. (2019). *New Flora of the British Isles, Fourth Edition.* C. & M. Floristics, Suffolk.

Staffordshire Wildlife Trust. (2016). *The State of Staffordshire's Nature. Technical Report.*

Stevenson, I. P. & Gaunt, G. D. (1971) *Geology of Country around Chapel-en-le-Frith.* Explanation of One-inch Geological Sheet 99, New Series. British Geological Survey Memoirs. HMSO.

Stroh, P. A., Leach, S. J., August, T. A., Walker, K. J., Pearman, D. A., Rumsey, F. J., Harrower, C. A., Fay, M. F., Martin, J. P., Pankhurst, T., Preston, C. D. & Taylor, I. (2014). *A Vascular Plant Red List for England.* Botanical Society of the British Isles.

Sudd, J. H. & Lodhi, A. Q. K. (1981). The distribution of foraging workers of the wood-ant *Formica lugubris* Zetterstedt (Hymenoptera: Formicidae) and their effect on the numbers and diversity of other arthropoda. *Biological Conservation* 20 (2), 133–145.

Tallis, J. H. (1964). Studies on Southern Pennine Peats: III. The Behaviour of Sphagnum. *Journal of Ecology* 52 (2), 345–353.

Tallis, J. H. (1997). The pollen record of *Empetrum nigrum* in Southern Pennine peats: implications for erosion and climate change. *Journal of Ecology* 85 (4), 455–465.

Tallis, J. H. (1991). Forest and Moorland in the South Pennine uplands in the mid-Flandrian Period: III. The spread of Moorland – local, regional and national. *Journal of Ecology* 79 (2), 401–415.

Tallis, J. H. (1997a). The Southern Pennines experience: an overview of blanket mire degradation. In: *Blanket Mire Degradation, Causes, Consequences and Challenges.* (Eds. Tallis, J. H., Meade, R. & Hulme, P. D.), 7–15. Conference Proceedings, University of Manchester, British Ecological Society Mires Research Group.

Tallis, J. H. & Switsur, V. R. (1973). Studies on Southern Pennine Peats: VI. A Radiocarbon-dated Pollen Diagram from Featherbed Moss, Derbyshire. *Journal of Ecology* 61 (3), 743–751.

Tallis, J. H. & Switsur, V. R. (1990). Forest and Moorland in the South Pennines Uplands in the Mid-Flandrian Period: the Hillslope Forests. *Journal of Ecology* 78 (4), 857–883.

Tattersfield, P. (1990). Terrestrial Mollusc Faunas from some South Pennine Woodlands. *Journal of Conchology* 33, 355–374.

Taylor, D. M., Griffiths, H. I., Pedley, H. M. & Prince, I. (1994). Radiocarbon-dated Holocene pollen and ostracod sequences from barrage tufa-dammed fluvial systems in the White Peak, Derbyshire, UK. *The Holocene* 4 (4) 356–364.

Thomas, C. D., Hill, J. K., Anderson, B. J., Bailey, S., Beale, C. M., Bradbury, R. B., Bulman, C. R., Crick, H. Q. P., Eigenbrod, F., Griffiths, H. M., Kunin, W. E., Oliver, T. H., Walmsley, C. A., Watts, K., Worsfold, N. T. & Yardley, T. (2011). A framework for assessing threats and benefits to species responding to climate change. *Methods in Ecology and Evolution* 2, 125–142.

Thomas, I. & Cooper, M. (2008). The Geology of Chatsworth House, Derbyshire. *Mercian Geologist* 17 (1), 27–42.

Thompson, D. (2008). *Carbon Management by Land and Marine Managers.* Natural England Research Report NERR026.

Tomlinson, M. L. & Perrow, M. R. (2003). *Ecology of Bullhead.* Conserving Natura 2000 Rivers Ecology Series no. 4. English Nature, Peterborough.

Trinder, S., Askew, A. & Whitlock, R. (2020). Climate-driven evolutionary change in reproductive and early-acting life-history traits in the perennial grass *Festuca ovina*. *Journal of Ecology* 108 (4), 1398–1410, https://doi. org/10. 1111/1365–2745. 13304.

Vera, F. W. M. (2000). *Grazing Ecology and Forest History*. CABI Publishing.

Virtue, A. (1976). *Memoirs of a Derbyshire Hill Farmer*. Virtue & Co Ltd, London & Coulsdon.

Visit Peak District. (Undated). '*The lands that time forgot*' Leaflet on Peak District Geology from Visit Peak District's website.

Wallace, I. (2004). The Sedge of the Water. *Salmo Trutta* 7, 70–73.

Walton, I. (1653). *The Compleat Angler*. Reprinted with an Introduction by Andrew Lang, Everyman's Library, 1906.

Ward, S., Smart, S. M., Quirk, H., Tallowin, J. R. B., Mortimer, S. R., Shiel, R. S., Wilby, A. & Bardgett, R. (2016). Legacy effects of grassland management on soil carbon to depth. *Global Change Biology* 22, 292902938 doi: 10. 1111/gcb. 13246.

Warren, P., Rotherham, I. D., Eades, P. Wright, S. & Howe, P. (1996). Invertebrate and Macrophyte communities of dewponds in the Peak District, with particular reference to the method of pond construction. *Peak District Journal of Natural History and Archaeology*. 1, 27–33.

Waters, C. N. (Undated) *Carboniferous Geology of Northern England*. British Geological Survey, Nottingham. http://nora. nerc. ac. uk/id/eprint/10713/1/Waters_Final. pdf

Webb, J & Brown, A. (2016). The Conservation Status of British Invertebrates. *British Wildlife* 27 (6), 410–421.

Westaway, R. (2019). Derwent Valley Caves: evolution of Peak District tributary gorges, dating and valley incision, 173. In: *The Quaternary Fluvial Archives of the major British Rivers*. (Eds. Bridgland, D. R., Briant, R. M., Allen, P., Brown, E. J. & White, T. S.) Quaternary Research Association, London.

Wheater, C. P. & Cullen, R. (1997). The Flora and Invertebrate Fauna of Abandoned Limestone Quarries in Derbyshire, United Kingdom. *Restoration Ecology* 5 (1), 77–84.

Whitehouse, N. J. & Smith, D. (2010). How fragmented was the British Holocene wildwood? Perspectives on the "Vera" grazing debate from the fossil record. *Quaternary Science Reviews* 29 (3–4), 539–553.

Whiteley, D. (1992). An Atlas of Sheffield Area Butterflies. *Sorby Record* 29, 19–56. Whiteley, D. (1997). *Reptiles and Amphibians of the Sheffield Area and North Derbyshire*. Sorby Record Special Series no 11.

Whiteley, D. (2004). The Slender Ground Hopper in the Sorby Area. *Sorby Record*, 40, 32–35.

Whitely, D. (2011). Oil Beetles in the Sorby Area. *Sorby Record* 47, 77–78.

Whiteley. D. (2011a). The 'Lead Rake Robberfly' in the Sorby Area. *Sorby Record* 47, 75–76.

Willmore, G. T. D., Lunn, J. & Rodwell, J. S. (2011). *The South Yorkshire Plant Atlas*. Yorkshire Naturalists' Union, Yorkshire & the Humber Ecological Data Trust.

Willmot, A. & Moyes, N. (2015). *The Flora of Derbyshire*. Pisces Publication.

Willmot, A. (1980). The woody species of hedges with special reference to age in the Church Broughton Parish, Derbyshire. *Journal of Ecology* 68 (1), 269–285.

Wiltshire, P. E. J. (Undated) *Palynological Analysis of Lismore Fields, Buxton, Derbyshire*. Ancient Monuments Laboratory Report 18. 91

Wood, D & Hill, R. Eds. (2013). *Breeding Birds of the Sheffield Area, including the North-east Peak District*. Sheffield Bird Study Group.

Yalden, D. W. & Albarella, U. (2009). *The History of British Birds*. Oxford University Press, Oxford.

Yalden, D. W. (1981). The occurrences of the Pigmy Shrew *Sorex minutus* on moorland, and the implications for its presence in Ireland. *Journal Zoology London* 195, 147–156.

Yalden, D. W. (1985). Diet, food availability and habitat selection of breeding Common Sandpipers *Actitis hypoleucos. Ibis* 128, 23–36.

Yalden, D. W. (1986). The Distribution of Newts, *Triturus* spp., in the Peak District, England. *Herpetological Journal* 1, 97–101.

Yalden, D. W. (1992). The influence of recreational disturbance on common sandpipers *Actitis hypoleucos* breeding by an upland reservoir, in England. *Biological Conservation* 61, 41–49.

Yalden, D. W. (1999). *The History of British Mammals.* Poyser, London.

Yalden, D. W. (2001). Fallow Deer *Dama dama* in the S. E. Peak District. *Sorby Record* 37, 31–33.

Yalden, D. W. (2001). Red Deer *Cervus elaphus* in the S. W. Peak District. *Sorby Record* 37, 25–31.

Yalden, D. W. (2013). The end of feral wallabies in the Peak District. *British Wildlife* 24 (3), 169–178.

Yalden, P. E. (1982). Pollen collected by the bumblebee *Bombus monticola* Smith in the Peak District, England. *Journal of Natural History* 16, 823–832.

Yalden, P. E. (1983). Foraging population size and distribution of *Bombus monticola* in the Peak District, England. *Naturalist* 108, 139–47.

Yeloff, D., Labadz, J. C. & Hunt, C. O. (2006). Causes of degradation and erosion of a blanket mire in the southern Pennines, UK. *Mires and Peat* 1 Article 4. 1–18, http://www.mires-and-peat. net

Zasada, K. A. (1981). Ephemeroptera, Mayflies. *Sorby Record Special Series* 4, 35–41. Zasada, K. A. (1981a). Plecoptera, Stoneflies. *Sorby Record Special Series* 4, 42–47.

Index

GENERAL INDEX